Intelligent
Freight
Transportation

AUTOMATION AND CONTROL ENGINEERING

A Series of Reference Books and Textbooks

Editor

FRANK L. LEWIS, PH.D.

Professor
Automation and Robotics Research Institute
The University of Texas at Arlington

Co-Editor

SHUZHI SAM GE, PH.D.

The National University of Singapore

Intelligent Freight Transportation

Edited by

Petros A. Ioannou
University of Southern California
Los Angeles, California, U.S.A.

CRC Press
Taylor & Francis Group
Boca Raton London New York

CRC Press is an imprint of the
Taylor & Francis Group, an **informa** business

Published 2008 by CRC Press
Taylor & Francis Group
6000 Broken Sound Parkway NW, Suite 300
Boca Raton, FL 33487-2742

© 2008 by Taylor & Francis Group, LLC
CRC Press is an imprint of Taylor & Francis Group, an Informa business

First issued in paperback 2019

No claim to original U.S. Government works

ISBN-13: 978-0-367-45268-1 (pbk)
ISBN-13: 978-0-8493-0770-6 (hbk)

Visit the Taylor & Francis Web site at
http://www.taylorandfrancis.com

and the CRC Press Web site at
http://www.crcpress.com

Library of Congress Cataloging-in-Publication Data

Ioannou, P. A. (Petros A.), 1953-
    Intelligent freight transportation / Petros A. Ioannou.
       p. cm. -- (Automation and control engineering)
    Includes bibliographical references and index.
    ISBN 978-0-8493-0770-6 (hardback : alk. paper)
       1. Intelligent Vehicle Highway Systems. 2. Freight and freightage. I. Title. II.
Series.

TE228.3.I63 2008
388'.044--dc22   2007049293

Contents

Preface

Freight transportation is vital to the functioning of societies. With globalization and relaxation of international trade barriers, the transportation chain has expanded considerably. The demand for goods has been increasing due to rising living standards, competition that lowers cost to consumers, and many other factors. This rising demand puts considerable pressure on all transportation nodes and links such as ports, ships, trucks, trains, airports, warehouses, roadway systems, and so on to improve their practices in order to keep pace with higher demand. The pressure comes at a time where in places such as metropolitan areas the freight transportation infrastructure is already saturated. Cost, social, and environmental constraints as well as lack of adequate space make it difficult, if at all possible, to expand the infrastructure using the traditional approach of "more land more capacity." The advances during the last three decades in electronics, computers, information technologies, and Internet as well as in optimization and software tools opened the way for alternative ways of improving the freight transportation system in order to keep pace with rising demand while meeting the constraints of the twenty-first century.

Intelligent freight transportation (IFT) is an area that deals with the use of advanced technologies and intelligent decision making in order to improve the freight transportation system by making the existing infrastructure more efficient. IFT is a multidisciplinary area with many complex problems, which in addition to technology include policy, socioeconomic, political, environmental, and human and other issues.

The purpose of this book is to bring together a number of experts working in different areas or disciplines to present some of their latest approaches and future research directions in the area of IFT. The field is so diverse that it is impossible for the seventeen chapters of the book to cover every single aspect of the area. The complexity of the problems involved, the issues raised, and proposed solutions will give the reader a clearer picture of the complexity of the problem as well as provide knowledge of the areas where progress could be made. The freight transportation system is an integrated system with many interconnected subsystems. While solving problems on the overall system level may be difficult, solutions on the subsystem level may be feasible. Such subsystem-level solutions are provided in several of the chapters of the book.

The book consists of seventeen chapters dealing with automation of container terminals, modeling of cross-border land transportation, port choice and competition, inland ports and alternative container transport systems, optimization techniques for efficient cargo movement, labor and environmental issues, and solutions.

The chapters present an indicative spectrum of the different areas and disciplines that can be classed as IFT. Many other important areas and aspects are omitted due to lack of space. The chapters are written by academicians, practitioners, and industry and labor union experts, giving the reader a very wide perspective of the issues and solutions that arise in IFT. The diversity of the topics and that of the authors is a unique feature of the book, making it very useful to a very wide range of readers. Policy makers, researchers, practitioners, economists, environmentalists, and so on could benefit from the book, as they will be able to read about diverse yet very relevant topics in a single book. The book in itself emphasizes with its structure and content the multidisciplinary nature of IFT, whose understanding could lead to solutions of freight transportation problems that are so vital to society.

I thank all the authors for their valuable time and efforts in putting together this book, for their hard work, and for sharing their experiences so readily. I also thank the reviewers for their valuable comments in enhancing the contents of this book. Last but not least I thank my colleagues at the University of Southern California (USC) and California State University at Long Beach (CSULB) for their interaction that enriched my knowledge by indirectly letting me know how little I know of the complexity of freight transportation. Special thanks go to Professors James Moore, Genevieve Giuliano, Maged Dessouky, Randy Hall, Hossein Jula, Elliott Axelband, Berok Khoshnevis, and Anastasios Chassiakos; Marianne Venieris; Stan Whitley; Dr. Isaac Maya; and others. I also acknowledge the financial support of the transportation centers, CCDoTT at CSULB, METRANS at USC/CSULB, and CATT at USC, as well as of the National Science Foundation, which made it possible to do research in IFT and become more familiar with the field. Furthermore, the valuable help of several former Ph.D. students, which include Drs. Hossein Jula, Chin Liu, Jianlong Zhang, Baris Fidan, and Marios Lestas, as well as current Ph.D. students Yun Wang, Hwan Chang, Ying Huo, Matthew Kuipers, Nazli Kahveci, and Jason Levin, and past associate researchers Professors Katerina Vukadinovic and Elias Kosmatopoulos, and industry collaborator Edmond Dougherty, Jr., made the preparation of this book a much easier task.

Petros Ioannou

Editor

Petros Ioannou received MS and PhD degrees from the University of Illinois in 1980 and 1982, respectively. He is currently a professor at the Department of Electrical Engineering-Systems, University of Southern California, director of the Center of Advanced Transportation Technologies, and Associate director for research of METRANS. He has been an associate editor for the *IEEE Transactions on Automatic Control*, the *International Journal of Control, Automatica*, and *IEEE*  *Transactions on Intelligent Transportation Systems*. Ioannou is currently associate editor-at-large of the *IEEE Transactions on Automatic Control* and chairman of the IFAC Technical Committee on Transportation Systems. His research interests are in the areas of adaptive control, neural networks, nonlinear systems, vehicle dynamics and control, intelligent transportation systems (ITS), and intelligent flight control. He is a fellow of IEEE, fellow of the International Federation of Automatic Control (IFAC), and author/coauthor of 8 books and over 150 research papers in the area of controls, neural networks, nonlinear dynamical systems, and ITS.

Contributors

Athanasios Ballis
National Technical University of Athens
Athens, Greece

Hwan Chang
University of Southern California
Los Angeles, California

Anastasios Chassiakos
California State University
Long Beach, California

Raymond K. Cheung
Hong Kong University of Science
  and Technology
Hong Kong, China

Carlos Daganzo
University of California at Berkeley
Berkeley, California

Edmond Dougherty
Ablaze Development Corporation
Villanova, Pennsylvania

Alan L. Erera
Georgia Institute of Technology
Atlanta, Georgia

Anne Goodchild
University of Washington
Seattle, Washington

Lindy Helfman
California State University
Long Beach, California

Petros Ioannou
University of Southern California
Los Angeles, California

Kenneth A. James
California State University
Long Beach, California

Hossein Jula
Pennsylvania State University–
  Harrisburg
Middletown, Pennsylvania

Kap Hwan Kim
Pusan National University
Busan, Korea

Rutger Kroon
JF Hillebrand
Haarlem, The Netherlands

Antonis Michail
Cardiff University
Cardiff, United Kingdom

Domenick Miretti
East Los Angeles College
Monterey Park, California

Kristen Monaco
California State University
Long Beach, California

Christine Ann Mulcahy
California State University
Long Beach, California

Edwin Savacool
Enterprise Management
  Systems, LLC
Manassas, Virginia

Karen R. Smilowitz
Northwestern University
Evanston, Illinois

Sotiris Theofanis
Rutgers University
Piscataway, New Jersey

Iris F. A. Vis
Vrije Universiteit Amsterdam
Amsterdam, The Netherlands

Thomas H. Wakeman
Stevens Institute of Technology
Hoboken, New Jersey

Christopher F. Wooldridge
Cardiff University
Cardiff, United Kingdom

1 Introduction to Intelligent Freight Transportation

Petros Ioannou

CONTENTS

1.1 INTRODUCTION/BACKGROUND

Economic restructuring and globalization have vastly increased the volume of commodity flows by all transport modes. Increased freight flows have had significant impact on metropolitan areas. Traffic at major freight generators (ports, airports, rail yards, warehouse/distribution nodes) has greatly increased, contributing to congestion, environmental pollution, and highway crashes. Economic forecasts are unanimous in predicting continued increases in international trade, with expectations that containerized cargo volume will increase considerably, leading to a commensurate increase in traffic at major import/export and distribution nodes. The U.S. highway, rail system, and ports are already struggling to keep pace with rising demand and deal with increased levels of congestion, environmental, and other issues. Increasing capacity at ports and goods movement supply chain in general, while satisfying environmental, economic, political, labor union, and other constraints, is a challenging problem that needs to be addressed.

Adding capacity by building new infrastructure is not a feasible short- or even medium-term alternative for managing increased cargo volumes, especially in metropolitan areas. Building new infrastructure faces numerous constraints: lack of funding, land scarcity, environmental concerns, community opposition, and others. The planning and review process for major projects can take a decade or more. Solving urban freight transportation problems will require better utilization of existing infrastructure and more efficient flows throughout the goods movement supply chain. Advances in information technology, telecommunications, data management, and computation tools, together with recent research in systems optimization and

1

control, make possible new approaches to the freight transportation problem. The term intelligent freight transportation represents an area where the use of advanced technologies and intelligent decision making can be integrated in order to come up with solutions that make better use of the existing infrastructure or introduce new transport ways in an effort to improve capacity and efficiency under environmental, economic, and other constraints.

The purpose of this book is to bring together experts from industry, academia, and other stakeholders to address freight transportation problems by raising issues, proposing solutions, discussing obstacles, and so forth from different points of view. In the following subsections we present the various areas covered by the chapters of the book.

1.2 AUTOMATION OF CONTAINER TERMINALS

Most terminals in metropolitan areas cannot meet increasing demand due to limited space and inefficient operations. Increasing capacity by using additional land is often a costly proposition due to the scarcity of land or its high cost. Advanced technologies and automation is an attractive way to increase capacity by replacing manual and often inefficient operations with automated ones that are optimized for efficiency.

In chapter 2, several automated container terminal (ACT) concepts are designed, analyzed, and evaluated. These concepts include the use of automated guided vehicles (AGVs) and different configurations of the terminals in an effort to save land while meeting throughput demand. Future demand scenarios are used to design the characteristics of each terminal in terms of configuration, equipment, and operations. Simulation models are developed to simulate each terminal system and evaluate its performance. In addition to performance, cost considerations are used to compare the different concepts. The results obtained demonstrate that automation could improve the performance of conventional terminals substantially and at a much lower cost.

Chapter 3 introduces a number of proposed container automation concepts and examines why some of them have been implemented and others not. The lessons learned from past successes and failures could lead to many more successful automated container terminals in the future. It points out that technology is often not the issue in implementing automation. Perceived need, funding, technical risk versus operational reward, and timing have been the key factors regarding terminal automation.

The trend in terminal operations is to minimize the time container vessels spend in ports. This can be achieved by improving the productivity of the various container handling operations by modernizing container terminals through the use of automated container handling facilities. Such changes introduce various new handling facilities and generate new research topics for efficient operations of terminals. Chapter 4 discusses the related decision-making problems that need to be explored by researchers.

1.3 MODELING OF CROSS-BORDER LAND TRANSPORTATION

The land transportation of containers between a container terminal and the origins or destinations of the containers may represent a very small portion of the global distribution network in terms of distance, but it could account for a significant portion of the total transportation cost due to cross-border issues such as having different regulatory policies and information flow giving rise to shipment delays. The modeling of such a transportation problem is not trivial. In chapter 5, several modeling perspectives are introduced and used to formulate cross-border land transportation problems under different situations. These situations depend on the level of policy restrictions that govern cross-border activities and the level of information available for decision making. A number of models, ranging from coupling drivers and tractors to matching resources with transportation requests in a dynamic, stochastic environment, are reviewed. The case of Hong Kong is used as an example to illustrate the challenges of managing cross-border container transportation.

1.4 PORT CHOICE AND COMPETITION

The containerized trade market is growing rapidly due to globalization and increased international trade with the Far East giving rise to port competition among the European ports in the Le Havre–Hamburg range for more capacity. The port of Amsterdam in an effort to meet such competition developed the Ceres Paragon Terminal in 2002. Characterized by a revolutionary concept known as an indented berth, served simultaneously by nine ultramodern post-Panamax gantry cranes, high productivity levels and low turnaround times can be obtained. Although the odds seemed favorable for the new terminal, enthusiasm was replaced by vexation as the terminal experienced a dramatically slow start. For years it was barely operational with only an incidental test run and some feeder and barge movements. Finally the first carriers were contracted in July 2005. In chapter 6 relevant main port choice and port performance criteria identified in literature are studied. Some of these criteria are applied to the port of Amsterdam in order to study the port's and terminal's chances for structural establishment in the competitive West European port arena.

1.5 INLAND PORTS AND ALTERNATIVE TRANSPORT SYSTEMS

The lack of sufficient storage capacity in terminals due to lack of adequate land as well as the increasing level of congestion associated with cargo movement within and outside the terminals motivates different approaches to deal with the situation. The concept of inland port, where containers are handled at a remote place where land is cheaper and the traffic network less congested before transferred from/to the terminal is one way to ease the pressure on terminals for more capacity. The use of alternative methods of cargo movement without disturbing the traffic network, especially in metropolitan areas, is another way of managing congestion and improving efficiency.

In chapter 7 existing inland port operations and port planning are studied and used to predict future development trends. Inland port operations are classified by the type and method of cargo handling to provide transportation and distribution planners sufficient information to begin the process. A review of concepts being considered for future inland port development is provided. The emerging efforts to develop an integrated inland port system within the United States are evaluated and summarized.

In chapter 8 the basic operational concepts reflected by the existing terminal types, mainly in terms of logistic activities, are studied. Furthermore, a number of innovative technological concepts proposed for the enhancement of inland terminal performance are presented.

Chapter 9 deals with a new approach of moving containers based on a magnetic levitation technology. The system is referred to as the Maglev Freight Conveyor System. This approach utilizes a proven Maglev "conveyor belt" technology that shows promise for both short-haul urban freight movement and interstate-bound containers. The application of this technology to container freight movement inside the port and beyond its confines is expected to reduce both highway congestion and pollution throughout the Los Angeles area.

1.6 OPTIMIZATION TECHNIQUES FOR EFFICIENT OPERATIONS

The application of advanced technologies, especially information technologies and automation, will open the way for the use of advanced optimization techniques in order to optimize performance and improve efficiency. Furthermore, the way practices are carried out could be changed or modified to make full use of technological developments that would lead to additional capacity, less congestion, and many other benefits.

In chapter 10, the time window appointment system, which has recently been introduced as a way to reduce congestion at the terminal gates, is investigated. With this system the trucks are assigned a window of time to show up at the terminal gates to be served. Given that trucks have to perform other delivery/pickup tasks associated with warehouses, customers, and so on using the road network, the overall problem can be formulated as an optimization problem with limits where trucks have to complete their tasks with minimum cost subject to time window and other constraints. In this chapter, the container movement by trucks in metropolitan areas with time constraints at origins and destinations is modeled as an asymmetric multi-traveling salesman problem with time windows (m-TSPTW) with social constraints. Different variations of the m-TSPTW are studied, and solution methods are reviewed and evaluated.

Intermodal drayage truck routing and scheduling problems represent a special class of vehicle routing problems called full *truckload pickup and delivery problems*. Feasible routes in such problems are primarily constrained by time restrictions. In chapter 11, methods to improve container drayage truck routing and scheduling practices through the use of advanced systematic scheduling approaches based on information technology are presented and analyzed.

Loading ships as they are unloaded (double-cycling) can improve the efficiency of a quay crane and therefore container port. In chapter 12 the double-cycling problem is investigated and solution algorithms to the sequencing problem are developed. Furthermore, a simple formula is used to estimate benefits. The objective in this case is to reduce the turnaround time of the ship by completing the loading/unloading process as fast as possible. Several optimization techniques, which include the greedy algorithm based on the physical properties of the problem and the formulation of the problem as a scheduling problem and its optimum solution using Johnson's rule, are used and compared. The results demonstrate that double-cycling can create significant efficiency gains.

Empty container repositioning is probably the single largest contributor to the congestion at and around marine ports. The reason is that terminals are used as storage and place of reference when it comes to empty containers leading to truck trips, which can be avoided if information and optimization technologies are used to facilitate exchange of empty containers between vendors without having to use the terminal for intermediate storage. With a huge number of empty containers at stake at many terminals, a small percentage reduction in empty repositioning traffic can be reflected in huge congestion reduction and improved operational cost. In chapter 13, the empty container reuse concept that facilitates the interchange of empty containers outside container terminals is studied. In particular, the depot-direct and street-turn methodologies are investigated, and variants of the empty container reuse problem are considered. These variants are modeled analytically, and optimization techniques are reviewed and discussed.

1.7   LABOR ISSUES

The labor unions are viewed by many as the main obstacle to the use of advanced technologies and automation at ports. Automation, while it may generate jobs in different areas, does take current jobs away, creating a social problem. The International Longshore and Warehouse Union (ILWU) has been a strong opponent of any change in ports that would lead to loss of jobs, becoming one of the most powerful unions in transportation. Competition, however, and the need for more capacity and efficiency in order to keep pace with rising demand put pressure on the ports and unions to do something about it. Jobs could be lost by business moving away to other places or countries even in the lack of any technology improvement. As was the case with containerization, the use of advanced technologies to meet rising demand is inevitable, and it is a matter of time when such technologies will be applied gradually. Chapters 14 and 15 deal with these issues from an academic point of view as well as a point of view of an expert associated with ILWU.

In chapter 14, a cross-country analysis of how labor was affected by changes in the structure of water transportation is presented, using a case study of West Coast port workers to contrast the outcomes under different economic, legal, and structural regimes. A discussion of current issues facing European and Asian ports is used to hypothesize their potential impacts on labor markets.

In chapter 15 the involvement of members of the International Longshore and Warehouse Union (ILWU) with goods movement technologies in West Coast seaports are studied. The union's involvement in three technological phases, the events of which have been instrumental in molding, coloring, and tempering its view of changing goods movement practices, is discussed in order to better understand how future introduction of intelligent transportation technologies could make seaports more productive while at the same time maintaining the integrity of the longshore workforce. The three technological phases include the union's formative years and the break-bulk era circa 1930–1959, the mechanization and modernization circa 1960–2001, and the years of expanding technology and the globalization of trade circa 2002 to the present. The findings of this study are based on a review of primary and secondary sources, related mainly to the longshore industry and a review of the oral history of the ILWU. Personal interviews were also conducted with key union and industry leaders. From a rather unique perspective, study findings and conclusions are drawn from personal observations and experiences gained of the author of the chapter as a member of both ILWU Locals 13 and 63 for more than five decades and as the union's senior liaison to the ports of Los Angeles/Long Beach for more than two decades.

1.8 ENVIRONMENTAL ISSUES

The environment is increasingly playing a bigger role in decisions affecting transportation practices. Trucks are highly polluting, and they have not been as regulated as passenger cars. Ports are major pollutant generators as heavy machinery, ships, and trucks are operating in a small area.

In chapter 16, the interaction between transport systems and the environment, and the significance of modal choices is investigated. Taking a European perspective, the extent to which policies of environmental protection are actually implemented is discussed with regard to the European transport system. In addition, the responses of the major players and operators in intermodal transport chains as they face the challenge of sustainable chain operation are identified. This includes related policies, recent initiatives, and examples of good practice. Finally, the focus is centered on the role of the major logistic nodes operators with special reference to seaports.

The complexity of components that constitute the apparently discrete areas of port and shipping environmental management and what makes the case for practicable integration of their implementation strategies through collaborative effort to mutual advantage of their commercial interests and the environment as a whole is presented in chapter 17. The global, strategic significance of the wide range of activities and operations facilitated through the shipping industry and port sector is established in terms of world trade, importance to the nation-state, and the increasingly significant, timely, and topical environmental imperative. With examples drawn from the United States and Europe, the major issues are identified and the various management response options are examined in terms of policy, environmental management systems, implementation, and case studies. The legislative and regulatory regimes are summarized; stakeholder pressure is identified by interest area, and the benefits of a proactive response to environmental liabilities and responsibilities are tabulated. The challenge of filling the gap between policy and actual delivery of continuous improvement of environmental quality is addressed through examination of the role and implementation of environmental management systems. Baseline and benchmark performance is established with reference to surveys of port sector achievements over the last 10 years of collaborative research and development of practicable tools and methodologies. Finally, an overview of options, instruments, and approaches for encouraging and facilitating further advances in the effective environmental management of port and shipping activities is set out for consideration.

2 Automated Container Terminal Concepts

Petros Ioannou and Hossein Jula

CONTENTS

ABSTRACT

In this chapter, we design, analyze, simulate, and evaluate several automated container terminal (ACT) concepts that have a strong potential for increasing the terminal capacity. In particular, the following four ACT concepts are considered and evaluated:

1. *The AGV-based ACT (AGV-ACT).* In this concept, we consider a terminal configuration that is similar to that of conventional terminals, but instead of using manually operated equipment we use automated guided vehicles (AGVs), to transfer containers within the yard, and automated cranes, for loading and unloading.
2. *The linear motor conveyance system (LMCS)-based ACT (LMCS-ACT).* This terminal is the same as the AGV-ACT one, with the exception that instead of the AGVs, automated shuttles driven by linear motors are used to transfer containers within the terminal. The shuttles play the role of AGVs, but unlike AGVs, their paths are fixed by the guide ways of the LMCS.
3. *The overhead grid rail (GR)-based ACT (GR-ACT).* This system is obtained by replacing the storage yard, in the general layout of the ACT, with a number of GR units. A GR unit or module consists of an overhead rail system with shuttles that travel over stacks of tightly packed containers in the yard, retrieve them, and carry them to the GR unit buffers or from the GR buffers to locations in the yard. AGVs are used to transfer containers between the GR buffers and the ship/truck/train interface buffer.
4. *The AS/RS-based ACT (AS/RS-ACT).* The GR units in the GR-ACT system are replaced by an Automated Storage/Retrieval System (AS/RS) that provides the same storage capacity as other ACT concepts. As in the GR-ACT system, AGVs are used to transfer containers between the AS/RS units and the ship/truck/train interface buffers. The AS/RS concept utilizes much less land than the other concepts as containers are stored in a high structure with multiple floors.

Each ACT concept is designed to meet a future expected demand. A cost analysis is then performed to compute the average cost to move a container through the terminal, which becomes a measure for comparison in addition to other performance characteristics.

2.1 INTRODUCTION

Globalization and economic boom have vastly increased the volume of commodity flows in all transport modes. Worldwide container trade is growing at a 9.5% annual rate, and the U.S. growth rate is around 6%. It is anticipated that the growth in containerized trade will continue as more and more cargo is transferred from break-bulk to containers.[1-3] By 2010, it is expected that 90% of all liner freight will be shipped in containers.[1-3] Every major port is expected to double and possibly triple its container traffic by 2020. The first generation of container ships had an average capacity of 1,700 twenty-foot equivalent units (TEUs), and in 1970 the industry predicted that ship capacity would peak at 3,250 TEUs. The largest container ships operating today have a capacity of about 8,000 TEUs, and there are predictions of 12,000 to 15,000 "megaships" in the future.[2] Larger-capacity ships have influenced shipping patterns, concentrating volumes on those locations with sufficiently deep channels and terminal capacity to serve large ships.

It is evident from the above discussion that container terminals, especially in metropolitan areas, have to make significant changes in order to keep pace with increasing demand under environmental, congestion, labor force, land costs, and other constraints. Adding terminal capacity by building additional facilities faces numerous constraints: land scarcity or high land cost is the dominant obstacle in metropolitan areas, whereas environmental concerns, community opposition, and others can be equally important. One important way of increasing terminal capacity is to make existing terminals more efficient by using advanced technologies and automation. Advances in information technology, telecommunications, data management, and computation tools and robotics, together with recent research in systems optimization and control, open the way for new approaches to container handling that includes partial or full automation of some or all container operations within the terminal. Automation removes a lot of randomness that is due to manual operations and human decision making and allows the use of optimization to maximize efficiency and increase capacity. Machines and computers do not get tired, have a predictable behavior, and do not suffer from the usual safety constraints human-operated machines do.

High-density automated container terminals are potential candidates for improving the performance of container terminals and meeting the challenges of the future in marine transportation. Recent advances in electronics, sensors, information technologies, and automation make the development of fully automated terminals technically feasible. The Port of Rotterdam is already operating a fully automated terminal using automated guided vehicles (AGVs) and automated yard cranes to handle containers, and the Port of Singapore, Thamesport of England, and the Port of Hamburg[1] experimented with similar ideas. Sea-Land at the Port of Hong Kong implemented the Grid Rail system referred to as the GRAIL,[15] designed by Sea-Land/August Design, Inc., a high-density manually operated terminal. A good account of the state of the art in the area of terminal operations can be found in several survey papers and books.[4–14]

In the United States, labor unions strongly oppose any type of automation that is viewed as a threat to current jobs, making it difficult, if at all possible, for terminals to advance in this area at the same speed as the European and other overseas counterparts. It is envisioned that competition in the global market will begin to put pressure on all sides involved to cooperate in order to improve productivity and reduce cost through the use of advanced technologies and automation.

In this chapter, we address the design, modeling, simulation, and evaluation of several automated container terminals. These include an ACT that employs automated guided vehicles (AGVs) and an ACT with a linear motor conveyance system (LMCS). The configuration of ACT is such that the storage area could be replaced with storage modules based on different concepts leaving the interfaces the same. This allows the comparison of different concepts for the same operational scenario. A concept based on an overhead grid rail (GR) system and one based on an Automated Storage/Retrieval System (AS/RS) are also considered and compared with the AGV-ACT and LMCS-ACT systems. Each ACT system is designed to meet the same demand based on future projections made by several ports on the expected size of ships and container volumes. A model is developed that is used to simulate all the operations of the ACT down to the finest detail of the characteristics of each piece of equipment. The model is exercised for each ACT system based on the same operational scenario, that is, based on the same incoming and outgoing traffic of containers at the interfaces. Performance criteria that include throughput in moves per hour per quay crane, throughput per acre, ship turnaround time, truck turnaround time, container dwell time, and idle rate of equipment are used to evaluate each system and make comparisons. A cost model is developed and used to calculate the average cost for moving a container through the ACT. The performance and cost criteria are used to compare the pros and cons of each ACT system and make recommendations.

2.2 AUTOMATED CONTAINER CONCEPTS: DESIGN, PERFORMANCE, AND COST CONSIDERATIONS

In the following subsections we present the general layout, design, performance, and cost considerations on which the design and evaluation of the proposed ACT concepts will be based.

2.2.1 ACT GENERAL LAYOUT

Figure 2.1 shows the general layout of the ACT considered in this chapter. The layout consists of interfaces of the gate, the train, and quay crane buffers with the storage yard. In the case of the AGV-ACT, the storage yard is a collection of stacks separated by roads where the containers are stacked and served by yard cranes. AGVs are used to transfer containers within the terminal and the storage yard. In the case of the LMCS-ACT the storage yard is the same as in the AGV-ACT system. The only difference is that shuttles driven on a linear motor conveyance system are used for the transport of containers. For the GR-ACT the storage yard in figure 2.1 is replaced with a number of GR units that provide the same storage capacity as in the other concepts under consideration. A GR unit consists of an overhead rail system with shuttles that can move above the storage yard and transfer containers between the GR buffers and the storage yard. AGVs in this case are used to transfer the containers between the GR buffers and the gate, train, and quay crane buffers. The AS/RS-ACT is similar to the GR-ACT system in the sense that the GR units forming the storage yard are replaced with AS/RS modules that form a high-rise structure for

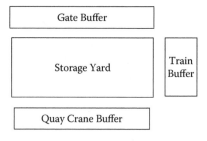

FIGURE 2.1 General layout of an automated container terminal.

container storage. As in the case of the GR-ACT system, AGVs are used to transfer containers between the AS/RS buffers and the gate/train/quay crane buffers.

The gate buffer is designed to interface between the manual operations (inland side) and the automated ones (internal terminal side). It provides a physical separation between the manual and automated operations for safety reasons and also for efficiency. It helps reduce the turnaround time for trucks by providing a temporary storage area for the export containers. The train buffer is the area next to the train where loading and unloading between the AGVs and the train take place.

The size of the ACT and its characteristics, such as storage capacity, number of lanes at gate, number of berths, number of quay cranes, and so on, depend on many parameters. These include the expected volume of containers the ACT has to process per day, the ship/truck/train arrival rates, and the volume of containers they carry. These considerations together with the type of equipment that is available could be used to specify and design the components of the ACT system in general and for each concept in particular.

2.2.2 DESIGN CONSIDERATIONS

The design characteristics for each ACT are based on future projections regarding expected demand.

Design Consideration 1. It is assumed that the ACT serves megaships capable of carrying 8,000 TEUs. The ships arrive every 24 hours 85% loaded and should be served in less than 24 hours. In our design, we assume a desired ship turnaround time of about 16 hours. This design consideration is adopted according to the plan for the Port of Rotterdam, the northwest terminal.[2,16]

Design Consideration 2. It is assumed that trucks deliver/pickup 60% of containers while the other 40% are carried by rail. These figures are based on the estimated projection at the Port of Los Angeles and Port of Long Beach for the year 2020.[17]

Design Consideration 3. It is assumed that the export container arrival pattern relative to the ship they are bound to is 0.2, 0.5, and 0.3. In other words, 20% of containers arrive during the second day before the ship arrives, 50% arrive during the first day before the ship arrives, and 30% arrive the same day and early enough to be loaded while the ship is at the berth. The arrival pattern is adopted according to Evers and Boonstra[16] based on the arrival pattern at the Port of Rotterdam.

It should be noted that the expected export container arrival patterns at many terminals can be determined from prior experience with land transport (road and rail) carriers. These patterns, however, are found to vary considerably from one port to another. Taleb-Ibrahimi[18] and De Castilho[19] present various container arrival and retrieval patterns.

Design Consideration 4. It is assumed that the import containers are retrieved during 3 days, with retrieval rates 0.5, 0.3, and 0.2, meaning 50% of the containers are taken away by trucks and trains during the day the ship was served, 30% the second day, and 20% during the third day. Out of the 50% of the containers that are taken away the same day, 30% are taken away directly without any intermediate storage and 20% are temporarily stored in the yard before being taken away. Similarly, the retrieval pattern is adopted according to Evers and Boonstra.[16]

Design Consideration 5. It is assumed that the trucks and trains at the ACT will operate in cycles of 24 hours. Although trucks operate in cycles of less than 24 hours in today's ports, there is a trend to increase the time to close to 24 hours in order to meet the demand and avoid traffic delays in the inland transportation system. This could be proven crucial in metropolitan areas where highways and surface streets during peak areas are highly congested.

2.2.3 CHARACTERISTICS OF THE ACT SYSTEM

The above design considerations are used to determine the characteristics of the ACT system including the storage capacity, number of berths, number of yard and quay cranes, and number of lanes at the gate. These characteristics are derived and specified in Liu et al.[20] and are summarized in tables 2.1 to 2.4.

In order to come up with the number of equipment that is needed to meet the demand, we assume a certain pattern of incoming and outgoing flow of containers. This operational scenario will be used to simulate and evaluate different ACT systems and is summarized below. All containers are assumed to be 40-foot containers.

2.2.4 PERFORMANCE CRITERIA

A container terminal is a complex system that involves a variety of entities (e.g., berths, cranes, gates, etc.) and various constituencies that play a role in operations of combinations of these entities (e.g., the port, the shipping line, the truckers, etc.). This complexity limits the possibility that a single measure of productivity can reflect the efficiency of all operations and their interaction in a meaningful way. Therefore, *measures* for marine container terminal productivity are required. A list of these measures is provided in National Research Council.[21] The most often used

---

TABLE 2.1
Summary of the General Characteristics of ACT Systems

Storage capacity of ACT	Around 22,000 TEUs
Berth and quay cranes	1 berth with 5 cranes. The average speed of each crane for combined loading/unloading is 42 moves per hour with 15% variance.
Number of lanes for inbound gate	9 (24-hour operation, processing time 3 minutes per truck)
Number of lanes for outbound gate	6 (24-hour operation, processing time 2 minutes per truck)
Number of yard cranes at gate buffer	6 with an average speed of 34 moves per hour per crane with variance of 10% of the average value
Number of yard cranes at train buffer	2 with an average speed of 28.3 moves per hour per crane with variance of 10% of the average value

TABLE 2.2

Arrival and Departure Rates of Containers

Ship arrival rate	One ship every 24 hours to be unloaded and loaded with 6,800 TEUs in less than 24 hours (desired ship turnaround time 16 hours)
Container arrival/departure rate by trucks	85 containers per hour
Container arrival/departure rate by trains	56.67 containers per hour

measure of performance of loading/unloading equipment is the average cycle time expressed in moves per hour. Moves per hour can be used either to evaluate the performance of single loading/unloading equipment or to evaluate the productivity of the terminal. Since the throughput of a terminal cannot exceed the best quay crane performance, a good measure of the *terminal throughput* is the average number of moves per hour per quay crane. A terminal can maintain a high throughput but could be utilizing a lot of land. To include the amount of land used in our measurement, we use the *throughput per acre*, or throughput measured in moves per hour per quay crane per acre. In many ports, such as Port of Long Beach, a similar measure defined as the number of processed TEUs per acre per year is often used. The time a ship spends at the berth for the purpose of loading or unloading is referred to as the *ship turnaround time*. The ship turnaround time is well recognized as an important factor in the overall transportation cost of containers, and its reduction to a minimum possible is one of the main priorities for shippers and terminal operators. This is easy to understand given that modern container ships may cost around $30,000 to $40,000 to operate per day.[22] In our design considerations for the ACT systems we chose a desired ship turnaround time of 16 hours. Since in practice the actual ship turnaround time may vary due to randomness of the performance characteristics of equipment, and so on, the ship turnaround time may be different from the desired one. Therefore, the ship turnaround time is another good measure for evaluating the performance of the proposed ACT systems. The typical external truck cost used by the trucking industry is $75 for each hour the truck is in use. This cost includes maintenance and labor.[23,24] The ability of the terminal to serve the trucks in short time will translate to cost reduction for the truckers and will make the terminal more attractive to do business with. Therefore, another useful measure of performance is the average time a truck spends in the terminal for the loading/unloading process

TABLE 2.3

Number of Export Containers, Bound for One Ship, Arrived by Trucks and Trains

Container Arrival Times	2nd Day Before Ship Arrival	1st Day Before Ship Arrival	Same Day for Direct Loading
Number of containers arrived by trucks	816 TEUs	2,040 TEUs	1,224 TEUs
Number of containers arrived by trains	544 TEUs	1,360 TEUs	816 TEUs

TABLE 2.4

Number of Import Containers, Unloaded from One Ship and Retrieved by Trucks and Trains

	Same Day while the Ship is at Berth (Direct Transfer)	Same Day after Ship Left the Berth	1st Day after Ship Departed	2nd Day after Ship Departed
Number of containers retrieved by trucks	1,224 TEUs	816 TEUs	1,224 TEUs	816 TEUs
Number of containers retrieved by trains	816 TEUs	544 TEUs	816 TEUs	544 TEUs

and the waiting time in queues to be processed by the gate. This time is referred to as the *truck turnaround time* and does not include the actual processing time at the gates. A secondary measure that affects the truck turnaround time is the *gate utilization* expressed in percentage of time the gate spends serving the incoming and outgoing container traffic. A low gate utilization for certain arrival and departure container rates shows that the gate is underutilized and could meet the demand with less number of lanes and people. On the other hand, if the gate utilization is high (close to 100%), then small changes in the container rates might cause congestion at the gate that may propagate into the terminal. The time that a container stays in the terminal before being taken away is referred to as the *container dwell time*. A high container dwell time could affect the transportation cost and the time to reach its destination in an adverse way. In addition, a high dwell time raises the required storage capacity of the yard since containers stay longer in the yard before being taken away. An efficient terminal would keep the dwell time as low as possible.

A cost-effective terminal is the one that keeps the amount of equipment to the minimum possible that is necessary to meet the expected demand. Since demand may vary with time, a good measure as to how effectively the equipment is utilized is the *idle rate of the equipment.* This is measured in the percentage of time the equipment is idle. Low idle rates indicate an efficient utilization of the equipment, whereas higher idle rates indicate that the equipment is underutilized. Underutilization may suggest design changes, reduction of the number of machines used, or improvement of the management of operations, and so on, in order to save costs and improve productivity. Table 2.5 summarizes the performance criteria that are used in this chapter to evaluate and compare different ACT systems.

2.2.5 COST MODEL

The average cost per container (ACC) processed through a terminal is among the most important cost measures considered by port authorities.[22] Though ACC does not express pricing, revenues, or terminal profits, it provides a basis for economic evaluation of container terminal operations. In this chapter, we adopted this measure in order to evaluate and compare the cost associated with each proposed ACT system. The ACC of a terminal is defined as

TABLE 2.5
Performance Criteria

Throughput	The number of moves per hour per quay crane
Throughput per acre	The throughput per acre
Annual throughput per acre	Number of TEUs processed per acre per year
Ship turnaround time	The time it takes for the ship to get loaded/unloaded in hours
Truck turnaround time	The average time it takes for the truck to enter the gate, get served, and exit the gate minus the actual processing time at the gate
Gate utilization	Percent of time the gate is serving the incoming and outgoing container traffic
Container dwell time	Average time a container spends in the container terminal before being taken away from the terminal
Idle rate of equipment	Percent of time the equipment is idle

$$ACC = \frac{C_{lct} + C_{lnd} + C_{eqp} + C_{lbr}}{V} \tag{2.1}$$

where V is the total annual number of containers that are processed by the terminal and C_{lct}, C_{lnd}, C_{eqp}, and C_{lbr} are costs of locations, land, equipment, and labor, respectively. These costs are briefly discussed below. Interested readers are referred to Ioannou et al.[25] for in-depth discussions.

1. **Cost of locations:** This is the cost of locations where activities (operations) take place. This includes buildings and facilities such as gates, berth, storage yard, customs, maintenance area, and central controller.

   The total annual cost of locations can be further classified into *fixed* and *variable* costs. The variable cost is mainly due to consumption cost of electricity, while the fixed cost is mainly due to capital investment. The life of capital investment is assumed to be 25 years except for the central controller, whose life is 10 years. The total investment for a location is depreciated within this period and is calculated based on a straight-line depreciation method.[26] Other fixed costs are assumed to be 3% for repair, 1% for insurance, and 10% for interest per year. More precisely, the fixed cost per year for locations is calculated by

$$\text{Fixed Cost} = \frac{\text{investment}}{\text{accounting life}} + \text{investment} \times (\text{repair} + \text{insurance} + \text{interest})$$

$$\tag{2.2}$$

2. **Cost of land:** This includes the capital investment for land in different areas, for example, berth area, storage area, and so forth. The land cost is considered to be investment only. It is calculated based on the area of each part (in acre) multiplied by the land cost per acre. The inflation rate is

assumed to be 5% per year, and the interest rate 10% annually. Based on the above assumptions, the annual land cost can be calculated as follows:[27]

$$A = P \times R \times \left(\frac{(1+R)^n}{(1+R)^n - 1} \right)$$

where P is the initial land investment, R is the inflation rate, and n is the accounting life, which is assumed to be 25 years. The total annual land cost is then computed as

$$\text{Total Annual Land Cost} = P \times IR + A$$

where IR is the average (over 25 years) annual interest rate that represents lost investment opportunity. In our cost model, IR is taken to be equal to 10%.

3. **Cost of equipment:** This is the cost of yard equipment (e.g., yard cranes), quay cranes, AGVs, management infrastructure (software/hardware system), and so forth. The cost of equipment depends on the type of the ACT system under consideration.

    Similar to the cost of locations, this cost can be classified into *fixed* and *variable* costs. The cost associated with energy consumption by each piece of equipment is considered to be the variable cost, while the capital investment to purchase each piece of equipment is the fixed cost. We assume that the total investment for the equipment is depreciated over a period of 15 years and is calculated based on the straight-line depreciation method similar to number 2. Other fixed costs parameters assumed are 10% for repair, 1% for insurance, and 10% for interest per equipment per year. The total cost value for equipment is calculated by adding the fixed costs and variable costs of all equipment.

4. **Cost of labor:** This is the total cost of labor, which is calculated for all employees at the facility for the hours they are physically present (scheduled to work) at the terminal. The employee's regular working week is assumed to be 40 hours/week (2,080 hours/year). The employees may get paid overtime, if they are scheduled to work more than a shift a day. The overtime pay is assumed to be 1.5 times the base pay.

2.2.6 SIMULATION/COST MODEL VALIDATION

The performance of the proposed ACT systems is studied using microscopic simulation models developed using the software packages Matlab, Simulink, and Stateflow developed by MathWorks, Inc. The simulation models simulate the characteristics and movements of every piece of equipment and vehicle in the terminal in detail as well as their interactions with each other and with the incoming/outgoing traffic. The models are time based and can be used to simulate different yard configurations and characteristics, different operating scenarios, different strategies and optimization techniques for cargo handling, and so forth.

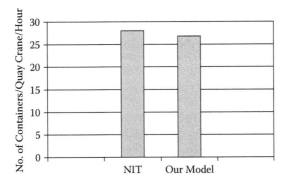

FIGURE 2.2 A comparison of a simulated throughput and actual one measured at NIT terminal.

The developed simulation model is validated using real data from Norfolk International Terminal (NIT) presented in Larsen and Moses[28] and Towery et al.[29] We use the characteristics of NIT and its equipment presented in Larsen and Moses[28] and Towery et al.[29] as inputs to the simulation model. We then exercise the model for the same operational scenario and compare the throughput generated by the model with that obtained from real data. The results are shown in figure 2.2. The throughput generated by the simulation model is 26.8 moves per hour per quay crane versus the measured one of 28 moves per hour per quay crane, a difference of –4.3%, which is a good approximation.

To validate our cost model, we use the developed model to calculate the ACC value for terminal operations where vehicles and equipment are manually operated. We use the simulation model described above to calculate the number of vehicles and cranes with the same characteristics as those in NIT that generate the same throughput as the NIT. Given the throughput of each quay crane and assuming that 5 quay cranes are working in parallel, 16 hours a day (2 working shifts), 365 days a year, the total projected annual volume of the yard is equal to 1,635,000 TEUs. These values, together with all the other data generated by the simulation model, are fed into the cost model. The results of the cost model are obtained based on the 2000 figures and are summarized in table 2.6.[25]

TABLE 2.6
Cost Data for Manual Operations

Annual projected volume	1,635,200 TEUs
Annual variable cost	$25,371,000
Annual fixed cost	$22,571,000
Annual land cost	$7,930,000
Annual labor cost	$61,602,000
Total annual cost	$117,475,000
Cost per container	$143.7

The calculated ACC value for the hypothetical terminal that has a throughput similar to that of many current conventional terminals is found to be $143.7. This cost value is found to be within the range of values reported in the literature for current terminal operations.[23]

2.3   ACT USING AGVS

An automated guided vehicle (AGV) is driven by an automatic control system that serves the role of the driver. The AGV system consists of the vehicle, onboard controller, management system, communication system, and navigation system. Due largely to the repetitive nature of movements within the container terminals, AGVs are very well suited to be deployed for terminal operations. The promise of deploying AGVs in container terminals lies in their capability of achieving high container throughput, round-the-clock continuous operation, and reduced operational costs.

In this section, we consider the AGV-based ACT (AGV-ACT) system proposed in Liu et al.[20] Figure 2.3 shows the basic configuration of the proposed AGV-ACT system.

In this concept, AGVs are used to transfer containers between the gate, train, and quay crane buffers and the storage yard. More precisely, the following three tasks, as shown in figure 2.4, are performed by AGVs:

Task 1: Containers are transferred between the quay crane and gate buffers/storage area/train buffers.

Task 2: Containers are transferred between the gate buffer and the storage area.

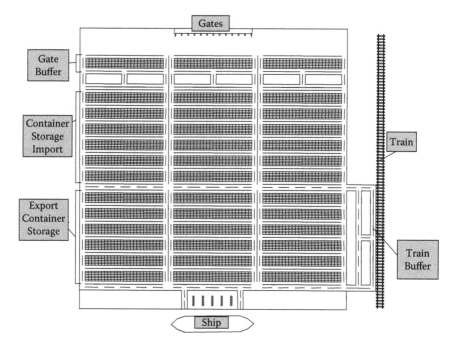

FIGURE 2.3 AGV-based automated container terminal (AGV-ACT) layout.

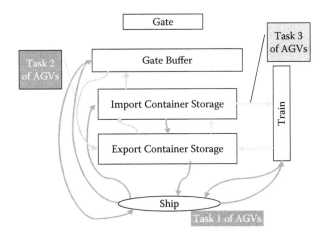

FIGURE 2.4 Different tasks assigned to AGVs.

Task 3: Containers are transferred between the train buffer and the storage
area.

The main characteristics of the AGV-ACT system are the same as those of the
general ACT described earlier and summarized in table 2.1. Specific to the AGV-
ACT system are the number of yard cranes needed in the storage yard and the number of AGVs to perform the various tasks in order to meet the expected container
volume as described in section 2.2.3. Table 2.7 summarizes the characteristics of the
AGV-ACT system according to the findings in Ioannou et al.[25]

TABLE 2.7
AGV-ACT: Summary of the Physical Characteristics of the Terminal

Size of the terminal	1,633 × 1,875 ft² (70.29 acres)
Storage capacity	22,464 TEUs
No. of berths	1
Capacity of quay cranes	42 moves per hour (combined loading and unloading)
No. of quay cranes	5
Gates service time	3 minutes inbound gate, 2 minutes outbound gate
No. of gate lanes	9 inbound, 6 outbound
Capacity of yard cranes at buffers	Yard crane's speed is 5 mph, takes 15 seconds to line up with the container, and an average time of 65 seconds to unload/load an AGV
No. of yard cranes at gate buffer	6
No. of yard cranes at train buffer	2
Capacity of yard cranes at storage yard	Yard crane's speed is 5 mph, takes 15 seconds to line up with the container, and an average time of 45 seconds to unload or load an AGV
No. of yard cranes at import and export storage yard	36
Speed of AGVs	10 mph for empty, 5 mph for loaded AGVs
No. of AGVs	85 (48 for task 1, 26 for task 2, 6 for task 3, plus 5 spare)

TABLE 2.8
AGV-ACT: Performance Results for 1-Day (24-hour) Simulation

Ship turnaround time	16.81 hours
Throughput	40.45 containers/quay crane/hour
Throughput per acre	0.576 containers/quay crane/acre/hour
Annual throughput per acre	35,310 TEUs/acre/year
Gate utilization	66.03%
Truck turnaround time (does not include time at the gate)	126.75 seconds
Throughput (train crane)	29.42 containers/hour/crane
Throughput (buffer crane)	33.7 containers/hour/crane
Idle rate of AGVs over 24 hours	36.3%
Idle rate of yard cranes over 24 hours	70.2%
Idle rate of buffer cranes over 24 hours	12.7%
Idle rate of train cranes over 24 hours	23.0%
Idle rate of quay cranes over 24 hours	31.7%
Container dwell time	19.1 hours

In our simulation model a variance of 10% is assumed in all values associated with speeds and time with the exception of the speed of the quay cranes, where a variance of 15% is assumed.

2.3.1 PERFORMANCE ANALYSIS

The characteristics of the AGV-ACT system are used as inputs to the simulation model together with the arrival/departure patterns of containers brought in and taken out by ships/trucks/trains, as shown in tables 2.2 to 2.4. We assume that the patterns of container arrivals and departures to or from the terminal by ship, trucks, and train are repeated every 24 hours so that a 24-hour simulation was sufficient to make projections about annual productivity. This assumption may not be valid today due to the randomness that exists in the system. The use of automation and information technologies, however, coupled with optimum dispatching and scheduling techniques will lead to scenarios that are very close to the assumed one. The results of a 1-day (24-hour) simulation are shown in table 2.8.

The ship turnaround time obtained from the simulations is 16.81 hours, which is close to the desired 16 hours. We should note that for a maximum speed of forty-two moves per hour per crane, the best possible ship turnaround time is 16.2 hours. The difference between the simulated and best possible ship turnaround time is mainly due to the variance introduced in simulations for the characteristics of the quay cranes and other equipment.

It should be noted that the idle rate of the cranes is calculated over a period of 24 hours. Since the ship was at the berth for only 16.81 hours, the quay cranes were idled for 24 − 16.81 = 7.19 hours, which is 30.0% of the time, close to the 31.73% obtained from simulations, indicating that while the ship was at the berth the quay cranes were operating very close to maximum capacity. Similarly, after the ship is served, the AGVs responsible for the task of serving the ship will be idle until the next ship arrives about 7 hours later. This accounts for most of the 36.3% idle rate for the AGVs.

TABLE 2.9
AGV-ACT: Cost Results

Annual projected volume	2,482,000 TEUs
Annual variable cost	$28,408,000
Annual fixed cost	$39,046,000
Annual land cost	$7,930,000
Annual labor cost	$20,113,000
Total annual cost	$95,498,000
Average cost per container (ACC)	$77.0

The throughput of the terminal is close to the maximum possible, indicating that the AGVs met the service demand imposed by the quay cranes' speeds.

The idle rate of the yard cranes was found to be high. This is due to two reasons: First, reshuffling has not been considered in our simulation, which implies that the cranes had it easy—something that may not be true in a real situation. However, one could argue that the use of automation could improve yard planning to the point that the number of reshufflings or unproductive moves is negligible. The second reason, which was stated earlier, is that the number of yard cranes has not been optimized. Instead, one yard crane was assumed for each stack in order to simplify the operations and the control logic of the AGVs. A smaller number of yard cranes could be used to achieve the same throughput. This can be achieved by having yard cranes serving two stacks instead of one. In such case, the crane may have to cross roads used by the AGVs, and therefore their motion relative to that of the AGVs has to be controlled and synchronized in order to avoid collisions, delays, and deadlocks. Another way is to change the configuration of the stacks so that a single crane can serve more than one stack without crossing roads used by the AGVs. These possibilities have not been explored in this study.

2.3.2 Cost Analysis

The simulation results obtained in section 2.3.1 together with the characteristics of the terminal listed in table 2.7 are used to calculate the average cost of moving a container through the terminal, that is, the ACC value, by exercising the cost model for the AGV-ACT system. In addition to these data the model is fed with several other parameters and data that are necessary for the operation of the terminal. These include number of people, salaries, and cost data regarding equipment, land, facilities, and so forth. Most of the cost data are collected from the open literature[22,23] and modified after discussions with experts in the field such as terminal operators.[30–32] Table 2.9 summarizes the results of the cost model based on the data obtained in the year 2000.

2.4 ACT USING LMCS

Linear motor conveyance systems (LMCS) are among the technologies that have recently been considered for cargo handling. The promise of employing linear motor technologies lies in its very high positioning accuracy, reliability, and robustness of equipment. In this section, we focus on the ACT that uses LMCS to transfer

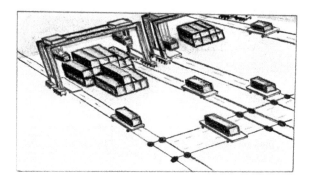

FIGURE 2.5 Transfer of containers in a yard using LMCS.

containers between the gate, train, and quay crane buffers and the storage yard. Figure 2.5 shows part of a conceptual container yard using LMCS.

We considered the LMCS yard layout to be identical to that of the AGV-ACT system in figure 2.3. The only difference is that in the LMCS-ACT all paths are pre-built guide ways. For instance, a road in the AGV-ACT system becomes a guide way track that allows shuttles to travel in opposite directions. In other words, in figure 2.3 the AGVs are replaced with shuttles that are moving on the linear motors conveyance system. Thus, the shuttles can be considered as AGVs moving on a fixed path.

The characteristics of equipment used for the LMCS-ACT system to meet the demand are the same as those of the general ACT described in section 2.2.3. The characteristics and the number of yard cranes are the same as in the AGV-ACT system. The speeds of empty shuttles and loaded shuttles are assumed to be the same as those in AGVS given in table 2.7. We assumed that at each corner of the guide way, it takes 5 seconds for the shuttle to change its direction of movement. Despite this change, the number of shuttles needed to meet the demand was calculated to be the same as the number of AGVs used in the AGV-ACT system. Table 2.10 summarizes the characteristics of the LMCS-ACT system.

2.4.1 PERFORMANCE ANALYSIS

A simulation model for the LMCS-ACT system is developed and used to simulate the terminal based on the operation scenario given in section 2.2.3. The results are shown in table 2.11.

Since the terminal yard layout, control logic of vehicles, speed of vehicles, and characteristics of the yard equipment are exactly the same for both AGV-ACT and LMCS-ACT systems, the performance of the two terminals is almost identical. The difference is that AGVs are moving freely in the yard, while LMCS shuttles are traveling on fixed guide paths.

2.4.2 COST ANALYSIS

The main difference between the AGV-ACT and LMCS-ACT systems is in the cost, as the cost of implementing the LMCS-ACT concept is much higher than that of the

TABLE 2.10

LMCS-ACT: Summary of the Physical Characteristics of the Terminal

Size of the terminal	$1,633 \times 1,875$ ft^2 (70.29 acres)
Storage capacity	22,464 TEUs
No. of berths	1
Capacity of quay cranes	42 moves per hour (combined loading and unloading)
No. of quay cranes	5
Gates service time	3 minutes inbound gate, 2 minutes outbound gate
No. of gate lanes	9 inbound, 6 outbound
Capacity of yard cranes at buffers	Yard crane's speed is 5 mph, takes 15 seconds to line up with the container, and an average time of 65 seconds to unload/load an AGV
No. of yard cranes at gate buffers	6
No. of yard cranes at train buffer	2
Capacity of yard cranes at storage yard	Yard crane's speed is 5 mph, takes 15 seconds to line up with the container, and an average time of 45 seconds to unload/load an AGV
No. of yard cranes at import and export storage yard	36
Speed of shuttles	10 mph empty, 5 mph loaded
No. of shuttles	82 (48 for task 1, 26 for task 2, 6 for task 3, plus 2 spare)

AGV-ACT. The cost data were estimated and obtained from Dougherty[30] as no such system of the scale under consideration is in operation. The results obtained from cost model and cost analysis for the LMCS-ACT system, based on the 2000 figures, are summarized in table 2.12.

TABLE 2.11

LMCS-ACT: Performance Results for 1-Day (24-hour) Simulation

Ship turnaround time	16.83 hours
Throughput	40.40 containers/quay crane/hour
Throughput per acre	0.575 containers/quay crane/acre/hour
Annual throughput per acre	35,310 TEUs/acre/year
Gate utilization	66.03%
Truck turnaround time (does not include time at the gate)	126.8 seconds
Throughput (train crane)	29.42 containers/hour/crane
Throughput (buffer crane)	33.7 containers/hour/crane
Idle rate of shuttles over 24 hours	36.2%
Idle rate of yard cranes over 24 hours	70.2%
Idle rate of buffer cranes over 24 hours	12.7%
Idle rate of train cranes over 24 hours	23.0%
Idle rate of quay cranes over 24 hours	31.8%
Container dwell time	19.1 hours

TABLE 2.12
LMCS-ACT: Cost Results

Annual projected volume	2,482,000 TEUs
Annual variable cost	$30,008,000
Annual fixed cost	$124,486,000
Annual land cost	$7,930,000
Annual labor cost	$20,113,000
Total annual cost	$182,539,000
Average cost per container (ACC)	$147.1

2.5   ACT USING GR SYSTEM

The concept of loading and unloading containers in the yard using overhead rail and shuttles is another attractive way of utilizing yard space more efficiently. Figure 2.6 shows an example of this concept, known as GRAIL.[15] It uses linear induction motors, located on overhead shuttles that move along a monorail above the terminal. The containers are stacked beneath the monorail and can be accessed and brought to the ship as needed.

The concept of loading and unloading containers in the yard using overhead rail and shuttles was the subject of an in-depth study in Ioannou et al.[15] We note that the grid rail (GR) system is modular in the sense that it is a collection of grid rail units. In Ioannou et al.[15] each unit is optimized by using a new dispatching algorithm for assigning shuttles to containers within the unit. The developed GR systems in Ioannou et al.[15] are used in Liu et al.[20] to propose an ACT using GR. The GR-ACT system is shown in figure 2.7 and is similar to that of the AGV-ACT system, with the only difference that the storage yard is replaced with eight GR units. The use of several GR units instead of a large one is done for robustness and reliability purposes as well as for simplifying the operations, as explained in Ioannou et al.[15] This number of units is chosen so that the storage capacity of the GR-ACT system meets the required level.

FIGURE 2.6 An example of a GRAIL system (courtesy of Horizon Lines, Inc.).

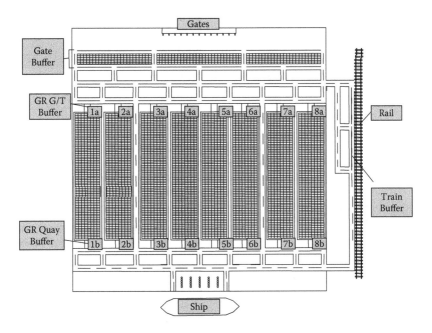

FIGURE 2.7 The GR automated container terminal.

The tasks to be performed by the AGVs in the GR-ACT system are the same as task 1 to task 3 discussed in section 2.3. The only difference, as shown in figure 2.8, is that the GR units are replacing the storage yard.

Table 2.13 summarizes the characteristics of the GR-ACT designed in Liu et al.[20] It should be pointed out that due to the high density of the GR units, less land is needed to obtain the same storage capacity as in the AGV-ACT and LMCS-ACT systems.

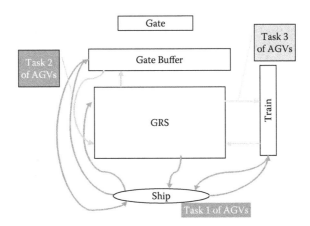

FIGURE 2.8 The tasks assigned to AGVs in the GR-ACT system.

TABLE 2.13

GR-ACT: Summary of the Physical Characteristics of the Terminal

Size of the terminal	$1{,}472 \times 1{,}875$ ft² (63.36 acres)
Storage capacity	22,464 TEUs
No. of berths	1
Capacity of quay cranes	42 moves per hour
No. of quay cranes	5
Gates service time	3 minutes inbound gate, 2 minutes outbound gate
No. of lanes at the gate	9 inbound, 6 outbound
Capacity of yard cranes at buffers	Yard crane's speed is 5 mph, takes 15 seconds to line up with the container, and an average time of 65 seconds to unload/load an AGV
No. of yard cranes at gate buffer	6
No. of yard cranes at train buffer	2
Average service time for loading and unloading an AGV at the GR buffers	30 seconds
No. of shuttles in each GR unit	15
No. of GR units	8
Speed of AGVs	5 mph
No. of AGVs	72 (42 for task 1, 21 for task 2, 6 for task 3, plus 3 spare AGVs)

2.5.1 PERFORMANCE ANALYSIS

The characteristics of the GR-ACT system given in table 2.13 together with those for each GR unit developed in Ioannou et al.[15] are fed into the simulation model for the GR-ACT system and simulated for the operational scenario described in section 2.2.3. In this simulation, we assume that each GR unit performed as designed in the sense that in the case of outgoing containers from the GR units, the GR buffers were always ready to deliver a container to an AGV, and in the case of incoming containers to the GR units, the GR buffer was always ready to receive a container. This property of the GR units was made possible in Ioannou et al.[15] by choosing an optimum number of shuttles and using a new dispatching algorithm to assign containers to shuttles within the unit and control their motion. The results of the simulation are shown in table 2.14.

The simulation results indicate that the GR-ACT system performs efficiently by having the quay cranes operate close to maximum capacity and keeping the ship turn-around time close to the desired one. Similarly, the yard cranes at the train and gate buffer worked close to maximum capacity. The idle rate of the quay cranes is over a 24-hour period. This means that 31.38% of the time the quay cranes were idle because the ship was not at the berth. The same goes for the AGVs dealing with task 1.

2.5.2 COST ANALYSIS

The performance characteristics of the GR-ACT system together with cost data are used as input to the cost model that calculates the ACC value for the

TABLE 2.14
GR-ACT: Performance Results for 1-Day (24-hour) Simulation

Ship turnaround time	16.47 hours
Throughput	41.68 containers/quay crane/hour
Throughput per acre	0.652 containers/quay crane/acre/hour
Annual throughput per acre	39,173 TEUs/acre/year
Gate utilization	65.7%
Truck turnaround time (does not include time at the gate)	120 seconds
Throughput (train crane)	28.6 containers/hour/crane
Throughput (buffer crane)	36.7 containers/hour/crane
Idle rate of AGVs over 24 hours	31.8%
Idle rate of buffer cranes over 24 hours	10.8%
Idle rate of train cranes over 24 hours	31.9%
Idle rate of quay cranes over 24 hours	31.8%
Container dwell time	19.0 hours

proposed system. The cost data generated by the cost model of the GR-ACT system, based on the 2000 figures, are summarized in table 2.15.

2.6 ACT USING AS/RS

Shown in figure 2.9 is an Automated Storage/Retrieval System (AS/RS) module with four major components: the storage and retrieval machine (SRM), the rack structure, the horizontal material handling system, and the planning and control system. The SRM simultaneously moves horizontally and vertically to reach a certain location in the rack structure. It travels on floor-mounted rails guided by electrical signals. In each AS/RS module, served by a single SRM, six rack structures are built to hold and store containers. The SRM is designed to move from one set of two racks to another within the module. Each module has two buffers, one on each side. Each buffer has two slots, one for outgoing containers to be picked up by AGVs and one for incoming containers brought in by AGVs. (See Ioannou et al.[25,33] and Khoshnevis and Asef-Vaziri[34] for more details.)

TABLE 2.15
GR-ACT: Cost Results

Annual projected volume	2,482,000
Annual variable cost	$36,152,000
Annual fixed cost	$47,880,000
Annual land cost	$7,338,000
Annual labor cost	$20,000,000
Total annual cost	$111,370,000
Average cost per container (ACC)	$89.7

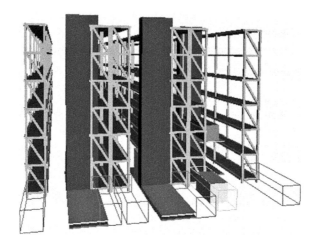

FIGURE 2.9 Automated Storage/Retrieval System (AS/RS) module.

The promise of the high productivity of the AS/RS lies in its capability to permit access to any container within the storage structure randomly (random access), without having to reshuffle containers. This high-productivity property together with the ability to have a high-storage capacity makes the AS/RS concept very attractive in places where land is very limited or costly.

The control logic that dictates the motion of the AGVs within the AS/RS-ACT system is exactly the same as in the case of the GR-ACT system. Similarly, the tasks performed by the AGVs are the same as indicated in figure 2.8.

TABLE 2.16

AS/RS-ACT: Summary of the Physical Characteristics of the Terminal

Size of the terminal	$1,265 \times 1,875$ ft^2 (54.45 acres)
Storage capacity	23,328 TEUs
No. of berth	1
Capacity of quay cranes	42 moves per hour
No. of quay cranes	5
Gates service time	3 minutes inbound gate, 2 minutes outbound gate
No. of lanes at the gate	9 inbound, 6 outbound
Capacity of yard cranes at buffers	Yard crane's speed is 5 mph, takes 15 seconds to line up with the container, and an average time of 65 seconds to unload/load an AGV
No. of yard cranes at gate buffer	6
No. of yard cranes at train buffer	2
Average service time at AS/RS buffers	45 seconds
No. of AS/RS modules	15
Average service time for SRM	110 seconds (double move), 80 seconds (single move)
Speed of AGVs	5 mph
No. of AGVs	58 (36 for task 1, 14 for task 2, 5 for task 3, plus 3 spare)

TABLE 2.17
AS/RS-ACT: Performance Results for 1-Day (24-hour) Simulation

Ship turnaround time	16.24 hours
Throughput	41.7 containers/quay crane/hour
Throughput per acre	0.767 containers/quay crane/acre/hour
Annual throughput per acre	45,583 TEUs/acre/year
Gate utilization	66.4%
Truck turnaround time (does not include time at the gate)	110.75 seconds
Throughput (train crane)	30.6 containers/hour/crane
Throughput (buffer crane)	38.32 containers/hour/crane
Idle rate of AGVs over 24 hours	30.9%
Idle rate of buffer cranes over 24 hours	6.8%
Idle rate of train cranes over 24 hours	27.86%
Idle rate of quay cranes over 24 hours	32.33%
Container dwell time	18.9 hours

Table 2.16 summarizes the characteristics of the AS/RS-ACT system used in the simulation model.

2.6.1 PERFORMANCE ANALYSIS

The characteristics of the AS/RS-ACT system summarized in table 2.16 are fed into the simulation model, which was then exercised for the operational scenario presented in section 2.2.3. The results of the simulation are shown in table 2.17.

The performance of the AS/RS-ACT system is comparable with that obtained with the other concepts. The throughput per acre, however, is higher due to the system's requirement for less land.

2.6.2 COST ANALYSIS

The performance characteristics of the AS/RS-ACT system generated by the simulation model as well as cost data specific to the AS/RS structure are used to perform a cost analysis. The results of the cost analysis are summarized in table 2.18.[25]

TABLE 2.18
AS/RS-ACT: Cost Results

Annual projected volume	2,482,000 TEUs
Annual variable cost	$25,806,000
Annual fixed cost	$82,427,000
Annual land cost	$6,576,000
Annual labor cost	$11,718,000
Total annual cost	$126,528,000
Average cost per container (ACC)	$101.96

2.7 SUMMARY OF SIMULATED CONCEPTS

The performance results for each proposed ACT system in sections 2.3 through 2.6 are summarized in table 2.19. Figure 2.10 shows the layout of the AS/RS automated container system.

Since the amount of equipment and number of vehicles in each ACT system are chosen so that the ACT system can meet the same demand, it is not surprising that the performance for each system is almost identical for all measures with the exception of the throughput per acre. The highest throughput per acre was obtained for the AS/RS-ACT system since it requires less land to be implemented for the same storage capacity. Next comes the GR-ACT system, which also requires less land for the same storage capacity. All the ACT systems operated close to the maximum possible capacity of the quay cranes, which was assumed to be forty-two moves per hour per crane for combined loading/unloading. This is much higher than the average of about twenty-eight moves per hour measured in many of today's conventional terminals. The simulation model when exercised for a hypothetical conventional terminal of the same layout as the ACT with characteristics of equipment and operations based on data from an actual terminal generated a throughput of about twenty-seven moves per hour per quay crane, which is very close to the value of twenty-eight that was actually measured in the terminal from where the data were obtained.

The significant difference between the various systems is the average cost per container. The LMCS-ACT was found to be the most expensive due to the high infrastructure cost associated with the LMCS. The second most expensive system is the

TABLE 2.19
Performance and Cost Results

	AGV-ACT	LMCS-ACT	GR-ACT	AS/RS-ACT
Ship turnaround time (hours)	16.81	16.83	16.47	16.24
Throughput, while the ship is at berth (moves/quay crane/hour)	40.45	40.40	41.68	41.7
Throughput per acre, while the ship is at berth (moves/quay crane/acre/hour)	0.579	0.575	0.652	0.767
Annual throughput per acre (TEUs/acre/year)	35,310	35,310	39,173	45,583
Gate utilization	65.7%	66.03%	65.7%	66.4%
Truck turnaround time (seconds)	127	127	120	110.75
Throughput (train crane) (moves/hour/crane)	29.4	29.4	28.6	30.6
Throughput (buffer crane) (moves/hour/crane)	33.7	33.7	35.7	38.32
Idle rate of AGVs over 24 hours	36.3%	36.2%	31.81%	30.9%
Idle rate of gate buffer cranes over 24 hours	12.7%	12.7%	10.8%	6.8%
Idle rate of train cranes over 24 hours	23.0%	23.0%	31.9%	27.86%
Idle rate of quay cranes over 24 hours	31.7%	31.8%	31.8%	32.33%
Container dwell time (hours)	19.1	19.1	19	18.9
Average cost per container (U.S.$)	77.0	147.4	89.7	102.0

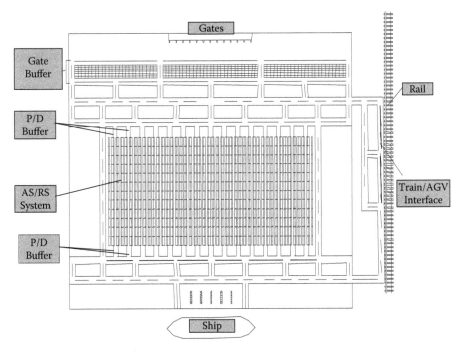

FIGURE 2.10 Automated terminal yard layout using AS/RS.

AS/RS-ACT, due to the infrastructure cost of the AS/RS structure. The AGV-ACT system was found to be the most cost effective, followed by the GR-ACT. As the cost of land increases, however, our model shows that after a certain land cost the AS/RS-ACT becomes more attractive.

REFERENCES

1. Ioannou, P. A., Kosmatopoulos, E. B., Jula, H., Collinge, A., Liu, C.-I., Asef-Vaziri, A., and Dougherty, E. 2000. *Cargo handling technologies*. CCDoTT technical report, Center for Advanced Transportation Technologies, University of Southern California.
2. Ryan, N. K. 1998. The future of maritime facility designs and operations. In *Proceedings of the 1998 Simulation Conference*, Washington, DC: IEEE Press, 1223–27.
3. Haveman, J. D., and Hummels, D. 2004. *California's global gateways: Trends and issues*. San Francisco: Public Policy Institute of California.
4. Preussag Noell GmbH. 1998. A fully automated container yard: Linear motor based transfer technology. http://www.noell.de/actuell_L/Pros-km.
5. Gould, L. 1996. AGVs: A bigger hit in other places—by far. *Modern Materials Handling*, April, 13.
6. HHLA Container Terminal Altenwerder. 1999. Latest technologies and simulation development for optimizing container terminal layout, logistics and equipment. Paper presented at IIR Conferences.
7. Traffic jam at the dock. 1996. *Traffic World* 28:27–28.

8. Personal communication with Captain Thomas Lombard from American Presidents Lines (APL).

9. Gunther, H.-O., and Kim, K. H., eds. 2005. *Container terminals and automated transport systems: Logistics control issues and quantitative decision support.* Berlin: Springer-Verlag.

10. Duinkerken, M. B., Evers, J. J. M., and Ottjes, J. A. 2002. Improving quay transport on automated container terminals. In *Proceedings of IASTED International Conference on Applied Simulation and Modelling (ASM 2002)* ed. L. Uberti. Calgary, Canada: Acta Press.

11. Evers, J. J. M., and Koppers, S. A. J. 1996. Automated guided vehicle traffic control at a container terminal. *Transportation Research A* 30:21–34.

12. Qiu, L., Hsu, W. J., Huang, S. Y., and Wang, H. 2002. Scheduling and routing algorithms for AGVs: A survey. *International Journal of Production Research* 40:745–60.

13. Vis, I. F. A., and Harika, I. 2004. Comparison of vehicle types at an automated container terminal. *OR Spectrum* 26:117–43.

14. Vis, I. F. A., and de Koster, R. Transshipment of containers at a container terminal: An overview. *European Journal of Operational Research* 147:1–16.

15. Ioannou, P., Kosmatopoulos, E. B., Vukadinovic, K., Liu, C. I., Pourmohammadi, H., and Dougherty, E. 2000. *Real time testing and verification of loading and unloading algorithms using grid rail (GR).* CCDoTT technical report, Center for Advanced Transportation Technologies, University of Southern California.

16. Evers, J. J. M., and Boonstra, H. 1996. Coaster express: An option for large-scale coastal container feedering. In *Proceedings of the Third European Research Roundtable Conference on Shortsea Shipping*, ed. F. A. J. Waals, 185–205. Bergen, Norway: The Netherlands: Delft University Press.

17. Meyer, Mohaddes Associates, Inc. 1996. *Gateway cities trucking study.* Gateway Cities Council of Governments, Southeast Los Angeles County.

18. Taleb-Ibrahimi, M. 1989. Modeling and analysis of container handling in ports. Ph.D. thesis, Department of Industrial Engineering and Operation Research, University of California, Berkeley.

19. De Castilho, B. 1992. High-container terminals: Technical and economic analysis of a new direct-transfer system. Ph.D. thesis, Department of Civil Engineering, University of California, Berkeley.

20. Liu, C. I., Jula, H., and Ioannou, P. 2002. Design, simulation, and evaluation of automated container terminals. *IEEE Transactions on Intelligent Transportation Systems* 3:12–26.

21. National Research Council, Committee on Productivity of Marine Terminals. 1986. *Improving productivity in U.S. marine terminals.* Washington, DC: National Academy Press.

22. Toth, T. 1999. Analysis of a simulated container port. MS thesis, University of Delaware.

23. Jones, E. G. 1996. Managing containers in marine terminals: An application of intelligent transportation systems technology to intermodal freight transportation. Ph.D. thesis, University of Texas at Austin.

24. Frankel, E. G. 1987. *Port planning and development.* New York: Wiley Publication.

25. Ioannou, P., Jula, H., Liu, C. I., Vukadinovic, K., Pourmohammadi, H., and Dougherty, E. 2001. *Advanced material handling: Automated guided vehicles in agile ports.* CCDoTT technical report, Center for Advanced Transportation Technologies, University of Southern California.

26. Sullivan, W., and Wicks, E. 2000. *Engineering economy.* Englewood Cliffs, NJ: Prentice Hall.

27. Park, C. S., and Park, M. 1997. *Contemporary engineering economics.* Reading, MA: Addison Wesley Publication Company.
28. Larsen, R., and Moses, J. 1998. AVCS for ports: An automation study for Norfolk International Terminals.
29. Towery, S. A., et al. 1996. *Planning for maximum efficiency at Norfolk International Terminals.* JWD report, AAPA, Tampa, FL.
30. Personal communication with Ed Dougherty, August Design, Inc., 2000.
31. Personal communication with Peter Ford from Sea-Land Mearsk, Port of Long Beach.
32. Personal communication with Mr. Philip Wright from Hanjin terminal, Port of Long Beach.
33. Ioannou, P., Kosmatopoulos, E. B., Jula, H., Collinge, A., Liu, C. I., Asef-Vaziri, A., and Dougherty, E. 2000. *Cargo handling technologies.* CCDoTT technical report, Center for Advanced Transportation Technologies, University of Southern California.
34. Khoshnevis, B., and Asef-Vaziri, A., 2000. *3D virtual and physical simulation of automated container terminal and analysis of impact on in land transportation.* METRANS technical report, University of Southern California.

3 Automated Container Terminals

Lessons Learned for Future Successes

Edmond Dougherty

CONTENTS

ABSTRACT

Since marine containers were first introduced by McLean in the 1950s, improving the efficiency of container terminals has captured the imagination of many, from top-level corporate management to stevedores, from PhDs in systems engineering to home brew inventors. Thousands of ideas big and small have been proposed to automate container terminals, but to date few have been fully implemented successfully. Why? Automation technologies that have been perfected in high-production manufacturing plants provide the basis of many of the automated containerized terminal concepts. Automated guided vehicles, machine vision, linear induction motors, crane stabilization systems, optical character recognition, radio frequency (RF) ID tags, artificial intelligence, stereovision, laser guidance, autonomous overhead crane systems, and

Stewart platforms are all examples of technologies that have been successfully used in manufacturing plants and hold promise for automated container terminals. However, many technologies that work well in structured environments have found difficulty when moving from the controlled environments of the factory. In an efficient automated factory the temperature, humidity, and lighting are typically well controlled. Positioning and timing of product movement within an automated factory can be precise and crisp, allowing computers to control material handing devices effectively. In a container terminal, automated systems are exposed to sunlight, rain, snow, and temperature extremes. Containers, truckers, and vessels arrive to and depart from a terminal on a schedule, but the schedule is far from precise, often unpredictable. Container movements within a terminal are often chaotic, responding to the positioning of other cargo and the asynchronous arrival of vessels and trains, and depending on the availability of container handling equipment.

But technology is often not the issue. Perceived need, funding, technical risk versus operational reward, and timing have been the key factors regarding terminal automation.

This chapter will introduce a number of proposed container automation concepts to examine why they were or were not implemented. It is hoped that the lessons learned from past successes and failures will lead to many more successful automated container terminals in the future.

3.1   INTRODUCTION

The worldwide container shipping industry is an amazing machine. Massive ships, imposing trucks, and miles of long intermodal trains move vast fortunes around the world on a daily basis in hundreds of thousands of containers. It is all performed with machinelike precision and efficiency. However, the universal corporate need to minimize transportation costs while maximizing profits drives the shipping industry to totally eliminate any flaws and maximize benefits.

Because of the need for this robotic-like perfection, new ideas are always welcomed in the container terminal industry. But to be successfully implemented they must be affordable, not disrupt existing operations, have low technical risks, and have significant measurable rewards. The ideas can be big or small, and they will be considered, but it is a difficult task to move from being considered to being adopted in busy container terminals.

3.2   IDEAS, BIG AND SMALL

Ideas in themselves are wonderful. They can fill our imaginations and raise our hopes and spirits. But in the world of container terminals poetry and patents have little value unless they can help move real-world cargo efficiently. Small or big, ideas can only help if they result in actual operating systems.

The phrases "small idea" and "big idea" will be used often in the paragraphs below, not to describe the physical size or even value of a concept, but rather its cost and scope. A small idea in this context is one that can be developed and tested at fairly low cost with little risk of disrupting existing operations. A big idea is one that could provide great rewards, but is much more costly, and has a much higher risk of negatively disrupting existing operations.

From a corporate standpoint a big idea is one that will make your company the industry leader, or at least make the company the one to watch. From a personal standpoint, a big idea is one that could make or break an entire career. You get it right, and you will be widely recognized as a visionary who can positively change the industry. You get it wrong, and you will probably need to switch careers.

An example of a small idea is a vision system that can provide guidance to hostler drivers to help them align chassis quickly and accurately under the hook of a ship crane.[1] It might cut just a few seconds per container loading cycle. But those few seconds add up to large savings on an annual basis. A rule of thumb often used in the industry is if a quay crane can increase its average container moves per hour by a single container, that crane will bring in an additional million dollars of revenue per year to the terminal. As with any rule of thumb it is not exact, but it emphasizes the fact that in this industry seconds saved can add up to millions earned. In addition, small ideas can normally be implemented under a small budget, with small risk, and small, if any, installation disruption to normal operations. As a result, small ideas are usually welcomed into terminals for at least a trail test, and often remain in place as permanent installations.

In contrast to small ideas, big ideas often have a difficult time getting to the stage where they are actually in operation in a terminal. For example, a big idea such as a totally automated container warehousing system that requires a very small footprint, increases throughput, reduces operating costs, reduces damage, and reduces trucker dwell time would certainly be *considered* by corporate management. However, it will only be seriously considered if it is affordable, if it does not disrupt the terminal during construction of the automation, and if it has been proven to work somewhere else. Those are very big "ifs." But they make sense. An existing container terminal that is already working to its capacity would have great difficulty shutting down part of the terminal for new construction, no matter how promising the new automated system may be. No one in such a low-profit-margin, high-capital-cost business really wants to be the first to try out a new technology—especially if the technology costs tens and maybe hundreds of millions of dollars.

3.3 HOW CAN THE BIG IDEAS BECOME REAL?

So if they can make or break careers, and affect the positioning of entire companies within an industry, how can big ideas for terminal automation ever be put into practice?

If the terminal is new and starts from a green field, there is no issue of disruption of existing operations other than drawing resources from other operations. Quite often a new terminal wants to be perceived as moving the industry forward with advanced technology, so a big idea has a better chance in new construction versus renovating an existing working terminal.

But for an unproven, first-time system, there are still the issues of cost and technical risk—both major obstacles. Of course, though rare, there are situations where the high cost and high technical risk are acceptable to management, such as the automated guided vehicles and automated cranes at Shimizu Container Terminal and the ECT terminal in Rotterdam.[2] In such cases, such big-idea systems are developed, installed, and put into operation.

But what of the majority of situations where cost, risk, and even temporary disruption of service are not acceptable options? How can big ideas be implemented?

The answer is to start small and, based on success, grow larger. Divide the big idea into smaller ones and prove the small ideas work. Eventually the big idea can be implemented based on the proven small ideas.

3.4   BIG IDEA THAT WORKED

For example, though the following cannot literally be considered a terminal automation project, it provides good lessons. In the 1980s Sea-Land Services and their partners ETC in Hong Kong realized that since a significant part of their business was LTL (less-than-full trailer load) cargo, a large consolidated freight station (CFS) on port would be of great benefit. However, they did not want to give up the cargo footprint at the port to have that additional warehouse space for loading and unloading LTL containers. They hypothesized that a CFS "on stilts" above the stacked container yard would give them the best of all worlds. While the idea was seen as very worthwhile and low technical risk, it would be costly to implement and could be disruptive to normal operations at one of the busiest ports in the world. So they went "small" in a number of ways.

First, they set about creating a number of simulations to look at the idea and its benefits without the cost and disruption associated with actually building it. Greatly encouraged by the results of the simulation, they then looked at dividing the implementation into acceptable phases.

In this situation, "acceptable phases" meant the capital costs to implement each phase was considered a reasonable investment for the duration of the phase, and the disruption to the normal operations was also considered acceptable. The disruption was considered acceptable because, based on the results of the simulations, they were able to increase the density of the terminal areas that were not going to be affected by construction, by converting parked areas to stacked areas, and maintain production during the construction phases.

But the most important part of going small was that each phase would result in a workable CFS section, so that when the first phase was completed it could go into operation on its own, bringing in revenue even though the entire system was not yet built. In this way the company also had the option of checking the operation and productivity of the concept without building the other phases. As it turns out, it worked very well, and not only were several additional phases completed in the same piecewise manner, but most of the terminals in Hong Kong adopted the same concept. Today, the Hong Kong port looks like a group of incredibly large CFS warehouses.

A lesson to be learned is that very large projects can be often be divided into smaller projects to reduce cost, disruption, and risk. But more importantly, each smaller project, or phase, should ideally be able to provide benefit to the terminal on its own. This will allow the terminal to examine the real-life benefits and problems with the technology before fully committing to the full project. If any phase of the project were implemented and did not provide the full benefits expected, the future phases need not be built.

In the following sections a number of automated terminal concepts that have been proposed will be discussed, in no particular order. Most have never been implemented. Though it is somewhat opinion based, some of the reasons the concepts were never fully implemented will be discussed.

3.5 GRAIL AUTOMATED OVERHEAD AUTOMATED CONTAINER TERMINAL

One of the best ideas never fully implemented was Sea-Land's GRAIL concept (figure 3.1) first conceived in the 1980s (U.S. Patent 5,511,923). The GRAIL (Grid Rail) concept used overhead shuttles that move above the terminal container stacks.[3] The stacked containers can be accessed and brought to the ship as needed via the overhead rails. Inexpensive switching mechanisms at the end of each rail allowed the shuttles to move from one rail to the next. Elevated automated platforms below the quay cranes allowed efficient access to the shuttles and allowed the quay cranes to pick/place cargo without worrying about the exact order of shuttle arrival. Moving at top speeds of 15 mph, several computer simulations and physical models demonstrated the viability of the system from a cost and operation standpoint.

The primary issues with the GRAIL were the cost of the steel in the overhead structure, the concern about operating costs, and, in spite of the working-scale model and computer simulations, the fact that the GRAIL was the first of its kind, raised the perception that it had a large technical and operational risk.

Recognizing the reluctance of upper management to take on such technical risks for such a big idea, the development team developed a much smaller GRAIL-like system concept that could be built upon. The concept was much less expensive and had much less technical risk, but would require the development of some of the same subsystems as would be used in the GRAIL, so that any development would be directly applicable to the full GRAIL system. The variation would have allowed a

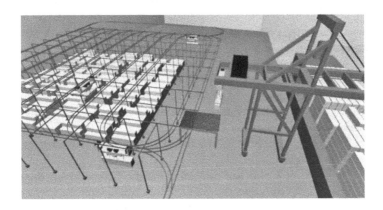

FIGURE 3.1 GRAIL concept (U.S. Patent 4,973,219).

significant amount of cargo storage to be located away from the quay, onto much less expensive land on the other side of the port highway. In this case, the GRAIL would have simply acted as an efficient overhead shuttle that would avoid highway traffic by moving cargo overhead, above the highway, shuttling between the main storage terminal and the supplemental storage terminal. This would likely have provided a great benefit to the port, and would have provided a means to prove the technology at much lower cost.

However, after having spent the previous few years convincing upper management that the full GRAIL concept was the goal, project management was reluctant to propose the much simpler system, even if it were seen as a stepping stone for the full system. And the full GRAIL system was seen as too expensive and too risky to upper management.

So, ironically, smaller portions of a GRAIL-like system were approved for development and implemented in Sea-Land's Hong Kong terminal. The basic system was a clever matrix-stacked storage system, installed underneath the multilevel CFS mentioned previously. The CFS was elevated so that the containers could be stacked beneath it. While the cranes/shuttles could move from track to track, they did not use the GRAIL switches, but rather more conventional switches that were operationally proven but much slower. Also, the installed minisystem did not include a shuttle track spur to the ship, and as a result, the cargo had to be delivered to hostlers/chassis to move the cargo out of the matrix storage system. Because of this, many of the predicted benefits of the system could not be fully realized.

One of the major development costs for the GRAIL was the creation of a safe, automated control system. As it turned out, though, the partial GRAIL system installed in Hong Kong was automatic, the cost of labor at the terminal was so low that the costly automatic controls were turned off within a few months, and the system was operated manually. If it were determined sooner that the full automation was not needed, the development costs could have been much less, the technical risk would have been seen as much lower, and perhaps the full GRAIL system would have been implemented.

What are the lessons learned? The development team recognized that the big idea of GRAIL needed to be made smaller; however, the timing was such that the concept of running the system over the highway to an inland terminal was not as attractive as placing the system under the elevated CFS. So the lesson is that even if the big idea is broken down into smaller ideas and implemented, they need to be the right small ideas that can highlight the benefits of the big idea.

In the case of the GRAIL, many millions were committed to the CFS, and having a GRAIL-like functionality as part of the CFS was seen as a relatively low-risk beneficial side project. The GRAIL-like system under the CFS did provide operational benefits, but the automation was turned off soon after the implementation and did not really highlight the benefits of a full GRAIL system like the over-the-highway system may have.

So the main problem was just bad timing. If the CFS in Hong Kong was not going to be built, it is likely that the simpler, very useful over-the-highway system would have been proposed and implemented. It would have provided a useful operating tool for the port and also laid the foundation for acceptance of the full GRAIL.

3.6 SPEEDPORT: THE RAIL EXTENDS ABOVE THE SHIP

Once described as "GRAIL on steroids," Speedport (figure 3.2) is a concept put forth by ACTA Maritime Development Corporation.[4] It takes the GRAIL concept further by extending the rails over the ship. In Speedport, straddle carrier-like devices, called spiders, bring cargo alongside the ship. The spiders along with their cargo attach themselves to overhead rails and move over the deck of the ship. In the concept, the hoisting mechanism in the spiders is able to pick and place cargo from the ship. Once the spiders reach the other side of the ship, they lower themselves to the ground where they are able to travel to the yard to pick up the next container.

Though much interest was expressed in the Speedport concept, the system was never built. Again, the main problem is the fact that the big-idea system is very expensive, has never been proven in a real-world application, and has not found a client for a small-idea version of the system. Therefore, though people can see the value in the system, they are reluctant to accept the combination of high cost and perception of high risk.

The technical issues with Speedport are concerns regarding its ability to accommodate various size and styles of vessel as well as the fact that the spiders must carry enough cable to reach deep in the cell of large container ships. The cables and winches would be substantial and, though not impossible, would be a great engineering feat to efficiently move such well-equipped agile spiders throughout the terminal. If coupled with the Cell Elevator (described below), the spider could be much lighter because it would only need to reach the deck rather than the full depth of the ship to access the first placed in the hold.

3.7 HIGH-RISE AUTOMATED CONTAINER WAREHOUSING SYSTEMS

The GRAIL and Speedport concepts involve storing containers in vertical stacks and accessing them from above. Such systems can be very efficient if each of the vertical columns/stacks of containers holds containers with essentially the same parameters (e.g., similar weight, same destination, same size). This is not difficult to do if the maximum stack height is limited to four or five containers. However, if the footprint of the storage system needs to be much smaller, the stack height would need to be

FIGURE 3.2 Speedport.

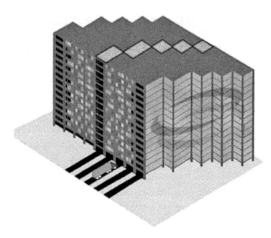

FIGURE 3.3 Computainer system.

higher and the throughput may drop due to additional juggling of the containers in the stacks, especially with inbound containers that would normally be called for by the customer by container number rather than parameter. To solve this issue, a number of concepts have been proposed that permit near-random access of the containers.

For example, two high-rise, or "pigeonhole," storage and retrieval systems have been proposed by Computainer Systems International[5] and Krupp (figure 3.4). Though at first glance each system looks similar, the detailed operations and techniques employed create their own capabilities and issues.

Through private investment and some government funding, Computainer Systems International has built a full-scale prototype of the handling system with a small number of storage bins. According to the Computainer website (www.computainer-systems.com), "The design and structure of the Computainer System enables fully

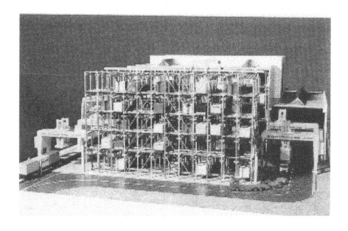

FIGURE 3.4 Krupp Fast Handling System.

loaded 20 and 40 foot containers to be stored and sorted vertically to a minimum of 10 and a maximum of 14 rows high. This allows a port to dramatically increase the number of containers it can sort and store per acre per year."

A variation has been proposed by Krupp. Krupp is a manufacturer of marine cranes, mineral processing, materials handling, and mining. It is located in Germany with offices throughout the world. The Krupp Fast Handling System is an automated system concept[6] that can be used in rail and marine terminals. Like Computainer, the Krupp system includes a multistory bin storage system. Krupp has a portion of the system installed as a prototype at the terminal in Duisburg-Rheinhausen.

A number of similar high-rise systems have been proposed as well as patents such as Robotic Container Handling Company's 5,511,923, Hagenzieker's 6,902,368, and Dobner's 6,698,990.

Though Computainer built a small-scale system, and thus showed how the large idea could be broken into a small idea, because of the high capital costs involved, the automated multistory system only shows its full benefit if it can store thousands of containers on a small parcel of land. Therefore, it cannot easily be shown that it is a concept that can start small and grow larger. So all high-rise automated warehouse systems are seen as expensive and high risk.

3.8   CELL ELEVATOR: A CONCEPT THAT CAN SUPPLEMENT OTHER AUTOMATIONS

The Cell Elevator (U.S. Patent 6,572,319) is intended to supplement existing quay cranes as well as such automated concepts as GRAIL and Speedport. The purpose of the device would be to raise the top container in each cell to deck level where the crane can easily access the container. Once the crane removes the container, the Cell Elevator would retrieve the next container and raise it above deck. The system would also work in reverse; that is, the crane operator could place a container above deck and the mechanism would stow the container below deck. The Cell Elevator would be located at the port or travel with the ship. The quay crane (or a Speedport spider or a GRAIL shuttle) would use its normal spreader bar to pick up the Cell Elevator and place it about a container cell, where it would be able to raise and lower containers into and out of the ship cell. The effect is that the cargo can be moved much more quickly and the task of the crane operator is much simplified. The system also lends itself well to operating with fully automated cranes.

It was estimated that a cell elevator would cost approximately the same as a top-of-the-line spreader bar. The design was such that it could be produced using the spreader bar manufacturing process, and so it could easily be a new product for a spreader bar manufacturer who could sell the product to its existing clients. The Cell Elevator, a relatively small idea, was fairly low cost, could produce large gains in productivity, and seemed to fit perfectly into the product line of spreader bar manufacturers.

Though a patent was issued, the concept was never developed beyond the idea phase. Why? Besides it being seen as a new concept (technical risk), the main reason it was never built is, again, a very common though unpredictable item—bad business timing. August Design, Inc., the company who created the Cell Elevator concept, was in the process of merging with another company, and the new business entity was

FIGURE 3.5 Cell Elevator.

about to focus on software rather than hardware development. In addition, the key potential licensee of the technology, a major spreader bar company, was also going through a number of worldwide corporate changes. Because of these two business timing factors, the Cell Elevator was placed on hold and, to date, never reactivated.

3.9   CARGO SENTRY: 100% CONTAINER INSPECTION

A similar situation occurred with another spreader-bar-related concept, the Cargo Sentry, U.S. Patent 6,998,617. The patent describes a means for a container inspection instrumentation package to be attached to and carried by any spreader bar. Though this device does not increase quay crane throughput at a port, it is intended to permit inspection of each container touched by the crane's spreader bar without slowing down the process. In the post–9/11 world, there is great concern that illicit materials could be hidden inside a container and brought into a country illegally. The Cargo Sentry (figure 3.6), equipped with the appropriate sensors, would prevent illicit cargo from ever being loaded onto the ship, as the container would be screened while being handled by the crane. If illicit materials were detected, the container would not be placed on the ship, but rather placed aside for further inspection.

Though, as mentioned, the Cargo Sentry does not increase productivity, it permits a means to perform 100% container inspection without reducing throughput, potentially solving a problem that has been expressed by many U.S. government officials.[7–9] To date the system has not yet been developed as a product, due to lack of

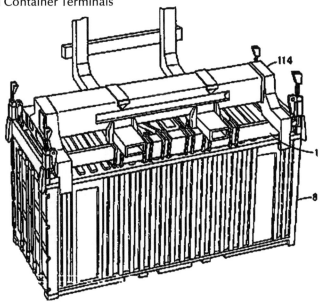

FIGURE 3.6 Cargo Sentry.

funding. The company who proposed the concept, Advent, Inc., primarily produces software, and the Cargo Sentry is a hardware system. It is unlikely that the U.S. ports would fund the Cargo Sentry because, though it would protect the United States and its ports, its primary use would be in foreign ports preventing cargo from being loaded onto ships bound for the United States. As a result, a potentially valuable concept remains only a concept.

3.10 DARTS: DIRECT ACQUISITION RAIL TO SHIP SPREADER

While the previous two spreader-bar-related technologies (Cell Elevator and Cargo Sentry) have not yet been built, a third system called the Direct Acquisition Rail to Ship (DARTS) spreader was designed, built, tested, and is available for commercial purchase.[10] The objective of the DARTS spreader (figure 3.7) was to produce an efficient means to move shipping containers from rail cars to a ship without the complications of intermediate handling. For outbound containers arriving on rail, the DARTS spreader eliminates the time- and labor-consuming tasks of moving containers from railroad cars to chassis to storage to the quay crane.

This system is very useful in ports that have rail on dock. Using DARTS, railcar containers can be rapidly picked and placed directly by a quay crane quickly. In addition to normal operations, the DARTS spreader can move parallel to the vessel and railcars in order to position the cargo accurately relative to the location of the railcar beneath the crane. DARTS can also simultaneously handle two 20-foot containers even when the containers are separated by distances up to 3 feet. The spreader portion of DARTS is available from Bromma.

Why was the DARTS concept brought to the point that it is a commercial product while the Cargo Sentry and Cell Elevator are still at the concept stage? In the

FIGURE 3.7 DARTS spreader.

case of the DARTS spreader, the Army and Navy identified the basic problem of rapidly moving cargo between ship and rail. Along with the Army and Navy, a team of LMI, August Design, CCDOTT, and Bromma were brought together to formulate a solution and build and test the concept. Bromma is a leading manufacturer of container spreaders. The Department of Defense (DOD) funded the initial effort, and at the end of the successful testing phase, Bromma made the solution available as a commercial product. So in the case of DARTS, a clear need was identified, a solution was proposed, funding was made available for development, intensive testing at a working port was conducted, and a leading manufacturer was involved in the process from the beginning.

The DOD played a major role in the development of the DARTS spreader as it clearly identified a need, provided funding, and conducted extensive field tests. The result was a commercially available product. However successful, the DARTS spreader was a relatively small idea, but being a small idea helped it become a commercial product.

3.11 AACTS: AUTOMATED ALL-WEATHER CARGO TRANSFER SYSTEM

The Automated All-Weather Cargo Transfer System (AACTS) is a big idea.[11] It had a similar foundation of DOD support as the DARTS spreader, but as a big idea it has still not reached commercial use.

The DOD originally identified a need to move containers from one vessel to another at sea and provided funding for concept development and prototyping. The AACTS concept was a large stabilized floating platform capable of loading and unloading two container ships simultaneously. While the full concept was seen as a good solution for at-sea container transfers, it was recognized that it was a big idea

FIGURE 3.8 AACTS wave tank testing.

that would take years and a great deal of funding to produce. In order to take positive, manageable steps toward reaching the final goal, it was decided to break the
concept down into a number of small ideas. It was determined that the most generally useful subsystem would be the AACTS crane with its six degrees of freedom
(DOF) spreader bar.

The AACTS crane (figure 3.8) is essentially a large-scale combination of a
SCARA robot, a rigid hoist, and a six DOF spreader bar[12] that can move in surge,
sway, heave, roll, pitch, and yaw.

Though radically different in design than existing quay cranes, it was believed that
the crane could find application in commercial ports because of its potential ability to
produce rapid cargo transfers, its fully automatic operation, and its operational flexibility due to its ability to reach multiple hatches, and its ability to move cargo directly
among railcars, chassis, and the ship. Coupled with the Cell Elevator, it was estimated
that the AACTS crane could nearly double existing quay crane throughput.

The lightweight SCARA arm is rigid horizontally, but articulated in two sections. It is able to rotate about the "shoulder" joint and "elbow" joint. The rigid
hoist transports the spreader bar vertically and eliminates swaying movement due
to cables. The six DOF spreader bar is essentially an inverted Stewart platform and
allows the spreader to move in a controlled fashion to follow the target container.
Since total automation was one of the goals of the AACTS, machine vision can be
employed to automatically track the moving target to guide the spreader into the
proper position to pick or place the container (U.S. Patent 5,943,476). In addition to
the automatic pick-and-place capability, the system also permits telerobotic control
to place the person in the control loop.

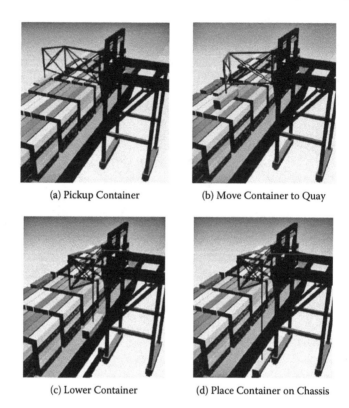

(a) Pickup Container   (b) Move Container to Quay

(c) Lower Container   (d) Place Container on Chassis

FIGURE 3.9 AACTS crane unload sequence.

To prove the concept in general and the technology in particular, a sixteenth-scale ACCTS crane was developed under DOD funding and very successfully tested in a wave tank. As a follow-up, a full-scale six DOF spreader bar was also developed and tested. As with the DARTS spreader development, Bromma, a leading manufacturer of spreader bars, was a team member in the design and fabrication of the spreader bar in the hope that it would become a commercial product.

While the total AACTS is clearly a big idea and the AACTS crane a small idea by comparison, building and testing a full-scale AACTS crane would cost about $10,000,000. So the AACTS crane is actually still a big idea. Since the crane was of such a radical design and was unproven in full scale, no ports or crane manufacturers were anxious to invest the millions needed to produce the first system. The DOD, though interested in seeing the crane developed, wanted a crane for offshore cargo handling and was not interested in funding a port crane (figure 3.9). Though the six DOF spreader bar was built and could be produced by Bromma, it had no crane to support it, so it had no customers.

In looking for lessons learned, the AACTS started out very well—the need was clear (military offshore container handling), initial funding was provided by the DOD, the system was divided into simpler components from a single big idea to a number of small ones, a leading manufacturer was part of the team, and the technology tested out—but a full-scale AACTS crane has not yet been built. Why? Besides the fact that the costs would be in the millions, the main flaw was taking technology that was to fit a specific need—offshore container handling—and trying to apply it to a different application—standard quay crane. While the AACTS crane would likely work well dockside in a commercial port, and could be a major improvement, it was still seen as a technical risk and not as a major need. The ports would love to have a same-cost replacement crane that could greatly improve productivity, but they are not willing to take the risk to be the first.

So while there was a desire to greatly improve commercial terminal operations, there was not a critical need to do so. Ports would rather make small improvements, at small cost with small technical risk. DOD's need was offshore, so it could not justify funding a land-based crane development. In retrospect, the focus should have remained on the needs of offshore applications, military or commercial.

3.12 CONCLUSION

Though timing can be an uncontrollable parameter that could work for or against terminal automation deployment, big ideas in terminal automation can be successful if the following conditions exist:

- A need is clearly identified.
- A funding source is clearly identified.
- The big idea can be reduced to a number of well-thought-out smaller ideas that can be the foundation of the big idea.
- The smaller ideas can independently provide significant benefits.
- The technical risk is commensurate with the rewards.

It should be understood that it is much easier to implement terminal automation big ideas in green field developments. It is also interesting to note that many of the terminal automation examples discussed in this chapter are still viable concepts that can be successfully implemented if the timing is right and the conditions described above are met.

REFERENCES

1. Wagner Associates. Container chassis positioning system. U.S. Patent 5,142,658 issued 1992.
2. Saanen, Y., et al. 2003. The design and assessment of next generation automated container terminals. Paper presented at Proceedings of the 15th European Simulation Symposium, Delft, The Netherlands.
3. Dougherty, E. J., and Bohlman, M. T. 1989. GRAIL: The container terminal of the future. In *Proceedings of the Fourteenth Ship Technology and Research Symposium*. Society of Naval Architects and Marine Engineers, New Orleans, LA.
4. Ioannou, P. A., et al. 2000. *Cargo handling technologies final report*, 66. Center for Commercial Deployment of Transportation Technologies, Long Beach, CA.

5. Computainer Systems, Inc. 1991. *Compact, high density storage of cargo containers*, WO/1991/013011. Invented B. J. Coatta, et al. World Intellectual Property Organization Publication, Vancouver, BC, Canada.

6. Woxenius, J. 1996. Development of new technologies for integrated transport chains in Europe. Paper presented at Trafikdage pa AUC '96, AAlbourg, Germany.

7. Menendez, R. Senate floor statement. 100 Percent Scanning Amendment H.R. 4954: The Port Security Bill. Wednesday, September 13, 2006, Washington, DC.

8. Cantwell, M., U.S. senator (D-WA). http://cantwell.senate.gov/news/record.cfm?id=262857&&days=30&, Washington, DC.

9. Port Security Act of 2006. Introduced by Senators Susan Collins (R-ME) and Patty Murray (D-WA) "to scan 100 percent of cargo."

10. Dougherty, E. J. 1999. *Direct Access Rail to Ship (DARTS): A new system for improving intermodal cargo transfers*. Marine Transport System R+D Conference, National Academy of Sciences, Washington, DC.

11. Dougherty, E. J. 1996. Automated All-Weather Cargo Transfer System (AACTS). *Mobility Times* 6:14–20.

12. Dougherty, E. J. 1998. Intelligent spreader bar for cargo handling. Paper presented at the second semiannual Mobile Offshore Base Technology Exchange Conference, Washington, DC.

4 Operational Issues in Modern Container Terminals

Kap Hwan Kim

CONTENTS

ABSTRACT

Because container vessels spend a significant amount of transportation time in ports, it is essential to improve the productivity of various handling activities at port container terminals. Further, in order to modernize container terminals, automated container handling facilities have been recently developed and installed. This trend has introduced various new handling facilities and opened up new research topics for the efficient operation of these terminals. This chapter discusses the related decision-making problems that need to be explored by researchers.

4.1   INTRODUCTION

Container terminals have played an important role in global manufacturing and businesses as multimodal interfaces between sea and land transport. Over the last 30 years, the marine container industry has grown dramatically. In order to increase the benefits of economies of scale, the size of containerships has steadily increased during the last decade. With increasing containerization, the number of container terminals has rapidly increased, and the competition among them for becoming a hub port has become severe. Owing to the request for a higher level of services by vessel carriers, issues pertaining to container terminal operations have recently gained the attention of the academic community. By installing automated facilities and utilizing advanced information technologies, many container terminals are attempting to increase their throughput and decrease the turnaround times of vessels and customers' trucks.

Depending on the circumstances under which a container terminal is constructed, the type of container handling facility varies. In North America, a large area of land is available for stacking the containers; however, in Asian countries, the yard space available is usually a bottleneck when attempting to increase the throughput of a container terminal. Thus, stacking facilities used in North America and European countries are for lower tiers of stacks, while containers in Asian countries are stacked in higher stacks. Because handling characteristics of different stacking facilities are different from each other, the characteristics of operational rules for these facilities would also be different.

In most existing container terminals, computers are employed to plan and control various handling operations. Because a container terminal is a complicated system with various interrelated components, operators or planners are required to make several complicated decisions. Further, since computer systems can store a large amount of data and analyze it in a short interval of time, they have been utilized to assist human experts during decision-making processes. This chapter introduces various handling systems in container terminals and decision-making issues for the efficient operation of container terminals. With regard to decision-making models, there are four similar review papers on this topic (Meersmans and Dekker, 2001; Steenken et al., 2004; Vis and de Koster, 2003; Kim, 2005).

The next section introduces the handling facilities and operational procedure in container terminals, and section 4.3 describes the operation planning problems that exist in container terminals. Section 4.4 introduces real-time decision-making problems. The last section provides concluding remarks.

4.2 OPERATIONS AND HANDLING FACILITIES IN PORT CONTAINER TERMINALS

There are many different types of quay cranes (QCs), such as single-trolley QC, double-trolley QC, dual-trolley QC, and dual-cycle elevator conveyor QC. The QCs that were developed after the single-trolley QC have the purpose of reducing the cycle time of the unloading and loading operations. The double-trolley QC has two trolleys with the same function. The water-side trolley of the dual-trolley QC delivers containers between the vessel and the platform of the QC, which is located between the two legs of the crane, while the land-side trolley delivers containers between the platform and transporters on the ground. The dual-cycle elevator conveyor QC has an elevator to deliver containers between the water-side and land-side trolleys. The part of the QCs that directly grasps a container is called a spreader. A twin-lift spreader can lift up to two 20-foot containers at a time, while a tandem spreader can lift up to four 20-foot containers or two 40-foot containers at the same time.

There are various types of yard cranes (YCs): transfer crane (TC), rail-mounted gantry crane (RMGC), automatic stacking crane (ASC), dual RMGC (DRMGC), and overhead bridge crane (OHBC). Yard blocks can be classified into two categories according to the positions where YCs transfer containers to or from transporters (transfer position): in the first, transfer positions are at the end of the blocks; in the second, they are at the sides of each block. Because in the yard layout of the former case blocks are usually laid out perpendicular to the direction of the berth, the layout is called the perpendicular layout, while the layout of the latter case is called the parallel layout. In East Asian countries, the parallel layout is usually applied, while in European countries, the perpendicular layout is more popular. In the parallel layout, YCs can move from a yard block to another, while in the perpendicular layout, YCs cannot move between blocks. Thus, in the parallel layout, more YCs can be installed additionally even after the terminal starts its operation. However, in the perpendicular layout, it is important to accurately estimate the productivity of YCs to be installed in a block as well as the throughput requirement of the block; further, it must also be checked whether the capacity of the selected YCs satisfies the throughput requirement.

There are various types of transporters such as yard truck (YT), straddle carrier (SC), multiload yard truck, automated guided vehicle (AGV), shuttle carrier, reach stacker, and forklift. YTs are the most popular transporters, and they are currently used in combination with YCs in many Asian countries. SCs have been used in many European countries, and they are used not only for transporting containers between the yard and apron but also for storing and retrieving containers into and from the yard. Automated SCs are successfully being used at the Patrick Container Terminal in Brisbane. In this terminal, all the transportation, storage, and retrieval operations are performed automatically. Multiload YTs are used at the Europe Container Terminal (ECT) in Rotterdam and at some terminals in Singapore. Because five to ten containers can be delivered by a tractor at the same time, they are used for moving containers between positions far away from each other, for example, between a berth and rail terminal or between two distant berths. AGVs are used at the ECT in Rotterdam and the Container Terminal Altenwerder (CTA) in Hamburg. They can move containers between the yard and apron by traveling on guide paths that are electronically stored within the memory of the supervisory control computer. Shuttle carriers are identical to SCs except for the fact that they can pass over stacks of only one tier with a container on its spreader, while SCs can pass over stacks of two or three tiers. Thus, shuttle carriers are used only for transporting containers from one place to another. Forklifts and reach stackers are used for storing or retrieving empty or light-weight containers.

The handling operations in container terminals are of three types: vessel operations associated with containerships, receiving/delivery operations for road trucks, and container handling and storage operations in the yard. Vessel operations include the discharging operation, during which containers in a vessel are unloaded from the vessel and stacked in a marshalling yard, and the loading operation, during which containers are handled in the reverse direction of the discharging operation. During the discharging operation, QCs transfer containers from a ship to a transporter. Then, in the case of the yard–crane relay system, the transporter delivers the inbound (import/discharging) container to a YC, which picks it up and stacks it into a position in a marshalling yard. For the loading operation, the process is carried out in the opposite direction. In case of the direct-transfer system, transporters move unloading containers to the yard and stack them onto the yard without requiring the use of YCs. For loading outbound containers, transporters pick them up by themselves and deliver them to QCs.

During receiving and delivery operations, when a container arrives at a container terminal by a road truck, the container is inspected at a gate to check whether all the required documents are ready and damages to the container are present. Further, at the gate, information on where an export container is to be stored and where an import container is located is provided to the road truck. When the road truck arrives at a transfer point of the yard, yard equipment that can be a YC or SC either receives a container from the truck, which is called the receiving operation, or transfers a container from the stack to the truck, which is called the delivery operation.

Important performance measures of container terminals are the vessel turnaround time and the road truck turnaround time. Because the maintenance cost of a vessel per day is very large, vessel carrier companies consider the vessel turnaround time to be the most important service measure. The road truck turnaround time is also important from the perspective of customer service. Thus, various activities of operation planning and real-time control for container terminals focus on the improvement in these two performance measures by using resources efficiently.

4.3 OPERATION PLANS IN CONTAINER TERMINALS

Before handling operations in container terminals actually happen, planners in the container terminal usually schedule the operations in advance with the goal of maximizing the efficiency of the operations. Target resources for the planning process are usually those whose capacities are limited; thus, priorities among handling activities that require the resources must be determined through the planning process. The target resources include berths, QCs, YCs, other handling equipment, yard spaces, and human operators. Expensive resources usually have limited capacities and thus become main target resources of planning. Thus, in container terminals, berths are considered to be the most critical resource, followed by QCs. Therefore, operation plans are usually formulated first for berths and then for QCs in a way of satisfying the requirements regarding important performance measures. In the case of the storage space not being enough, the usage of the storage space must be carefully planned before containers start to arrive at the yard.

4.3.1 BERTH PLANNING

The berth planning process consists of berth scheduling and QC deployment. During the process of berth scheduling, the berthing time and position of a containership, which may be the berth ID or the bitt number on the quay, are determined. QC deployment is the process of allocating QCs to each vessel so that the sum of allocated numbers of QCs does not exceed the total number of QCs available. This process determines the vessel that each QC will serve and the time during which the QC will serve the assigned vessel. Berth scheduling and QC deployment are

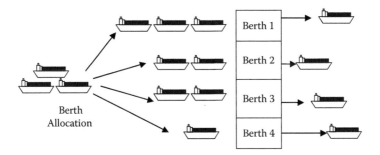

FIGURE 4.1 Berth allocation.

interrelated because the number of QCs to be assigned to a vessel affects the berthing duration of the vessel. Despite this interrelationship, because of the complexity of the integrated problem, most academic researchers have decomposed the problem into two independent parts, except in the study by Park and Kim (2003).

A quay is usually partitioned into several berths, each of which is assigned a unique ID. Many researchers have treated berths as discrete resources (discrete berths). Vessels are allocated to one of the berths when they arrive at the quay, as shown in figure 4.1. Many researchers have proposed methods for allocating vessels to discrete berths (Lai and Shih, 1992; Imai et al., 2001, 2003; Nishimura et al., 2001).

A quay is just a structure along the water, and thus it can be considered a continuous line (continuous quay) shared by vessels with limited lengths. Although all the above studies considered berths as discrete resources that can be allocated to vessels, some studies (Park and Kim, 2002; Moorthy and Teo, 2006) have considered the quay as a continuous line that multiple vessels can share with each other at the same time. Thus, when the quay is considered as a continuous line, more vessels can be served simultaneously at a quay of a particular length if the vessels are shorter in length. A berth schedule is illustrated in figure 4.2. However, when the

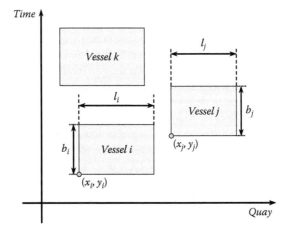

FIGURE 4.2 An example of a berth schedule.

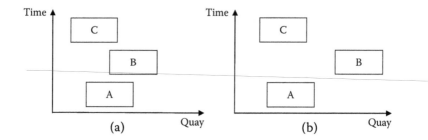

FIGURE 4.3 Two berth schedules for comparison.

quay is considered a collection of discrete berthing locations, the number of vessels to be served simultaneously is fixed regardless of their lengths. The vertical axis represents the time axis, while the horizontal axis represents positions on the quay. Thus, the solution space can be represented by a large box on which small rectangles, representing schedules for vessels, will be placed. That is, the horizontal side of a small rectangle represents the length of the vessel, and the position of the small rectangle on the horizontal axis corresponds to the berthing position of a vessel on the quay. For a berth schedule to be feasible, small rectangles must not overlap with each other.

The most important objective of berth scheduling is to complete ship operations within the time prespecified by a mutual agreement between the ship carrier and terminal operator. Also, because outbound containers for a vessel may already be stacked in the marshalling yard, there is a most preferable berthing position for a vessel. An illustrative objective function is to minimize the costs resulting from the delayed departures of vessels and the additional handling costs resulting from deviations of the berthing position from the best location. Figure 4.2 illustrates the relationships between variables and input data. Notations b_i, l_i, x_i, and y_i represent the ship operation time required for vessel i, length of vessel i, berthing position of vessel i (a decision variable), and berthing time of vessel i (a decision variable), respectively.

One more factor to be considered is the uncertainly of the arrival time or operation time. Consider two schedules as shown in figure 4.3. The schedule in figure 4.3(b) is preferred to that in figure 4.3(a) by planners in practice. When the arrival of vessel A or the ship operations for vessel A are delayed, the departures of vessels B and C in schedule 1 have to delayed, while those in schedule 2 do not. This factor was considered in the paper by Moorthy and Teo (2006).

Also, there are many different types of constraints that must be considered when determining the berthing positions of vessels. Examples are the depth of water along the quay and the maximum outreach of QCs installed at specific positions on the quay. If the depth of the water of a part of the quay is not enough, or the outreach of QCs installed at a part of the quay is shorter than that necessary, the corresponding vessel cannot be assigned to that part of the quay.

Although most previous research assumed that the quay is in the form of a straight line, there are many different types of quays in practice. Figure 4.4 illustrates

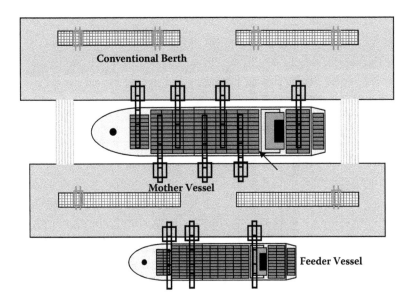

FIGURE 4.4 A conceptual model of a hybrid berth.

a conceptual model for the hybrid berth. The floating berth may be used to transfer containers between mother vessels and feeder vessels. By discharging and loading containers by using QCs at both sides of mother vessels, we can reduce the ship operation time significantly. The QC scheduling problem in this case is a new issue that must be solved. The container yard on the floating berth can be used for stacking transshipment containers temporarily. In this case, the synchronization of arrivals of feeder vessels with the arrival of a mother vessel is very important for the efficient use of the container yard on the floating berth. The operation of the container yard on the floating berth would be very dynamic because containers on the yard cannot stay for a long time as in the case of a conventional yard. Also, considering that more than one small vessel can berth between the floating berth and conventional berth, the berth scheduling problem would pose further challenging issues.

4.3.2 SHIP OPERATION PLANNING

The planning process of ship operations consists of QC scheduling (termed QC work scheduling in practice) and discharge and load sequencing. QCs are one of the critical resources in container terminals because of their high prices. Inbound containers of the same size and unloaded by the same ship are said to be included in the same container group. Likewise, outbound containers with the same destination port, of the same size, and to be loaded onto the same ship are said to be in the same container group. For the sake of efficiency in the discharging and loading operations, a collection of adjacent slots are usually allocated to containers of the same group in the stowage plan of a ship. In the stowage plan for loading, a cluster is defined to be a collection of adjacent slots in which containers of the same group are to be loaded.

Quay Crane Schedule

| | QC 1 (scheduled time: 09:00 ~ 12:00) | | | | | |
Operation Sequence	Cluster Number	Location of Task	Type of Task	Number of Containers	Start Time	Finish Time
1	6	1 Hold*	D**	47	09:00	09:47
2	1	1 Hold	L**	42	09:47	10:29
3	8	3 Hold	D	32	10:31	11:03
4	3	3 Deck	L	8	11:03	11:13
5	4	5 Hold	L	23	11:13	12:00

| | QC 2 (scheduled time: 09:00 ~ 12:00) | | | | | |
Operation Sequence	Cluster Number	Location of Task	Type of Task	Number of Containers	Start Time	Finish Time
1	7	3 Deck	D	39	09:00	09:39
2	9	5 Hold	D	46	09:41	10:27
3	10	7 Deck	D	24	10:27	10:51
4	5	7 Deck	L	5	10:51	11:15
5	2	3 Deck	L	24	11:15	12:00

* Hold of ship-bay 1   **D (discharging), L (loading)

FIGURE 4.5 An example of a quay crane schedule.

In the stowage plan for discharging, a cluster implies a collection of adjacent slots in which inbound containers of the same group are stacked.

For constructing a QC schedule (Daganzo, 1989; Park, 2003; Kim and Park, 2004), as illustrated in figure 4.5, planners are usually given information such as the stowage plan of the ship, as illustrated in figure 4.6, and the time interval in which each QC is available. The QC schedule in figure 4.5 shows a sequence of clusters that a QC will transfer. It also shows the location of the clusters, number of containers to be transferred, and time schedule of the operations. The stowage plan in figure 4.6 consists of four cross-sectional views, each corresponding to a ship bay and labeled with an odd number from 1 to 7. Small squares correspond to slots in which containers should be loaded in this container terminal. The shaded pattern in each slot represents a specific group of containers to be loaded or picked up from the corresponding slots. A set of adjacent slots with the same shade is called a cluster, as mentioned before. Figure 4.6 shows that four groups of containers should be discharged from five clusters of slots, and then four groups of containers should be loaded into five clusters of slots.

Once a QC starts to load (or discharge) containers into (from) a cluster of slots, it usually continues to do so until all the slots in the cluster become filled (empty). When discharging and loading operations must be performed at the same ship bay, the discharging operation must precede the loading operation. When the discharging operation is performed in a ship bay, containers on a deck must be transferred before containers in the hold of the same ship bay are unloaded. Further, the loading operation in a hold must precede the loading operation on the deck of the same ship bay. It should also be noted that QCs travel on the same track. Thus, certain pairs of clusters cannot be transferred simultaneously when the locations of the two clusters are too close to each other; this is because two adjacent QCs must be apart from each other by at least one or two ship bays so that they can simultaneously perform the transfer operations without interference. Moreover, if containers for any two clusters must be picked up at or delivered to the same location in a yard, the tasks for the two clusters

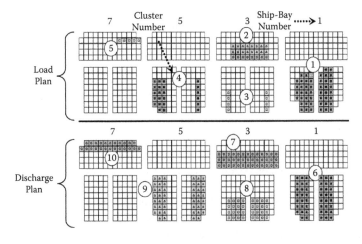

FIGURE 4.6 A partial example of a stowage plan.

TABLE 4.1

Transportation Matrix among Ports

(Number of Containers)

	Destination			
Source	2	3	4	5
1	1	4	4	1
2		2	0	1
3			4	2
4				4

cannot be performed simultaneously; this is because doing so will cause interference among the corresponding YCs.

Recently, new types of QCs have been developed and applied to several container terminals. Examples are twin-lift QCs whose spreader can pick up two 20-foot containers at once, and tandem QCs whose spreader can simultaneously pick up two 40-foot containers or four 20-foot containers. These new types of QCs have thrown open new challenging research issues regarding QC scheduling problems and the problems of sequencing discharging and loading operations, as described below.

The stowage plan, which is an important document for planning ship operations, is usually provided by vessel carriers. Stowage planning determines into which block (cluster) of slots in a ship bay a specific group of containers should be stacked. The relocations must be done when containers bound for a port are located in higher tiers in a vessel than containers bound for ports preceding the port. Further, various indices of the stability and strength of the containership must be checked. Consider an example of a containership calling four ports (Avriel and Penn, 1993). The ship has one bay consisting of three rows and four tiers. Table 4.1 shows the transportation requirement between ports. Table 4.2 shows the configurations of stacks when the ship departs ports 1, 2, 3, and 4. The notation a → b means that the container is located at port a and unloaded at port b. The asterisk in a → b* means that a relocation occurs at the corresponding port. The main issue in storage planning is this: what is the best way to stack containers on the ship in order to minimize the number of relocations? Note that the number of relocations in this example is four when containers are loaded onto the vessel as shown in table 4.2.

After constructing the QC schedule, the sequence of containers for discharging and loading operations is determined. Table 4.3 illustrates a load sequence list. The fourth column presents the storage location of a container before loading, and the fifth column mentions the slot in the vessel in which the container should be loaded. When the indirect-transfer system is used, the loading sequence of individual containers significantly influences the handling cost in the yard; on the other hand, in the direct-transfer system, the handling cost in the yard remains the same for different loading sequences. For unloading containers, because determining the discharging sequence is straightforward and determining the stacking locations of containers in the yard is done in real time, researchers have focused on the sequencing problem

TABLE 4.2

Configuration of Stacks When the Ship Departs Each Port

Departing Port	Tier	Row 1	Row 2	Row 3
1	4	1 → 2		
	3	1 → 4	1 → 3	1 → 3
	2	1 → 4	1 → 3	1 → 3
	1	1 → 5	1 → 4	1 → 4
2	4	2 → 3	2 → 3	2 → 5
	3	1 → 4	1 → 3	1 → 3
	2	1 → 4	1 → 3	1 → 3
	1	1 → 5	1 → 4	1 → 4
3	4	3 → 4	3 → 4	3 → 4
	3	1 → 4	3 → 4	3 → 5
	2	1 → 4	3 → 5	2 → 5*
	1	1 → 5	1 → 4	1 → 4
4	4			
	3	4 → 5	4 → 5	
	2	4 → 5	4 → 5	3 → 5*
	1	1 → 5	3 → 5*	2 → 5*

for loading operations rather than for discharging operations. In loading operations, containers to be loaded into slots in a vessel must satisfy various constraints on the slots prespecified by a stowage planner. Because the locations of outbound containers may be scattered over a wide area in a marshalling yard, the time required for loading operations depends not only on the transfer time of QCs, but also on that of YCs. Furthermore, the transfer time of a QC depends on the loading sequence of slots, while the transfer time of a YC is affected by the loading sequence of containers in the yard.

TABLE 4.3

An Example of a Load Sequence List (Kim et al., 2004)

QC Number	Sequence	Container Number	Location in Yard	Location in Vessel
101	1	MFU8408374	2C-06-01-03	05-07-01
101	2	DMU2975379	2C-06-01-02	05-08-01
101	3	DMU2979970	2C-06-01-01	05-07-02
101	4	OLU0071308	2C-06-02-03	05-08-02
101	5	MTU4015162	2C-06-02-02	05-07-03
…	…	…	…	…

Because the problem of load sequencing is highly complicated, most studies (Gifford, 1981; Cojeen and Dyke, 1976; Kim et al., 2004) have applied heuristic algorithms to solve it. The following typical objectives must be pursued and the following constraints (Kim et al., 2004) must be satisfied by the loading sequence. The objectives related to the operation of QCs are to (1) fill slots in the same hold, (2) stack containers onto the same tier on deck, and (3) stack containers in the same weight group as specified in the stowage plan. The objectives related to the operation of YCs are to (1) minimize the travel time of YCs, (2) minimize the number of rehandles, and (3) pick up containers in locations nearer to the transfer point earlier than those located farther from the transfer point. Constraints related to the operation of QCs are (1) to maintain the precedence relationships (according to the work schedules for QCs and the relative positions between slots in a ship bay) among slots, (2) not to violate the maximum allowed total weight of the stack on the deck, (3) not to violate the maximum allowed height of the stack of a hold, and (4) to load the same type of containers as specified in the stowage plan. Constraints related to the operation of YCs are to maintain a distance between adjacent YCs such that they can transfer containers without interference between each other.

4.3.3 YARD SPACE PLANNING AND ASSIGNMENT

One of the important factors that affect the turnaround time of vessels and road trucks is the method of allocating storage spaces for containers arriving at the marshalling yard; this is because the locations of containers significantly affect the efficiency of delivery and loading operations (Chen, 1999). The popular objectives of the space planning are to (1) minimize the travel distance of transporters, (2) minimize the travel distance of YCs, (3) minimize the congestion of YCs and transporters in the yard, and (4) minimize the possibility of relocations.

For the first objective, outbound containers are usually stacked at the positions close to the berthing position of the corresponding vessel. The strategy for the second objective is different between container terminals with a parallel layout and those with a perpendicular layout of blocks. In container terminals with a parallel layout of blocks, trucks park at the side of each block. Because outbound containers of the same group are usually loaded onto the vessel consecutively, the gantry travel of YCs can be minimized when the outbound containers of the same group are located at the same yard bay. However, when attempting to allocate a yard bay to a container group, an empty yard bay may not be available at a location close to the corresponding vessel.

Figure 4.7 illustrates the result of space allocation for outbound containers in a yard in the case of a parallel layout. Thus, in container terminals with a parallel layout of blocks, space must be preallocated before outbound containers start to arrive at the yard (Taleb-Ibrahimi et al., 1993). Different patterns in the layout represent different groups of outbound containers. Note that several adjacent yard bays are allocated to a container group to reduce the travel time of YCs.

However, in container yards with a perpendicular layout of blocks, because the transfer points are located at the end of each block, YCs must move between the storage location and the end of its block. Thus, there is no way to reduce the travel distance by stacking outbound containers of the same group close to each other. Instead, for reducing the travel distance of YCs during the retrieval operation,

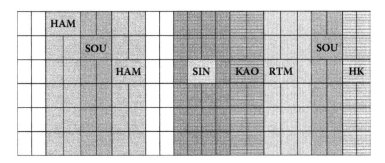

FIGURE 4.7 An illustration of the yard space plan.

outbound containers can be stacked at yard bays close to the water side, while inbound containers can be moved to yard bays close to the land side. In this case, storage space is not preallocated before containers start to arrive at the yard. Instead, the storage location for an arriving container is usually determined in real time by considering the various factors mentioned previously.

Congestion is another factor that lowers the productivity of the yard operation. Thus, it is better to spread the workload over many different blocks for reducing the congestion. This can be done by scattering incoming outbound containers and discharged inbound containers over many different blocks. Further, when the sequence of loading outbound containers is determined, the containers should not be picked up consecutively from the same block.

Another important objective of locating containers is to minimize the possibility of relocations during retrievals. When locating outbound containers, the weights of the containers must be taken into account. Because heavy containers are usually placed in low tiers in the holds of vessels for the stability of vessels, they are retrieved earlier than light containers from the yard. Thus, heavy containers must be stacked in higher tiers than light containers in the yard so that relocation can be avoided during retrievals of heavy containers. Suppose that outbound containers are classified into heavy, medium, and light containers according to their weights. Further, consider a bay of outbound containers as shown in figure 4.8(a). Assume that containers of different weights arrive at the bay randomly, heavier containers are picked up earlier than lighter ones, and the maximum number of tiers is four. Then, the problem is to determine the best position to locate an arriving container in each

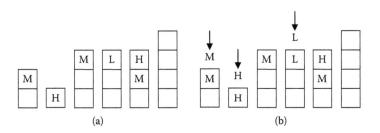

FIGURE 4.8 Example for locating outbound containers of different weights.

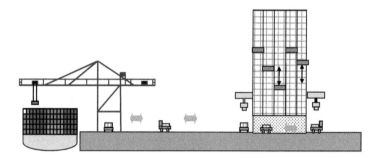

FIGURE 4.9 A conceptual design of the high-rise stacking system (HSS).

of the following cases: the arriving container is (1) heavy, (2) medium, or (3) light. The objective is to minimize the total expected number of relocations during the retrieval after the bay becomes full. Kim et al. (2000) found that the optimal location is as shown in figure 4.8(b).

It should be mentioned that the relocation problem in the yard for inbound containers is even more serious than that for outbound containers (Castilho and Daganzo, 1993); this is because the sequence of retrieval of inbound containers is extremely uncertain.

Even among yards of the same type, storage strategies can be different from each other. The storage strategy for yards in which OHBCs are used, which have a larger number of rows and tiers per bay, may not be the same as that for yards in which rubber-tired gantry cranes (RTGCs) are used, which run over stacks of smaller size.

Figure 4.9 shows the conceptual design of a high-rise stacking system (HSS) (Ioannou et al., 2000; Choi and Ha, 2005), which resembles the automated storage and retrieval system. The main advantage of this type of storage system is that the number of relocations can be minimized. Another basic idea behind this system is that when the size of vessels becomes larger (for example, over 10,000 TEUs), the current storage system may not be able to accommodate all the storage requirements. Thus, a storage system with a much higher storage capacity like the HSS would be necessary in the near future. There are many design options for the HSS, and new storage strategies for the HSS must also be developed.

4.4 REAL-TIME SCHEDULING FOR YARD CRANES AND TRANSPORTERS

The plans explained in the previous section are formulated for critical resources (berths, QCs, and storage spaces in some cases) and tasks (loading and unloading operations). However, these plans cannot cover all the details of handling activities in container terminals. Thus, for the remaining activities, decisions on the utilization of equipment and on the assignment of tasks to each piece of equipment are usually made on a real-time basis. Examples are the assignment of tasks to transporters and the assignment of tasks to YCs. Two reasons for those activities not being preplanned are the high uncertainties of the situation and the lower importance of the resources compared with berths or QCs. Although for decision making a schedule can be constructed for the events of the near future (less than 10 minutes into the future), the decision is basically made in response to an event that has just occurred. Further, even the decisions included in the various plans can be modified and updated during the implementation responding to the deviation of the situation from expectations or forecasts.

There are two hierarchical problems even in task assignment problems: the equipment deployment problem and task-scheduling problem. The equipment deployment problem involves the deployment of a certain group of equipment to a specific type of task. For example, a group of YCs may be dedicated to delivery and receiving tasks for a certain period (Zhang et al., 2002; Cheung et al., 2002; Linn and Zhang, 2003), and a group of YTs may be assigned to the task of delivering a group of containers from a block to another for a certain period. This type of decision must be made before the start of the real-time assignment of tasks to each piece of equipment.

The task-scheduling problem involves the assignment of tasks to each piece of equipment and the sequencing of the assigned tasks to be carried out (Kozan and Preston, 1999; Bish, 2003; Hartmann, 2004). For unloading and loading tasks, considerations for the scheduling are as follows: (1) Because the most important objective of the unloading and loading operations is to minimize the turnaround time, the scheduling must be done in such a way as to accomplish the objective. (2) The loading and unloading operations are performed by QCs, YCs, and transporters together. Thus, the activities of these types of equipment must be synchronized with each other. (3) The priority of the unloading and loading operations is higher than that of the receiving and delivery operations.

There are two types of strategies for assigning delivery tasks to transporters. One is a dedicated assignment strategy and the other is a pooled strategy (Kim and Bae, 2004). In the case of the former strategy, a group of transporters is assigned to a QC and they deliver containers only for that QC. In the latter strategy, all the transporters are shared by different QCs, and thus any transporter can deliver containers for any QC, which is a more flexible strategy for utilizing transporters.

According to the delivery cycle, the delivery task follows either the single-cycle strategy or the dual-cycle strategy. When the former strategy is applied, a transporter delivers a container to a block (or a QC) and returns empty. However, when the latter strategy is applied (Bish et al., 2005), a transporter delivers a container not only when it moves from the apron to the yard, but also when it moves from the yard to the apron. When the single-cycle strategy is used, the dedicated assignment strategy is usually applied.

Transporters can usually deliver two 20-foot containers. Transporters such as multiload yard trucks can deliver more than two containers at a time. In this case, we have to identify the two containers for pickup and the sequence of deliveries. The assignment of multiple containers to transporters and the sequencing of deliveries affect the efficiency of the operation significantly (Grunow et al., 2004, 2006). Further, YTs and AGVs can load or put down containers with the help of cranes, while SCs and shuttle carriers can not only deliver containers but also pick those up from the ground by themselves. Thus, although the containers can be transferred by a QC to a YT or AGV only if the YT or AGV is ready under the QC, the operation of SCs and shuttle carriers does not have to be synchronized, which results in a higher performance than YTs or AGVs (Vis and Harika, 2004; Yang et al., 2004). This difference between the two types of transporters requires operation methods different from each other.

When automated guided transporters are used, the following issues must be addressed to ensure the efficiency of operations:

1. The traffic control problem is a critical issue (Evers and Koppers, 1996). Because the number of transporters is very large (more than 150 AGVs are used in the ECT) and the size of each transporter is also large, special attention must be paid to prevent congestion and deadlocks (Kim et al., 2006; Lehmann et al., 2006).
2. Because a container can be transferred between a QC and AGV or between a YC and AGV only when both are ready for the transfer, the travel of AGVs must be carefully preplanned (Briskorn et al., 2006). Thus, the scheduling of flows of AGVs is another challenging issue.
3. Transporters in container terminals are free-ranging vehicles that can move to any position on the apron with the help of GPS, transponders, or microwave radars. Thus, the guide path network must be stored in the memory of the supervisory control computer. Once the guide path network is designed, the route for a travel order can be determined. The guide path network and the algorithm to determine the routes of transporters impact the performance of the transportation system significantly, which is another important issue that should be investigated by researchers.

The task scheduling for YCs must also be done in real time (Lai and Lam, 1994; Kim et al., 2003; Ng, 2005). The priorities or due times of tasks must be considered to sequence the tasks. Further, the interference between YCs must be considered to estimate the completion time of a task by a YC. Note that the handling movements differ between YCs according to their type, which must be considered for scheduling purposes.

Because the system of container terminals is highly complicated, many researchers have used a simulation approach to solve various practical problems. The main advantages of this approach are as follows:

1. In the design stage, simulation can be used to determine the type of handling equipment, number of pieces of equipment, layout of the yard, and specification of blocks (the numbers of bays, rows, and tiers).
2. To determine the operational rules such as rules for the operation of transporters and YCs, rules for locating containers, and rules for assigning tasks to equipment, simulation study is usually employed.
3. To test the traffic control algorithm of the equipment, an emulation program can be used. In the emulation program, the detailed movement of vehicles is represented in order to assess the possibility of collisions and deadlock among vehicles.

4.5   CONCLUSIONS

This chapter introduced various types of container handling systems in container terminals, which are designed for satisfying specific requirements of container handling logistics. These requirements may be different from one country to another. This chapter also attempted to outline various operational problems that researchers have addressed. The problems were classified into planning problems and real-time control problems, depending on the length of the planning horizon for each planning process.

Recently, much effort has been devoted to automate various operations in container terminals. Automation has been realized in some container terminals such as the ECT in Rotterdam, CTA in Hamburg, and Thames port in the United Kingdom. Automation requires detailed operation orders and decisions for equipment that have been made by human operators in conventional container terminals. Thus, operations researchers now face much more challenging problems in realizing the automation of container terminals.

The sizes of containerships are continuously increasing, and containerships with a capacity larger than 8,000 TEUs will become popular in the next decade. Thus, the loading and unloading speeds of container handling equipment in ports must be increased dramatically so that large-sized vessels can adhere to their voyage schedules. In addition to developing equipment with higher speed, more efficient operational decision-making algorithms must be developed and the computational times of the algorithms must be significantly shortened.

Until now, researchers have considered operational problems of container terminals to be isolated from outside logistic nodes (rail yards, feeder ports, inland depots, and so on). However, considering that a container terminal is only a node in a much larger logistics network, many new decision-making problems, resulting from the integration of functions of outside nodes to those of the container terminal, would be promising topics for future studies.

ACKNOWLEDGMENTS

This research was supported by the MIC (Ministry of Information and Communication), Korea, under the ITRC (Information Technology Research Center) support program supervised by the IITA (Institute of Information Technology Advancement) (IITA-2006-C1090-0602-0013).

REFERENCES

Avriel, M., and Penn, M. 1993. Exact and approximate solutions of the container ship stowage problem. *Computers and Industrial Engineering* 25:271–74.
Bish, E. K. 2003. A multiple-crane-constrained scheduling problem in a container terminal. *European Journal of Operational Research* 144:83–107.
Bish, E. K., Chen, F. Y., Leong, Y. T., Nelson, B. L., Ng, J. W. C., and Simchi-Levi, D. 2005. Dispatching vehicles in a mega container terminal. *OR Spectrum* 27:491–506.
Briskorn, D., Drexl, A., and Hartmann, S. 2006. Inventory-based dispatching of automated guided vehicles on container terminals. *OR Spectrum* 28:611–30.
Chen, T. 1999. Yard operations in the container terminal: A study in the "unproductive moves." *Maritime Policy and Management* 26:27–38.
Cheung, R. K., Li, C.-L., and Lin, W. 2002. Interblock crane deployment in container terminals. *Transportation Science* 36:79–93.
Choi, S. H., and Ha, T. Y. 2005. A study on high-efficiency yard handling system for next generation port. *Ocean Policy Research* 20:81–126.

Cojeen, H. P., and Dyke, P. V. 1976. The automatic planning and sequencing of containers for containership loading and unloading. In *Ship operation automation*, ed. Pitkin, Roche, and Williams, 415–23. North-Holland Publishing Co.

Daganzo, C. F. 1989. The crane scheduling problem. *Transportation Research* 23B:159–75.

Castilho, B. D., and Daganzo, C. F. 1993. Handling strategies for import containers at marine terminals. *Transportation Research* 27B:151–66.

Evers, J. J. M., and Koppers, S. A. J. 1996. Automated guided vehicle traffic control at a container terminal. *Transportation Research* 30A:21–34.

Gifford, L. A. 1981. Containership load planning heuristic for a transtainer-based container port. Unpublished MSc thesis, Oregon State University.

Grunow, M., Günther, H.-O., and Lehmann, M. 2004. Dispatching multi-load AGVs in highly automated seaport container terminals. *OR Spectrum* 26:211–35.

Grunow, M., Günther, H.-O., and Lehmann, M. 2006. Strategies for dispatching AGVs at automated seaport container terminals. *OR Spectrum* 28:587–610.

Hartmann, S. 2004. General framework for scheduling equipment and manpower on container terminals. *OR Spectrum* 26:51–74.

Imai, A., Nishimura, E., and Papadimitriou, S. 2001. The dynamic berth allocation problem for a container port. *Transportation Research* 35B:401–17.

Imai, A., Nishimura, E., and Papadimitriou, S. 2003. Berth allocation with service priority. *Transportation Research* 37B:437–57.

Ioannou, D. A., Kosmatopoulos, E. B., Jula, H., Collinge, A., and Dougherty, Jr., E. 2000. *Cargo handling technologies*. Final report, Center for Commercial Deployment of Transportation Technologies.

Kim, K. H. 2005. Models and methods for operations in port container terminals. In *Logistics systems: Design and optimization*, ed. A. Langevin and D. Riopel, 213–43. New York: Springer.

Kim, K. H., Jeon, S. M., and Ryu, K. R. 2006. Deadlock prevention for automated guided vehicles in automated container terminals. *OR Spectrum* 28:659–679.

Kim, K. H., and Bae, J.-W. 2004. A Look-ahead dispatching method for automated guided vehicles in automated port container terminals. *Transportation Science* 38:224–34.

Kim, K. H., Kang, J. S., and Ryu, K. R. 2004. A beam search algorithm for the load sequencing of outbound containers in port container terminals. *OR Spectrum* 26:93–116.

Kim, K. H., Lee, K. M., and Hwang, H. 2003. Sequencing delivery and receiving operations for yard cranes in port container terminals. *International Journal of Production Economics* 84:283–92.

Kim, K. H., and Park, Y.-M. 2004. A crane scheduling method for port container terminals. *European Journal of Operational Research* 156:752–68.

Kim, K. H., Park, Y.-M., and Ryu, K. R. 2000. Deriving decision rules to locate export containers in container yards. *European Journal of Operational Research* 124:89–101.

Kozan, E., and Preston, P. 1999. Genetic algorithms to schedule container transfers at multimodal terminals. *International Transactions in Operational Research* 6:311–29.

Lai, K. K., and Lam, K. 1994. A study of container yard equipment allocation strategy in Hong Kong. *International Journal of Modeling & Simulation* 14:134–38.

Lai, K. K., and Shih, K. 1992. A study of container berth allocation. *Journal of Advanced Transportation* 26:45–60.

Lehmann, M., Grunow, M., and Günther, H.-O. 2006. Deadlock handling for real-time control of AGVs at automated container terminals. *OR Spectrum* 28:631–58.

Linn, R. J., and Zhang, C.-Q. 2003. A heuristic for dynamic yard crane deployment in a container terminal. *IIE Transactions* 35:161–74.

Meersmans, P. J. M., and Dekker, R. 2001. *Operations research supports container handling*. Econometric Institute report EI 2001-22, Erasmus University.

Moorthy, R., and Teo, C.-P. 2006. Berth management in container terminals: The template design problem. *OR Spectrum* 28:495–518.

Ng, W. C. 2005. Crane scheduling in container yards. *European Journal of Operational Research* 164:64–78.

Nishimura, E., Imai, A., and Papadimitriou, S. 2001. Berth allocation planning in the public berth system by genetic algorithms. *European Journal of Operational Research* 131:282–92.

Park, K. T., and Kim, K. H. 2002. Berth scheduling for container terminals by using a subgradient optimization technique. *Journal of the Operational Research Society* 53:1049–54.

Park, Y.-M., and Kim, K. H. 2003. A scheduling method for berth and quay cranes. *OR Spectrum* 25:1–23.

Steenken, D., Voß, S., and Stahlbock, R. 2004. Container terminal operation and operations research: A classification and literature review. *OR Spectrum* 26:3–49.

Taleb-Ibrahimi, M., de Castilho, B., and Daganzo, C. F. 1993. Storage space vs handling work in container terminals. *Transportation Research* 27B:13–32.

Vis, I. F. A., and de Koster, R. 2003. Transshipment of containers at a container terminal: An overview. *European Journal of Operational Research* 147:1–16.

Vis, I. F. A., and Harika, I. 2004. Comparison of vehicle types at an automated container terminal. *OR Spectrum* 26:117–43.

Yang, C. H., Choi, Y. S., and Ha, T. Y. 2004. Simulation-based performance evaluation of transport vehicles at automated container terminals. *OR Spectrum* 26:149–70.

Zhang, C., Wan, Y.-W., Liu, J., and Linn, R. J. 2002. Dynamic crane deployment in container storage yards. *Transportation Research* 36B:537–55.

5 Models for Cross-Border Land Transportation of Ocean Containers

Raymond K. Cheung

CONTENTS

ABSTRACT

The land transportation of containers between a container terminal and the origins or destinations of the containers represents a very small portion of the global distribution network in terms of distance. However, it accounts for a significant portion of the total transportation cost. It can also cause shipment delays and disruptions in the global network. When the transportation involves cross-border issues such as having different regulatory policies for transportation and information flow, modeling the problem is not trivial. In this chapter, we introduce several modeling perspectives. These perspectives help to formulate cross-border land transportation problems under different situations. These situations depend on the level of policy restriction that governs cross-border activities and the level of information available for decision making. We review a number of models, ranging from coupling drivers and tractors to matching resources with transportation requests in a dynamic, stochastic environment. This chapter uses the case of Hong Kong as an example to illustrate the challenges of managing cross-border container transportation and the breadth and depth in research development.

5.1   INTRODUCTION

The intensifying globalization has fueled the rapid growth of the goods flow in international port cities. This is particularly true for the port cities in the Asia Pacific (see figure 5.1). To cope with the growth, much effort has been spent on improving terminal operations and planning (see Steenken et al.[1] for a review). The connecting land transportation service between a container terminal and the origin or destination of a container, known as the *drayage service*, has become the bottleneck of the whole transportation process. The situation is even worse when border crossing is involved. The cross-border drayage problem, however, has received little research attention. The key challenges of the problem come from the different policies applied to the different sides of the border and the different levels of available information for decision making. In this chapter, we present several modeling perspectives and possible solution strategies for various drayage problems.

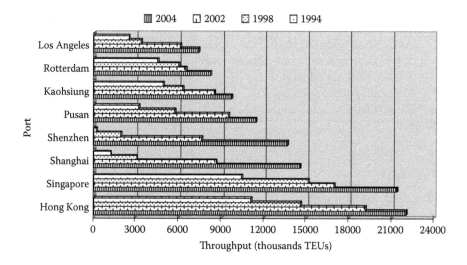

FIGURE 5.1 Throughput of the top eight container ports from 1994 to 2004. Source: Dubai Ports Authority, Port of Hamburg.

Take Hong Kong, the busiest port in the world, as an example. Over 80% of the truck trips to and from terminals are cross-border. It is because most containers passing through Hong Kong's port have either origins or destinations in the southern part of mainland China (PRC)–Pearl River Delta (PRD) region, which is within 100 km of Hong Kong. The cross-border drayage operations have very low productivity in terms of the driver's time, trip time, and tractor's time. Given the short distance between the factories in PRD and Hong Kong, drivers could make several round-trips per day. In practice, the average number of round-trips per day is only 1.2.[2] Moreover, in one of the two ways of a round-trip, an empty container is transported. Less than 40% of what it could be, the low productivity comes from the regulatory policies governing cross-border container trucking and the level of information on the availability of resources and possible movements. This low productivity has cost millions a day to shippers, affecting the competitiveness of the Hong Kong port.[3]

The drayage problem can be considered as a pickup and delivery problem,[4] since the container is picked up from a location and delivered to another location where no pickup is allowed until the previous delivery is done. It can also be considered as the short-haul version of the full-truckload management problem,[5] or, in a broader sense, a vehicle routing and scheduling problem.[6–8] Versions of these problems that consider uncertainty have appeared in literature, including the solution methods for stochastic vehicle routing problems[9] and stochastic dial-a-ride problems,[10] the modeling of the full-truckload problems with random demands,[11] and the dynamic stochastic approach to assigning drivers to loads for the land transportation of ocean containers.[12] If a driver can cover multiple requests, then the problem can be regarded as determining the tour of the driver. The length of the trip is typically constrained to the maximum number of hours a driver can spend on the road. To solve such a problem, tour construction and tour improvement procedures are used (see, for example, Ball et al.[13] and Desrosiers et al.[14]). The issues of policies for cross-border drayage activities and the impact of information availability on productivity have not yet been addressed in literature.

The contributions of this chapter are threefold. First, we show how the regulatory policies and level of available information affect drayage operations. Second, we provide new modeling perspectives and a new classification of constraints. Third, we present a range of models for different settings, illustrating how different modeling perspectives can be used in formulating the models. For the class of models that are quite new, we highlight directions for developing solution methods, rather than the details of their implementations and numerical experiments. This chapter is organized as follows. Section 5.2 discusses the problems and provides model categories and constraint classification. Sections 5.3 and 5.4 present models for handling policy relaxation and for different levels of information availability. Section 5.5 provides some concluding remarks.

5.2 MODEL ORIENTATIONS AND CONSTRAINT CLASSIFICATIONS

Regulatory policies and the level of available information are the key factors that affect cross-border drayage operations. In this section, we first look at their impact. Then, we illustrate the dependence of the modeling framework on the problem perspective, the availability of information, the levels of economies of scale, the scope, and the level of uncertainty. There is no one-size-fits-all model or solution algorithm. As our main objective is to present an illustration of the perspectives of different models and their key components, the notation used may not be as precise as it could be. We note that the terms *demand*, *task*, and *transportation request* are interchangeable in this chapter.

5.2.1 THE IMPACT OF REGULATORY POLICIES AND AVAILABLE INFORMATION

An empty container can be considered a commodity, rather than a piece of transportation equipment, when it crosses the border to enter another country or another region with different custom regulations. To avoid paying custom duty on the container, when the quadruple of driver-tractor-chassis-container crosses the border to take a loaded container to the destination, the exact same quadruple has to be together on the return trip. Moreover, the return trip must be made within a certain time limit. In the case of the Hong Kong–PRD border, this is called the *four-up-four-down* policy. Under this policy, the quadruple is considered a single resource rather than four individual resources. This policy severely constrains how resources are used. First, a driver who has taken a laden container crossing the border needs to wait hours for unloading. Second, the driver needs to take the empty container back on the return trip, thereby creating low utilization of the driver's time and the trip time.

Another common policy that has imposed strong constraints on resource management is the licensing requirement for cross-border drivers. In the case of the Hong Kong–PRD border, only Hong Kong drivers can operate the cross-border trucks. However, the license is issued to a cross-border driver for a particular truck. This is called the *one-driver-one-truck* policy. It is clear that this policy causes low utilization of truck time.

The level of available information also affects how to determine whether or not a request should be accepted and how to match resources to the accepted requests. Whenever a new request is known and when a decision needs to be made immediately, the decisions are likely made in a first-come-first-served manner. If we can delay the decision making until a batch of orders are known, then we can make a better optimized solution. Furthermore, if we know the orders for the whole day, then we can plan the trips of the drivers to deliver the orders. Suppose that the statistical information about the availability of the transportation requests (and the travel time) is known. We can then use stochastic models to help make the decisions.

5.2.2 MODELING PERSPECTIVES

The policy and information availability issues must be incorporated in the drayage problem where a set of drivers (and tractors) are to handle a set of transportation requests. Each request has an origin and a destination, and there is usually a time window for the request. Let

N = the set of locations,
L = the set of transportation requests or demands,
T = the set of time indices over a planning horizon of T periods, and
R = the set of resources (e.g., driver, tractors) to be managed.

5.2.2.1   Flow-Oriented Perspective

In terms of decisions, we can use different perspectives. The first one is the classical network flow model over a time-space network where resources are traveling through the arcs in the network. For each $t \in \mathsf{T}$, $i,j \in \mathsf{N}$, let

$c_{ij,t}^{L}$ = the contribution of moving a loaded container from i to j at time t,

$c_{ij,t}^{M}$ = the contribution of moving an empty container from i to j at time t,

$c_{ij,t}^{R}$ = the contribution of pure repositioning of a tractor from i to j at time t, and

t_{ij} = the travel time from i to j.

Notice that the travel time can depend on whether the tractor is taking a loaded or an empty container or a container simply undergoing empty repositioning. Thus, we have t_{ij}^{L}, t_{ij}^{M}, and t_{ij}^{R} for the three situations, respectively. The decision variables are the flow on the arcs:

$x_{ij,t}^{L}$ = the amount of flow (in units of loaded containers) from i to j at time t.

Similarly, we can define $x_{ij,t}^{M}$ and $x_{ij,t}^{R}$ for moving empty containers and empty repositioning, respectively. In such a flow-oriented perspective (or model), a transportation request for moving a laden container can be represented as the arc capacity because we know that the number of loaded movements cannot exceed the actual demand.

5.2.2.2   Matching-Oriented Perspective

When viewing the drayage problem as matching the drivers (or resources) to transportation requests, we can define

$c_{r\ell,t}$ as the contribution of assigning resource $r \in \mathsf{R}$ for handling request $\ell \in \mathsf{L}$.

Suppose that $\ell \in \mathsf{L}$ is a request for moving a loaded container from i to j, then $c_{r\ell,t}$ is equal to $c_{ij,t}^{L}$. The decision variable is defined as

$$x_{r\ell,t} : \begin{cases} 1 & \text{if resource } r \text{ is assigned to handle request } \ell \text{ at time } t, \\ 0 & \text{otherwise.} \end{cases}$$

This perspective allows us to embed some constraints in the definition of the decision variables. For example, if resource r' (e.g., a PRD driver) can never take request ℓ' (e.g., a load starting from Hong Kong), then $x_{r'\ell',t}$ is not defined.

5.2.2.3   Attribute-Decision-Oriented Perspective

What decision can be made on a resource not only depends on the type of resource, but also on the status of the resource. For example, if a driver has less than 2 hours of duty period left, the driver should not take a load across the border. Further, as we will see later, even the definition of a resource may change over time. To allow for a more general description of a resource, we can use an attribute vector to describe the characteristics of a resource at a particular time. For a given attribute vector, there is a set of decisions that can act on the resource. We let

A = the set of attribute vectors that describe the characteristics of a resource,

N_a = the set of decisions that can be made when the resource's attribute is $a \in$ A, and

c_{ad} = the contribution of making decision $d \in D_a$ on the resource with attribute $a \in$ A.

Then, the decision variable is written as

$$x_{ad} : \begin{cases} 1 & \text{if the resource with attribute } a \in \text{A takes decision } d \in D_a, \\ 0 & \text{otherwise.} \end{cases}$$

For the above example, let a be the attribute vector describing a driver who has less than 2 duty hours remaining. Then, the decision of taking a load cross-border will not be in D_a. Some possible elements in D_a may include returning to the depot, taking a non-border-crossing load, or simply doing nothing. We can see that this modeling perspective gives a flexible way to incorporate qualitative constraints.

5.2.3 CONSTRAINT CLASSIFICATIONS

In addition to classical system dynamics, a cross-border drayage problem also has policy-imposed constraints and information-availability-imposed constraints. These constraints are described as follows.

5.2.3.1 System Dynamics

The system dynamics include the physical laws that govern resources evolve over time and space. Constraints in this category include:

- Flow conservation constraints in which the inflow and outflow of resources at a location must be balanced
- Resource limitations (e.g., the number of trucks available, the duty hours of drivers)
- Demand-covering constraints (e.g., the minimum number of requests to be satisfied)
- Integrality constraints

For example, to ensure the flow conservation of resources in flow-oriented models, we write

$$\sum_{s \in \{L,M,R\}} \sum_{i} x^s_{ij,t-t^s_{ij}} - \sum_{s \in \{L,M,R\}} \sum_{k} x^s_{jk,t} = b_{j,t} \qquad \forall \; j \in \text{N}, t \in \text{T}, \tag{5.1}$$

where $b_{j,t}$ is the amount of net external resources arriving at location j at time t. In matching-oriented models and attribute-decision-oriented models, flow conservation means that each resource must take exactly one action, including the do-nothing option. We can write the following for flow-oriented models:

$$\sum_{\ell} x_{r\ell,t} = 1 \qquad \forall\, t \in \mathsf{T}, r \in \mathsf{R},$$

and the following for attribute-decision-oriented models:

$$\sum_{d \in D_a} x_{ad} = 1 \qquad \forall\, a \in \mathsf{A}.$$

5.2.3.2 Policy-Imposed Constraints

Other than physical laws, policies can constrain how the system behaves. Let us use the matching-oriented perspective to illustrate the difference between the four-up-four-down policy and two-up-two-down policy. Under the latter policy, the driver-tractor pair can leave the chassis-container pair at a location and then pick up another chassis-container pair. Consider a task (or request), $\ell \in \mathsf{L}$, that moves a laden container from Hong Kong to PRD. Under the four-up-four-down policy, after a resource, r (which is the driver-tractor pair), is assigned to this task, the combined unit has to be together throughout the entire round-trip. Denote $roundtrip^{\ell}$ as the time interval for the round-trip for this task. We know that if $x_{r\ell,t} = 1$ for any $t \in roundtrip^{\ell}$, then $x_{r\ell,t}$ must be 1 for all $t \in roundtrip^{\ell}$. Thus, the constraint is

$$\sum_{t \in roundtrip^{\ell}} x_{r\ell,t} = x_{r\ell,t} \cdot |\,roundtrip^{\ell}\,| \qquad \forall t \in roundtrip^{\ell}. \tag{5.2}$$

The last term gives the time required by the round-trip.

If the four-up-four-down policy is reduced to a two-up-two-down policy, then the resource, r, and task, ℓ, only need to be together for the shorter period, $oneway^{\ell}$, and constraint (5.2) becomes

$$\sum_{t \in oneway^{\ell}} x_{r\ell,t} = x_{r\ell,t} \cdot |\,oneway^{\ell}\,| \qquad \forall t \in oneway^{\ell}. \tag{5.3}$$

It is clear that constraint (5.3) is less restrictive than constraint (5.2). Notice that many policy constraints are difficult to express using a flow-oriented model. Matching-oriented and attribute-decision-oriented models are more convenient. For example, consider that the driver is a PRD driver, who can only handle requests whose origins and destinations are both within the PRD. To handle this situation in a flow-oriented model, we need to use the concept of multicommodity (here, commodity refers to the type of driver), increasing the number of decision variables. In matching-oriented models, we can simply define $x_{r\ell,t}$ only if resource r is eligible to handle task ℓ. This situation actually helps to reduce the number of decision variables.

5.2.3.3 Information-Imposed Constraints

The type of available information can significantly affect the types of decisions that are made. On the demand side (e.g., transportation requests), if there is no forecast information, then we can only match the resource with known demands. If we have some forecast information, then we can have a trip plan for covering several tasks. For example, consider that the demands are known with certainty up to time t^D; from time t^D to time t^S (where $t^D < t^S$), we only know the distribution of the demands. Using the flow-oriented perspective and supposing that we only consider the period for which we have deterministic information, the decision variables, $x_{ij,t}^L$, are only defined for $t \leq t^D$. If we also consider the period when the stochastic information is available, then $x_{ij,t}^L$ are defined for $t \leq t^S$. Since there are more decision variables (and thus a higher degree of freedom for making decisions), the total contribution of the expanded set of decisions should usually be higher. However, with stochastic information, the problem becomes more difficult to solve. To see that, let $u_{ij,t}^L$ and $\tilde{u}_{ij,t}^L$ be the demands for the deterministic period and the stochastic period, respectively. Then, the demand constraints are given as

$$x_{ij,t}^L \leq u_{ij,t}^L \qquad \forall t \leq t^D$$
$$x_{ij,t}^L \leq \tilde{u}_{ij,t}^L \qquad \forall t^D < t \leq t^S$$

Notice that the second capacity constraint is not a deterministic one. Thus, stochastic models are required to represent the problem.

As for the demand side, the availability of supply-side information can affect the set of constraints. This information includes the available time of drivers who will report for duty soon, the trip time left, and the possible outsourcing opportunity (e.g., in case that a preassigned driver for a request cannot make it eventually, then we can still use a subcontractor to cover the request by paying a premium).

5.2.3.4 A Generic Formulation

Using the simplest form of notation, define

$\mathbf{x}^{SD}$: as the set of solutions that satisfy the system dynamics constraints,
$\mathbf{x}^{P}$: as the set of solutions that satisfy the policy-imposed constraints, and
$\mathbf{x}^{I}$: as the set of solutions that satisfy the information-imposed constraints.

Then, the compact form of the cross-border drayage problem can be written as

$$\max \quad c^T \cdot x \tag{5.4}$$

subject to

$$x \in \mathbf{X}^{SD}, \tag{5.5}$$

$$x \in \mathbf{X}^{P}, \tag{5.6}$$

and

$$x \in \mathsf{X}^L. \tag{5.7}$$

The actual form of the objective function varies from model to model. The cost items in the objective function may include the contribution (or cost) for moving a container, the empty-repositioning cost, the outsourcing cost (if a subcontractor can be hired), the overtime cost, and the driver (labor) cost, which may be fixed. Notice that the contribution of having a loaded container movement must be carefully defined in the drayage problem. For example, even when the origin, the destination, and the start time of a loaded movement are fixed, the corresponding contribution for the loaded movement is not unique. This is because the movement may be moving one 20-foot container, two 20-foot containers, or one 40-foot container.

5.3   RESOURCE COUPLING AND POLICY RELAXATION

Regulatory policies are different for different types of drivers and tractors and depend on the current locations of the drivers and tractors. For example, the one-driver-one-tractor policy does not apply if the driver travels without crossing the border. However, if the driver crosses the border, the designated tractor has to be used. From this example, we can see that a driver can be a resource and a driver-tractor can also be a resource, and the decisions acting on the resources are of different types. This complexity is best handled by using the attribute-decision-modeling perspective.

In an attribute vector, a, that describes a resource, some elements are *static* (e.g., the identification number of a tractor)—will never change within the context of a model, but some are *dynamic* (e.g., location)—will change over time. A *combined* resource is formed by joining several individual resources. In the drayage problem, individual resource classes include the driver, tractor, and chassis, and the combined resource classes are the driver-tractor and the driver-tractor-chassis. The dimension of an attribute vector depends on the class of the resources.

There are three fundamental classes of decisions that can be made on individual resources and combined resources:

- *Couple* decisions that join different individual resources (e.g., combining a driver and a tractor) to form a combined resource
- *Uncouple* decisions that decompose a combined resource into individual resources
- *Adjust* decisions that change the resource's attributes such as moving the resource between locations

When a decision is made, the attribute vectors of the resources involved will be changed (including the values of its elements or even its dimensions). To tackle this type of attribute-decision model, we can build a candidate list, Q, that contains a list of attribute vectors. For a given attribute, a, we select and make the best decision among all decisions that can be applied. The decision, however, may result in some changes to the system such as making a previously made decision invalid, reducing or increasing the number of resources available, and changing the intrinsic values of the resources in the system. The major steps of this adaptive multiattribute labeling method follow:

Step 1: Initialization. Obtain the initial set of a, the candidate list Q, and the value of having a resource with attribute a.

Step 2: Making the decision. Remove an attribute vector, a, from Q, and select and execute the best decision, d^*, based on some cost evaluation (described below). Add new attribute vectors, a', to Q if needed.

Step 3: Decision adjustment. If the decision made in step 2 causes some previously made decisions to be invalid, then we need to adjust those affected decisions accordingly.

Step 4: Price and cost update. This includes reestimating the values of having resources with different attributes and the dual prices associated with the resources.

Step 5: Termination check. If the termination criteria are not met, then we repeat steps 2 to 4.

When making a decision in step 2, we consider three cost elements. The first is the direct contribution (or cost), c_{ad}, of the decision. The second is the estimated price, π_d, that we are prepared to pay for the resource used by decision d. The third is the future value that estimates the future benefit of having this new attribute, $v(a')$, of the attribute, a', that results from the decision just made. For a given attribute a, the best decision is then determined by maximizing the total contribution:

$$c_{ad} - \pi_d + v(a').$$

(5.8)

Obtaining $v(a)$ for all possible a is not practical in general, as the state space of the attribute vectors can be enormous. One approach is the use of aggregation. We can group a set of attribute vectors together and define a value function for the group as a whole. Alternatively, we can consider the use of predefined functional forms (e.g., polynomial functions) such that $v(a)$ can be expressed by a small number of parameters.

The price π_d measures the level of competition for getting the resource (if d is coupling a resource in order to meet a demand). The more intense the competition, the higher the price should be. One way to determine π_d is to use the concept of opportunity cost. Consider the best decision, d^*, and the second-best decision, d', that can be used on a. Let c^* and c' be the contributions associated with these two decisions. Then, we use $c^* - c'$ to increase the price of getting the resource to meet the demand associated with decision d^*. The rationale is that if another resource should be coupled to meet with the same demand, the benefit created by this new coupling should at least offset the opportunity cost.

One issue in this approach is how to maintain the candidate list, Q. For example, how should the elements be inserted and removed in Q (e.g., first-in-first-out or last-in-first-out) when these elements may affect the time complexity of the method? This certainly deserves more research effort.

Finally, using the attribute-decision perspective together with the solution approach, we can evaluate the impact of different types of policy relaxation. For example, let $D_a^{4\text{-up-4-down}}$, $D_a^{2\text{-up-2-down}}$, and D_a^{free} be the decision sets for attribution vector a under the four-up-four-down policy, two-up-two-down policy, and policy-free situations, respectively. Then, the total contributions obtained by using the corresponding models can provide an indication of the benefits of the policy relaxation.

5.4   IMPACT OF AVAILABLE INFORMATION ON MODELING

Suppose that the four-up-four-down policy is relaxed to a two-up-two-down policy and the one-driver-one-tractor policy remains intact. Thus, we can ignore the policy-imposed constraints (5.6) here. According to the availability of demand information and using the flow-oriented perspective, we can divide the decision models into four levels.

5.4.1   MYOPIC MODELS

The first level is whenever a transportation request, ℓ, is available, we immediately need to make the decision on whether a request is accepted or rejected and a decision on finding the best available driver to cover the request. This situation happens in spot markets or transportation bidding. With limited information, for request ℓ, we simply use the greedy rule to make the decision, that is,

$$r^* = \arg\max_r c_{r\ell}. \tag{5.9}$$

5.4.2   ASSIGNMENT MODELS

The second level is when we can make decisions in batches. That is, given a set of available resources and a set of available requests, we would like to match the resources and requests to maximize the total contributions. The situation usually happens when the information on requests is given in advance and the planner needs to schedule the resources in one stage (typically 2 to 3 hours) of a duty shift. The corresponding optimization problem is the classical assignment model written as max $c_{r\ell}x_{r\ell}$ subject to (5.5) and (5.7).

5.4.3   DETERMINISTIC LOOK-AHEAD MODELS

The third level is when we can make decisions in multiple stages. The demands in stage 1 are known and the average (or expected) amount of demands in the later stages is also available. Figure 5.2 shows a three-stage network. Using this framework, we can generate tours to cover the demands (the loaded container movement). A common approach is to use tour-generation techniques that have been applied widely in the crew-scheduling literature.

One limitation of this model is the assumption that vehicles are dispatched at the beginning of a stage. Suppose the start time of a dispatch is a decision variable. The model can be written as follows. Let

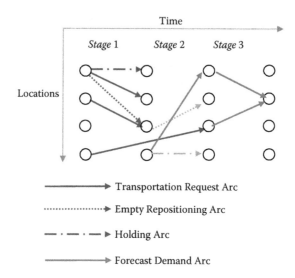

FIGURE 5.2 A three-stage network for the drayage problem.

S_i = the set of possible starting times of task i,
F_i = the set of possible finishing times of task i,
t_{ij} = the travel time from the finishing location of task i to the starting location of task j,
c_{ij} = the travel cost from the finishing location of task i to the starting location of task j,
r_i = the reward from task i,
τ_i = duration of task i,
c_{ij}^w = the minimum of the waiting cost per unit time at task i's finishing location and that at task j's starting location, and

$$x_{ij}^{fs} = \begin{cases} 1 & \text{if the driver who finishes task } i \text{ at time } f \in F_i \text{ starts task } j \text{ at time } s \in S_j, \\ 0 & \text{otherwise.} \end{cases}$$

For simplicity, define:

S_j = the starting time of task j (if it is covered) = $\Sigma_i \Sigma_{f \in Fi} \Sigma_{s \in S_j} s \cdot x_{ij}^{fs}$,
f_i = the finishing time of task i (if it is covered) = $s_i + \tau_i$, and
$x_{ij} = \Sigma_{f \in \mathcal{F}_i} \Sigma_{s \in S_j} x_{ij}^{fs}$.

Then, the constraints of the problem are

$$\sum_i x_{ij} \leq 1, \tag{5.10}$$

$$\sum_i x_{ij} = \sum_k x_{jk}, \tag{5.11}$$

and

$$f_i + t_{ij} - s_j \le (1 - x_{ij})M. \tag{5.12}$$

Constraint (5.10) ensures that each task is covered at most once. Constraint (5.11) is the flow conservation constraint: a driver who covers a task will be available later. Constraint (5.12) requires that a task's starting time be later than the time when the driver arrives at the task's starting location. In addition, we have to ensure that all the starting time windows of tasks are satisfied and the integrality requirement is met. The objective of the problem is to minimize the total assignment costs minus the total rewards. The assignment cost for the driver who finishes task i and then starts task j, denoted by $\bar{c}_{ij}$, consists of the travel cost and the waiting cost:

$$\bar{c}_{ij} = c_{ij} + \max\{s_j - (f_i + t_{ij}), 0\}c_{ij}^w .$$

Therefore, the objective function of the problem is

$$\min \ \sum_i \sum_j (\bar{c}_{ij} - r_j)x_{ij}.$$

Notice that the problem size grows quickly as the numbers of drivers, containers, and decision epochs increase. In this case, use of the attribute-decision-modeling perspective and the associated solution approach is helpful. For the special case of a single type of resource, Powell et al.[15] propose a labeling-based algorithm that works well for the situation when decisions are made on a rolling horizon basis. The algorithm is a specialized version of the adaptive multiattribute labeling method described in section 5.3 where a label can be viewed as a special type of attribute vector.

5.4.4 Stochastic Models

The fourth level is when we make decisions in multiple stages under uncertainty. One major source of uncertainty for the cross-border drayage problem is market demand. Notice that the number of loaded movements that are revenue generating is limited by the market demand. When using the flow-oriented perspective, market demands can be modeled as arc capacities of the loaded movement arcs (see figure 5.2). Assuming the distributions of the random demands in the later stages are known, we arrive at the multistage stochastic optimization model:

$$\min_{x_1 \in X_1} c_1^T x_1 + E\left\{ \min_{x_2 \in X_2} c_2^T x_2 + E\left\{ \min_{x_3 \in X_3} c_3^T x_3 + E\left\{ \cdots E\left\{ \min_{x_N \in X_N} c_N^T x_N \right\} \cdots \right\} \right\} \right\}, \tag{5.13}$$

where C_n, x_n, X_n represent the cost vector, the decision variable vector, and the constraint set in stage n, respectively. The element $x_{n,ij}$ of vector x_n is the number of vehicles moving from location i to location j in stage n. One solution strategy for such complex model is to use some simpler functions to approximate the objective function. Problem (5.10) can be rewritten as

$$\min_{x_1 \in X_1} c_1^T x_1 + EQ(s_2) \qquad (5.14)$$

and for $n = 2, 3, \ldots, N - 1$. Notice that we use N to represent the number of stages instead of T to avoid the confusion with the transpose operator T.

$$Q(s_n) = \min_{x_n \in X_n} c_n^T x_n + EQ(s_{n+1}), \qquad (5.15)$$

where s_n is the vector representing the available trucks at the locations at the beginning of stage n, that is, $[s_{n+1}]_j = \sum_i [x_n]_{ij}$.

In the approximation, the complex term $EQ(s_{n+1})$ in (5.15) is replaced by a more tractable function, $\hat{Q}_{n+1}(s_{n+1})$. Two functional forms are more promising as they facilitate the use of special problem structure. The first is the linear function:[16]

$$\hat{Q}_{n+1}(s_{n+1}) = \sum_j v_{n+1,j} [s_{n+1}]_j, \qquad (5.16)$$

where $v_{n+1,j}$ is the estimated average cost of having a vehicle arriving at location j in stage $n + 1$. The second is the separable, convex, piecewise linear function:

$$\hat{Q}_{n+1}(s_{n+1}) = \sum_j \sum_{k=1}^{[s_{n+1}]_j} v_{n+1,j}^k, \qquad (5.17)$$

where $v_{n+1,j}^k$ is the estimated marginal cost of having the kth vehicle arriving at location j in stage $n + 1$. The $v_{n+1,j}^k$ are increasing in k, reflecting the convexity (and the diminishing returns of having more vehicles arriving at a location). A decomposition method has been developed to estimate $v_{n+1,j}^k$ [17] that approximates the stage $n + 1$ problem by a set of stochastic trees that are easier to solve.[18]

5.5 CONCLUDING REMARKS

We have described the challenges in managing cross-border drayage operations, introduced new modeling perspectives and a constraint classification, and reviewed various models and their respective or potential solution methods. There are several likely developments in drayage operations. One of them is the use of relay centers that are close to the border, where containers can be exchanged. This allows a platform for better utilization of resources, if the restrictive regulatory policies are relaxed. Another is the use of barges in the last-mile connections. A barge can hold many containers and thus lead to a lower per-container transportation cost. These upcoming developments certainly create new challenges and thereby require more research efforts.

ACKNOWLEDGMENT

The author thanks the Research Grants Council of Hong Kong for supporting this research through grant HKUST612206.

REFERENCES

1. Steenken, D., S. Voß, and R. Stahlbock. 2004. Container terminal operation and operations research: A classification and literature review. *OR Spectrum* 26:3–49.
2. The Better Hong Kong Foundation. 2004. Restoring Hong Kong's competitiveness as a sea-trade logistics hub. News release of the Better Hong Kong Foundation.
3. Hong Kong Legislative Council. 2004. Minutes of meeting held in November, p. 11, part 36. http://www.legco.gov.hk/yr04-05/english/panels/es/minutes/es041122.pdf.
4. Yvan, D., D. Jacques, and S. Francois. 1991. Pickup and delivery problem with time windows. *Operational Research* 54:7–22.
5. Arunapuram, S., K. Mathur, and D. Solow. 2003. Vehicle routing and scheduling with full truckloads. *Transportation Science* 37:170–82.
6. Bodin, L., and B. Golden. 1981. Classification in vehicle routing and scheduling. *Networks* 11:97–108.
7. Desrosiers, J., F. Soumis, and M. Desrochers. 1984. Routing with time windows by column generation. *Networks* 14:545–65.
8. Fisher, M. 1995. Vehicle routing. In *Handbooks in operations research and management science*, Vol. 8, *Network routing*, ed. M. Ball et al., 1–33, Amsterdam: Elsevier Science.
9. Dror, M., G. Laporte, and P. Trudeau. 1998. Vehicle routing with stochastic demands: Properties and solution frameworks. *Transportation Science* 23:166–76.
10. Swihart, M., and J. Papastavrou. 1999. A stochastic and dynamic model for the single-vehicle pick-up and delivery problem. *European Journal of Operational Research* 114:447–64.
11. Powell, W. 1986. A stochastic model of the dynamic vehicle allocation problem. *Transportation Science* 20:117–29.
12. Cheung, R., and D. Hang. 2003. A time-window sliding procedure for driver-task assignment with random service times. *IIE Transactions* 35:433–44.
13. Ball, M., B. Golden, A. Assad, and L. Bodin. 1981. Planning for truck fleet size in the presence of a common carrier option. *Decision Sciences* 14:103–20.
14. Desrosiers, J., Y. Dumas, M. Solomon, and F. Soumis. 1995. Time constrained routing and scheduling. In *Handbooks in operations research and management science*, Vol. X, *Networks*, eds. C. Monma, T. Magnanti, and M. Ball. North Holland.
15. Powell, W., W. Snow, and R. Cheung. 2000. Adaptive labeling algorithms for the dynamic assignment problem. *Transportation Science* 34:50–66.

16. Frantzeskakis, L., and W. Powell. 1990. A successive linear approximation procedure for stochastic, dynamic vehicle allocation problems. *Transportation Science* 24:40–57.
17. Cheung, R., and W. Powell. 1996. An algorithm for multistage dynamic networks with random arc capacities, with an application to dynamic fleet management. *Operations Research* 44:951–63.
18. Powell, W., and R. Cheung. 1994. Stochastic programs over trees with random arc capacities. *Networks* 24:161–75.

6 Container Port Choice and Container Port Performance Criteria

A Case Study on the Ceres Paragon Terminal in Amsterdam

Rutger Kroon and Iris F. A. Vis

CONTENTS

ABSTRACT

Within maritime logistics the containerized trade market is growing rapidly with the uprising of the Far East. European container port competition among the ports in the Le Havre–Hamburg range is fierce as they are threatened by a shortage of terminal capacity. The port of Amsterdam identified this threat and realized a brand new container terminal, the Ceres Paragon Terminal, in 2002. Characterized by a revolutionary concept known as an indented berth, served simultaneously by nine ultramodern post-Panamax gantry cranes, high productivity levels and low turnaround times can be obtained. Although the odds seemed favorable for the new terminal, enthusiasm was replaced by vexation as the terminal experienced a dramatically slow start. For years it was barely operational with only an incidental test run and some feeder and barge movements. Finally the first carriers were contracted in July 2005.

The objective of this chapter is to study relevant main port choice and port performance criteria identified in literature. Reviewing these criteria for the port of Amsterdam gives us the opportunity to make an assumption on the port's and terminal's chances for structural establishment among the competitive West European port arena.

6.1   INTRODUCTION

The growth in seaborne trade of containerized cargo has outstripped the growth in world trade in general and world economic growth in particular since the introduction of the container during the 1950s on the west–east/east–west long-haul trades. As we write, volumes of containerized cargo are still growing relatively rapidly. More diverse cargo is being containerized and export and imports are increasing on a global scale. The relatively high growth rate for the global containerized trade is initiated by the strong uprising for the developing and transitory countries with respect to their trade volumes. Within these groups of countries Asia is responsible for the highest containerized cargo volumes in global trade, nowadays, and determines the containerized trade scene to a large extent. The containerized trade sector benefited especially from the strong growth of the Chinese economy. To keep up the pace and provide for the necessary capacity and tools to transship these massive volumes, container ports worldwide should be responsive and on guard in order to retain their levels of competitiveness.

The main stakeholders within the container transport chain are carriers, terminal operators, port authorities, regional and national authorities, transport companies, and, of course, the clients, i.e., the shippers of the containerized goods. Government organs set up a protocol for port authorities for granting port access and (partly) provide for infrastructural development. The actual port control on a daily basis lies with the port authority. This organ grants access to vessels and provides infrastructure for terminal areas. In return, they receive payments. The terminal operators run actual operations on a daily basis. A cash flow is generated through transshipment of containers, thereby serving the carriers. The terminal and carriers usually work according to contractual agreements. The containers find their way to and from the hinterlands through continental transport companies. In this chapter, the emphasis will be on the three stakeholders that are directly involved in the money flows: the port authority, terminal operators, and carriers.

In Europe, the leading container ports showed an increase in container throughput during the last years. Several of the ports located in the European Le Havre–Hamburg range are struggling with the enormous amounts of containerized cargo. Terminal capacity and throughput to the hinterlands are under strain on the short-term competition between the ports, with respect to future growth increases. Therefore, new projects with respect to capacity increases are deployed by established container ports as well as general and smaller ports and port areas, not recognized as container ports. One of these ports is the port of Amsterdam, established as an important bulk port but without any significant number of container transshipment.

In the middle of the 1990s the Greek-American stevedore Kritikos, chairman of Ceres Terminals Incorporated, appeared as the redeemer of the port of Amsterdam. He would establish the "container port" Amsterdam by giving the city the world's most beautiful container terminal, capable of handling at least 650,000 containers a year. He did so by realizing the Ceres Paragon Terminal in 2001.

Giants they are, the nine state-of-the-art cranes of the Ceres Paragon Terminal located on the quay at the "America" port in the western Amsterdam port area. They are rated by port technicians as the fastest, most efficient, and quietest cranes in the world, able to serve container vessels of all sizes, including the post-Panamax-plus container vessels. According to plans, approximately 300,000 containers should have been transshipped at the terminal in 2004, to eventually rise to 650,000 in 2005. However, up to the middle of 2005 the terminal had not handled a single container.

In September 2002 the Japanese company Nippon Yusen Kaisha (NYK) took a 50% share in the Ceres organization and the Ceres Paragon Terminal. Together with four other intercontinental container shipping companies (Hapag-Lloyd, Orient Overseas Container Line, Malaysia International Shipping Corporation Berhad, and P&O Nedlloyd), NYK operates in a consortium. As a member of this Grand Alliance, NYK initiated a lobby to encourage members to shift their actions to the port of Amsterdam. But all attempts seemed futile due to the fact that P&O Nedlloyd had a considerable interest in the Euromax terminal located at the port of Rotterdam. Through a billion-dollar takeover, Maersk Sealand recently became the full new owner of P&O Nedlloyd. Although the Grand Alliance has lost an important and powerful ally, this new situation offered the Ceres Paragon Terminal new chances. The Grand Alliance decided to operate two services to and from the port of Amsterdam and its terminal. Two vessels a week service their A-loop (Japan–Europe) and new F-loop (Far East–Europe). With an expected annual transshipment volume of 130,000 containers,[1] it was a meager start for such a sophisticated terminal. However, the terminal was finally offered an opportunity to prove its potential.

The container ambitions of the port of Amsterdam and the arrival of the Ceres Paragon Terminal were based on present market development for the containerized trade sector. At first sight the developments opened up major opportunities. But it became clear in the years following that there was more to it than just opportunities for the container terminal.

This chapter finds its objective in clarifying causes for the dramatically slow start of the Ceres Paragon Terminal. Container port choice criteria, favored by the main stakeholders identified earlier, create the basis for this chapter. We will introduce them in the next section. In section 6.3 we give an introduction on the port of Amsterdam and the Ceres Paragon Terminal in particular. In section 6.4 we utilize the port choice criteria as a tool to review the potential of Amsterdam in its favored role of container port by describing its strengths and weaknesses. The political opportunities and threats and the terminal's chances for structural establishment among the competitive West European port area are discussed in section 6.5. Will the terminal succeed to establish itself or is it offered false hope?

6.2 PORT STAKEHOLDERS' INTERESTS

This section offers a framework for the analysis on Amsterdam's container ambitions. We will start with the introduction on port choice criteria since these explain the initial choice for a particular port. These criteria do not guarantee the continuity of the port, however, as there is more to it than this initial choice. Actual port performance will be discussed in sections 6.2.2 and 6.2.3. Port performance will depend on terminal productivity and service degree to a large extent.

6.2.1 CONTAINER PORT CHOICE CRITERIA

The routing for a container transport is dependent on the choice of port(s) in the region of origin and the region of destination. Nowadays there is a considerably wide choice among container ports that are all within the perimeter of these regions. An example is formed by the ports in the European North Sea region (see figure 6.1).

The North Sea ports in the Le Havre–Hamburg range handle a large part of the container traffic to and from the European continent. The ports of Hamburg, Bremen, Rotterdam, and Antwerp are the largest ports in the region. Shippers and receivers located in, for example, Ludwigshaven (Germany) have the choice to transship their containers, to and from the Far East, via different combinations of:[3]

- Shipping line
- Port of call
- Inland transport mode (train, truck, barge)

With, say, 20 carriers, 4 ports of call, and 3 modes of inland transport (though not all combinations are relevant), the number of different routings serving a particular region easily exceeds 100. For routings including a hub port focusing on sea–sea transshipment, the number of options is even greater.

In literature a large set of port choice criteria has been presented.[4,5] These criteria can be classified in four main criteria corresponding to the port's physical and technical infrastructure, its geographical location, its management and administration perspective, and its carrier and terminal cost perspective. These four main criteria hold a number of subcriteria on port choice. Previous research[2,4–6] ranked these with respect to importance based on main stakeholders (i.e., carriers, port operators, port authorities, and shippers):

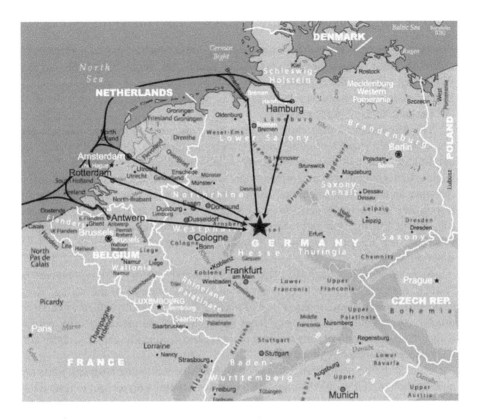

FIGURE 6.1 Different routing to one particular hinterland destination.[2] (Based 6 N Veldman and Buckmann (2003). http//www.johomaps.com/eu/germbenelux/germbenelux-ANN.html.)

- Handling cost of containers
- Geographical location:
  - Proximity to main navigation routes
  - Proximity to import/export areas
  - Proximity to feeder ports
- Basic infrastructural condition
- Intermodal links

The handling cost criterion is emphasized by the main stakeholders. The degree of importance for the rest of the criteria differs slightly among the main stakeholders. Intermodal links, for example, are particularly valued by the shippers. The basic infrastructural condition is an important port operator aspect. Carriers are more concerned with the proximity to main navigation routes as costs vary considerably with deviating routes.[2]

Extension of the discussion on port choice will eventually lead to a discussion on the port of choice's (and terminal's) performance. This provides a tool for measurement of the quality of the actual transshipment process. It provides a basis for remarks concerning the port's potential, stability, and continuity.

6.2.2 CONTAINER PORT PERFORMANCE

Productivity and efficiency are the two most important concepts in port performance, especially from the perspective of its main clients: the shippers and the carriers. Productivity of a producer can be loosely defined as the ratio of output(s) to input(s). Efficiency can be defined as relative productivity over time or space, or both.[7] Both are measurements of performance.

The discussion on port choice criteria has demonstrated that the location and cost criteria are important in the initial port choice process. Costs and location are not directly, although they are indirectly, related to a port's (and terminal's) performance. Criteria that are categorized under the port's physical and technical infrastructure and its management and administration focus on many variables directly responsible for the port's performance.[2] Many of these criteria, however, have been subordinated to the cost and location criteria identified by the main stakeholders. Examples are the port's technical structure and vessel's turnaround time, both subcriteria.

As a consequence, from the emphasis on cost and location criteria, "bigger" or "busier" does not always automatically mean "better," phrased by the saying "King of the hill doesn't always mean prince of ports."[8] Ports frequently called by carriers, usually corresponding to the very large ports, do not automatically offer performance levels substantially higher than do ports less frequently called, usually corresponding to smaller, regional ports.

Understanding performance is a concept fundamental to any business. Ports are no exception, and it is only by comparison between them, with respect to time intervals, cost structures, and service degree, that performance can be properly evaluated. Ports are, however, complex entities with many different sources of inputs and outputs, making direct comparison among apparently homogeneous ports difficult. Various port types[9] and port ownership and organizational structures, existing throughout the world, complicate it further. Since our research goal needs a positioning of the port of Amsterdam, comparisons between the port and its competitors will be made up to the terminal level in section 6.4. First, the relationship between a port's performance and vessel's waiting time will be introduced.

6.2.3 AVERAGE WAITING TIME FOR CONTAINER VESSELS

The performance of a port, from the perspective of the main stakeholders, directly corresponds to the average waiting times encountered by its clients. Important factors that contribute to waiting time are the delays in port transit and delays at the (terminal) quay.

The average waiting times encountered by vessels at a specific port are important for economic comparisons of the different situations, as time means money. Waiting times have a large impact on the total time of the vessel spent in the port. Frequent or increasing waiting times for vessels calling a specific port can have negative consequences for the number of vessels visiting that port, depending on the extent of the delay and the port's (economical) importance. Comparison of the present waiting times with forecasted waiting times might help to create a future perspective for the port.[10]

Costs for bulk transports that are the result of additional waiting times, caused by unexpected delays, are relatively low: 1.4 eurocent per ton per hour (t/h) for dry bulk and 4.5 eurocent per t/h for wet bulk. For container vessels, however, the costs accompanying waiting time are much higher since container carriers adapt a line services system: at least 8.2 eurocent per t/h.[11] Unexpected waiting times can cause maladjustments from sail schedules that can lead to negative consequences for later links in the transport chain. Moreover, the container transport often concerns (expensive) industrial goods. Bulk carriers do not adopt a line service, use vessels that are relatively cheap, and transport relatively cheap goods. In its shift from a bulk to a container port, Amsterdam should keep in mind these important differences between both cargo forms and the consequences they have for port operations.

6.2.4 RECONSIDERING PORT CHOICE

When a particular port of choice is not able to offer desired levels of efficiency and performance, or charges tariffs that are unsatisfactory, carriers and shippers will reconsider their port choice. The decision to continue or seize calling that particular port is a complex one, however. Costs accompanying a change of preferred port of call have to be offset by higher revenues generated with the new port of call.

Besides the cost-revenue balance, such a change process creates a considerable amount of uncertainty as well. If the new port of choice does not live up to the client's expectations, a smooth return to its initial base port will not be easy. As container terminal capacity is scarce, idle capacity will be seized immediately by others. Carriers that want to return to their former base port might end up waiting in line for available terminal capacity and risk the loss of past privileges. Carriers and shippers might therefore reconsider a possible port shift as uncertainties add up. The power of the established terminal operators therefore seems to be considerable.

With a theoretical basis consisting of port choice criteria, port performance, its correspondence with waiting times, and the power of an established port, the port of Amsterdam will be analyzed in the next section. The section starts with an introduction to the Amsterdam port area. From that point it will narrow down to the Ceres Paragon Terminal.

6.3 PORT OF AMSTERDAM

The Amsterdam port and its satellite ports (Beverwijk, Velsen/IJmuiden, and Zaanstad) are ranked sixth within the EU with respect to total throughput volume (tons).[12] The port of Amsterdam has established itself in the North Sea region as an important bulk port. Vast quantities of dry bulk are transshipped in the port, including goods processed into semimanufactured products like feed and grain products. Besides having these important cargo forms, Amsterdam is the largest cocoa port in the world. In the first half of 2006 Amsterdam showed a growth of 18% compared to the same period in 2005. This was the highest growth among the ports in the northwestern part of Europe.[13]

The port of Amsterdam is one of very few significant ports to be fully entered through a lock complex. One of the world's largest lock complexes separates the

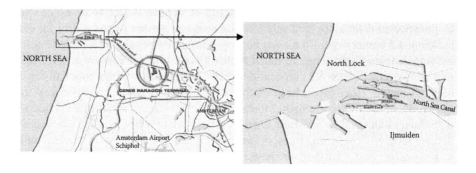

FIGURE 6.2 The Amsterdam seaport area and the sea lock complex at Ijmuiden.[14]

North Sea and the port's access canal. It comprises four locks and is operational 24 hours a day. Only the largest lock, the north lock, is compatible for the first generation of post-Panamax-plus container vessels (5,000–8,000 TEUs). Through the canal, the North Sea canal, the port area is fully accessible. The transit from lock entrance to berth at the Ceres Paragon Terminal takes, on average, 2.5 hours. Figure 6.2 shows the port entrance and lock situation for the port of Amsterdam.

Although established internationally as a bulk port, the port of Amsterdam hardly handled any containerized trade, however. With a trend of decreasing bulk cargoes at the expense of increasing amounts of containerized trade, the port of Amsterdam has made a disputed attempt, with the realization of the Ceres Paragon Terminal, to enter the container market.

6.3.1 Ceres Paragon Terminal

The Ceres Paragon Terminal was realized as a joint project of Ceres Terminal, Inc., and the Amsterdam Port Authority. Total investments were estimated to be €172 million, of which the Amsterdam Port Authority invested the larger part, €128.5 million in infrastructure and part of the cranes. The remaining part of the investment, € 43.5 million, was made by Ceres Terminal, Inc. As explained in the introduction, the Japanese shipping and transport company Nippon Yusen Kaisha (NYK), a member of the Grand Alliance consortium, has been a partial owner of the Ceres Paragon Terminal since 2002.

The terminal area covers 54 hectares, has a total quay length of 1,050 m, and an annual capacity of 1,000,000 TEUs. The terminal introduced a revolutionary and unique concept among container terminal facilities: a so-called indented berth enabled post-Panamax vessels to be serviced by a maximum of nine gantry cranes, having a reach up to twenty-two containers across deck from *both* sides of the vessel (see figure 6.3). This made it possible to enhance productivity to a high standard.

The indented berth has a length of 400 m and a width of 57 m and was designed for the latest generation of container vessels: the post-Panamax-plus category vessels. It also offered a classic quay with a length of 615 m where a maximum of five gantry cranes could be assigned to one vessel. The total berth time of the indented berth is accepted to be within 15 minutes.[14] The terminal thereby seems to be able

FIGURE 6.3 Docking situation at Ceres Paragon Terminal, Amsterdam.[14]

to offer a full package of services, and there is also space available to expand the terminal's activities in the future.

6.4 PORT ANALYSIS

Port and terminal comparison between the port of Amsterdam and its direct competitors, Rotterdam and Antwerp, although established as massive global container ports already, might be interesting to further explicate Amsterdam's competitive potential. We will begin in section 6.4.1 by comparing the main terminals in these container ports, forming the basis of the port's actual performance. Section 6.4.2 and further provide feedback on the port choice criteria, important in the initial port choice.

6.4.1 TERMINAL PERFORMANCE COMPARISON

If it is indeed as revolutionary as presumed, the indented berth might prove to be one of the terminal's strengths. The first reliable productivity figures cannot be reviewed yet, with the terminal only being fully operable since July 2005 and only functioning on a small percentage of its capacity (around 15%). Productivity gains from operating an indented berth are therefore not yet guaranteed. However, we can show some theoretical figures. With an average capacity of 25 movements per hour for one gantry crane, the capacity to load or discharge a container vessel with five cranes can theoretically lead to 125 movements per hour. With the introduction of the indented berth concept, productivity of the Ceres Paragon Terminal is to be doubled to at least 250 picks per hour. The cranes work independently from the transport process regarding supply and evacuation of containers, as straddle carriers are deployed. Turnaround times (and with that, costs) could be reduced to a considerable extent this way, estimated to be in the range of 30%–50%.

The Rotterdam ECT terminal is characterized by having a huge capacity. Productivity per berth, however, remains relatively low (70 units per hour per berth), with a maximum of five cranes per berth. Productivity of the cranes is also dependent on the productivity of the automated guided vehicles (AGVs) that the Delta terminal deploys. Antwerp, with its not yet (fully) automated container terminals, reaches a somewhat higher productivity per berth—110 units per hour per berth—one of the aspects that keep the port competitive, besides its favorable prices.

When productivity levels of the two massive terminals are offset against the productivity levels of the Ceres Paragon Terminal, the difference is striking. A marginal note must be made here. This productivity level for Ceres Paragon is reached with the indented berth fully serviced by nine cranes. With another vessel berthed simultaneously at the other, classic berth, no more cranes can be assigned to that particular ship. When the nine available cranes are divided over both vessels, respective productivity levels will be reached but will not match the massive number of 250–300 units per hour per berth. The other terminals operating the ports of Rotterdam and Antwerp have several berths that can be serviced simultaneously by a constant number of cranes. Overall productivity for the total number of berths can thereby be increased considerably. Productivity levels for Ceres Paragon remain striking, however. Numbers on terminal capabilities are summarized in table 6.1.

6.4.2 PORT CHOICE COMPARISON

Terminal comparison led to a satisfactory number on productivity for Amsterdam's Ceres Paragon Terminal. Container terminals are, however, dependent on a large set of port criteria in addition to their independent capabilities. A port comparison between the identified ports is therefore a necessity. Port comparison between these ports will be based on the main port choice criteria identified earlier.

The three criteria that correspond to the port's geographical location seem to be rated in basically the same way for the three ports. The ports are in each other's direct vicinity, with the distance between the two most remote ports, Amsterdam and Antwerp, being only a sheer 100 miles. International stakeholders' perception with regard to proximity to feeder ports, proximity to import/export areas, and proximity to main navigation routes seems not to differ to a large extent for both ports.

Although this statement is open to some ambiguity, it is the other important port choice criteria on which findings for the ports might deviate considerably: the ports' *infrastructural basis*, intermodal linkage, and handling cost of containers.

The port of Rotterdam, positioned among the largest (container) ports in the world, owes a considerable part of its status and rank to its favorable geographical

TABLE 6.1

Terminal Comparison between the Main Container Terminals in Amsterdam, Rotterdam, and Antwerp: Ceres Paragon Terminal, ECT Delta Terminal, and North Sea Terminal[14]

	Amsterdam		Rotterdam	Antwerp
	Paragon		ECT Delta	North Sea
	Ceres/NYK			
Terminal Operator	Indented	Marginal	ECT	Hesse-Noord Nation
Berth length (m)	400	612	4,500	1,200
Water depth (m)	13.7	13.7	16.6	14.3
Terminal area (hectares)	54		236	80
Reefer plugs (units)	433 (850 later)		2,696	660
Annual capacity	1,000,000		4,100,000	1,500,000
Main type of operation	Straddle carrier		AGV	Straddle carrier
Gantry cranes	9		25a	10a
Outreach	22 slots		16–18 slots	18–20 slots
Productivity per berth (units/hour)	250–300		70	110
Ship cargo operation	24 hours		24 hours	24 hours

[a] Being expanded.

location. It is centrally situated to the main trunks, either eastbound or westbound. It occupies a central position with regard to feeder ports throughout the (West) European continent. And it finds itself in a highly competitive position when it comes to its closeness and reach, e.g., its intermodal network, to the main European import and export areas.[2] The advantageous and competitive outcomes with respect to these three important main choice criteria are complemented by high ratings for the other identified choice criteria. The port's nautical access, as a basic infrastructural condition, is quite optimal. It offers an easy and continuous access through a more than sufficient depth. Its intermodal linkage is quite optimal as barge, rail, and road networks are extensive and of high quality. Container handling costs are competitive, especially since run-up costs to the port are kept low due to the favorable geographical and infrastructural situation. Since the port of Rotterdam seems to have acquired an ideal position with respect to the choice criteria, it will serve as a reference point.

6.4.3 Port's Infrastructural Condition

The nautical accessibilities for the three ports are quite different from each other. The port of Amsterdam is characterized as a lock system port, as is the port of Antwerp, partially. Because the main container operations with the port of Antwerp are executed in front of the lock complex, in tidal river waters, it will not be designated a lock port. The port of Rotterdam is directly accessible from deep water.

The nautical accessibility, as one of the main issues for the infrastructural basis,[4,5] will be discussed next for the port of Amsterdam and will be put in perspective with the other ports. Central in this discussion are the lock complex and the port's access canal, the North Sea canal.

6.4.3.1   Sea Lock Complex

A lock system has the advantage of ruling out tidal differences that might facilitate load and discharge operations. It offers a continuous 24-hour port access. It is the disadvantages that dominate the conversation, however. Load and discharge problems that accompany tidal differences are mainly caught up with by today's technology being applied in quay cranes and so forth. The disadvantages that accompany a lock complex are not ruled out that easily.

What follows is an overview of the bottlenecks that might be encountered with Amsterdam's lock complex in case of increasing inbound and outbound flows of cargo:

- The vulnerability of the lock system
  - The risk of encountering damage upon entering the large north lock depends on a combination of wind force and direction. Negative advice is given when wind forces exceed 6 Bft (i.e., wind speed) and wind blows from any other direction than parallel to the lock. Container vessels in particular are vulnerable to wind force and direction as their draft–height ratio is low.[15]
  - Unexpected jams in the lock complex, in particular jams in the north lock, which forms an obligatory link for the large vessels that have a destination beyond the lock.[16]
- The dimensions of the vessels calling the port of Amsterdam:[16]
  - The middle lock is characterized by a beam restriction, which can only handle vessels of well below Panamax dimensions, leaving the north lock as the only option for a majority of the deep-sea vessels calling at the port.
  - With the inevitable arrival of a new generation of vessels (very large container vessels [VLCVs], e.g., the *Emma Maerk*), the dimensions of these vessels, with respect to length, beam, and draft, might be beyond the dimensions of the north lock.
  - The outport at the seaside of the lock complex has been assessed too tight to grant an easy maneuvering for large vessels.
- The lock process
  - Liner vessels plan port calls as far as 3 months ahead. Bulk carriers usually make an unannounced call only 12 to 24 hours before arrival. This could lead to possible additional waiting times for liner shipping.[17]
  - The lockage times (including waiting times) at passing.[16]

At a certain point in time, heavily depending on the future development for the port of Amsterdam and the success of the Ceres Paragon Terminal, the maximum capacity of the lock complex will be reached. The waiting times at the lock complex might increase exponentially from that point:[16] estimates of total cargo flows for 2020 destined for the port of Amsterdam, under the current lock complex, vary from 64 to 73 million tons (including container traffic). The large difference (14%) between these estimates, however, might have a doubling effect on waiting times, from 165 to 330 minutes.[18] These times reflect total waiting times, including delays encountered with the lock processes, but also possible waiting times that stem from problematic passages at the North Sea canal, to be discussed later in this section.

As a result of congestion at the lock complex, liner vessels will take evasive measures. This will lead to stagnation of the volume growth for the Amsterdam port area. The first to make evasive maneuvers, involving extra costs, are the container carriers as they experience the highest costs that accompany waiting times. Moreover, a considerable amount of cargo might head for foreign destinations, through which additional loss of prosperity for the Netherlands might be encountered.

The outlined waiting time problem that might originate from extra pressure on the north lock and possibly lead to congestion seems to arise especially with an increasing number of container vessels. The construction of an extra lock is apparently an important edge condition for the future success of the container ambitions of Amsterdam and the Ceres Paragon Terminal.

Although it will not offer a solution to all the problems outlined above, an additional lock would have the following significant benefits:[19]

- It allows the port of Amsterdam to grow its traffic volume as cargo projections indicate negative growth with the current infrastructure.
- It increases the number of usable locks (i.e., usable by modern commercial deep-sea vessels) from two to three, which would provide sufficient capacity to accommodate increasing cargo volumes and associated vessel activity.
- It permits two or even three vessels to transit simultaneously in the new lock, thus helping to constrain lock operating costs, and further increase lock capacity.
- It makes the capture of significant container traffic volumes more feasible by reducing or even eliminating potential congestion at the locks.

In a research report prepared for the Dutch Ministry for Transport, Drewry Shipping Consultants[20] arrived at some interesting numbers when comparing the current lock situation with an additional lock; forecast developments with respect to container vessel size calling the port of Amsterdam in 2020 show an average size of 4,223 TEUs versus 5,698 TEUs for both scenarios, respectively.[19] The present lock has dimensions that allow the latest generation of container vessel. The margins are very small, however. The dimensions of the projected new generation of vessels will not be able to pass the lock. Therefore, stagnation of the average size of a container vessel will be encountered eventually without a new lock.

The assumption made here is that major customers are to be won with the introduction of a new lock. Dependent upon the lock infrastructure provided, new lines will be won that are active in the two European east–west trunks (i.e., to Asia and North America) and that there will also be some additional north–south trade traffic as well. The capture of a main Europe–Asia trade service would be a potential catalyst for a broadening of the services using the port, due to the generation of interline possibilities.

A decision concerning a new lock will ultimately be taken in 2008. Costs are high, as they are estimated to be between 450 and 550 million euros. The central government, one of the main stakeholders with respect to this issue, is yet convinced of the necessity of a new lock.[20] As a result, the construction of a new lock will not be started on a short notice. In addition to this infrastructural bottleneck, another might present itself when the number of ship movements to and from the Amsterdam port area rises. It has to do with its access canal, the North Sea canal.

6.4.3.2 North Sea Canal

As the canal has limited dimensions, the influence of wind, water displacement by the vessels, and the (few) turns in the canal might lead to problematic ship passages. This can lead to additional waiting times. The number of problematic passages nowadays is still acceptable, being restricted to some meetings per week.[10] With the realization of the Ceres Paragon Terminal, container vessels categorized in the largest vessel section were also expected to call the port on a frequent basis. If the new lock is approved, more and larger vessels should be able to call the port. For this reason, it is expected that the probability of difficult, or even impossible, passages on the North Sea canal would increase in the future.

Some suggestions to control or overcome these North Sea canal limitations have been put forward:[10]

- Broadening of the canal: The North Sea canal has a bank relating 1 to 3 that confiscates 100 m of the width of the waterway, since the total surface width is 270 m and the canal's fairway comprises only 170 m. If dam walls are created at the edge of the banks and the banks are dug off to a depth of 15 m or beyond, a waterway arises that is 270 m wide instead of the current 170 m. This could be applied particularly at "turning points" of the canal. The disadvantage of dam walls is the reflection of waves that might have negative consequences for the vessels sailing the canal.
- Limitation of wind: Since the vessels sailing the North Sea canal seem sensitive to wind, a solution might be found in the placement of wind awnings alongside the canal, again, particularly at the turnings. The transverse winds will have less influence, in the form of drifting, on the larger vessels.

If no measures are taken, serious nautical access obstructions for sea shipping are expected by the year 2010. These will definitely have consequences for the international competitive position of the port as waiting times will be raised to an unacceptable level.

To offer an acceptable port waiting and transit time, the Amsterdam port area should consider its options. The disadvantages that accompany the nautical accessibility stand in strident contrast to a (tidal) port such as Rotterdam. A direct, deepwater connection to sea, however, does not always mean a win-win situation, as is the case for the port of Rotterdam. The port of Antwerp, for instance, experiences

TABLE 6.2
Navigation Comparison between the Ports of Amsterdam, Rotterdam, and Antwerp[14]

	Amsterdam	Rotterdam	Antwerp
	Paragon		
	Indented Berth	ECT Delta	North Sea
Terminal Operator	Ceres/NYK	ECT	Hesse-Noord Nation
Length of waterway			
Pilot station–lock (km)	16	19	122
Lock–berth (km)	12		
Width (m)	275.0		>300
Acceptable draft (m)	13.7	22.5	14.5
Water depth (m)	13.7	24.8	14.0
Time (average, hours)			
Pilot station–lock	1.0	2.0	6.0
Lock transit	0.75		
Lock–berth	1.75		
Total duration			
Pilot station–berth	3.5	2.0	6.0

severe problems when it comes to tidal differences as the port's nautical entrance, the Westerschelde, is characterized by shallow waters at low tide. Vessels have to follow strict procedures to enter the port of Antwerp to prevent running aground, thereby experiencing additional waiting times. Transit times may add up to 6 hours before a container vessel berths. Table 6.2 gives an overview of numbers with respect to nautical accessibility, as it forms the main aspect with respect to the port's infrastructural basis, and the accompanying vessel transit times.

The infrastructural bottlenecks discussed for the port of Amsterdam, however, can not be held solely responsible for the slow start of the Ceres Paragon Terminal. Current lock capacity is still more than sufficient, and problematic passages were expected in only extreme cases and with higher numbers on traffic. Container vessels can safely call the port of Amsterdam the first years. When realistic predictions of future problems arise, instead of the current scenarios sketched, enough time will be granted to decide on further developments of the nautical accessibility. Lock capacity, however, needs primacy here, as the presented problems with respect to the canal are based on intangible simulation studies. They need not be rejected or neglected for that, though.

6.4.4 INTERMODAL LINKAGE

The lock process and the North Sea canal transit create some levels of uncertainty with respect to transit times. These are, however, not fully insuperable. So there must be other reasons that have led to the failure of the Ceres Paragon Terminal. Some doubt the capabilities of the port of Amsterdam with respect to hinterland connectivity.

6.4.4.1   Truck

The port area is situated right on the A10 and A9 motorways, an area that is characterized by congestion on a regular basis. These congestions are not solely applicable to the Amsterdam area. A large part of the Benelux's road network is characterized by heavy traffic and congestion, as is also the case for the Rotterdam and Antwerp port areas. Some carriers are foreseeing difficulties, however, for the road connections between Amsterdam and Rotterdam, characterized by frequent and major congestions.[21]

Plans have been presented for improvement of the road system in the Amsterdam port area. The Westrand road will be a new connecting road between the Coen tunnel (A10) and the A4/A9 highways. The first part of the road (designated A5) opened November 2003 and connects the A4, along Schiphol Airport, with the A9. With the construction of the remaining part, planned for the Westrand road (between the A9 and the port area), Schiphol Airport Amsterdam will be accessible within 15 minutes. The second part is planned to be delivered in 2008. The Coen tunnel, a known bottleneck, is to be widened. The number of lanes will increase from four to eight. Two of those will be rush-hour "exchangeable" lanes with the possibility of offering five lanes in the busiest direction. The Coen tunnel will be finished in 2010.[12,17] As other port areas are also characterized by heavy traffic and frequent congestion, hinterland connectivity by road is therefore not directly a factor that stands in the way of success for the Ceres Paragon Terminal.

6.4.4.2   Train

The port of Amsterdam is properly served with railway connections. These make the European hinterlands fully accessible. The port has its own marshalling yards and connections to main railway systems. There is a Rotterdam–Amsterdam shuttle with an international rail connection to Belgium, France, Switzerland, and beyond. A certain form of dependency on Rotterdam remains, however. Some of the container transports originating in Amsterdam are only indirectly connected to their final international destination as transports head for Rotterdam first. Amsterdam will shortly connect to the Betuwe line at Geldermalsen, however. The Betuwe line is a freight-only rail shuttle link between Rotterdam and Amsterdam in the west (independent of each other), and the German border in the east. Hinterland connectivity by rail is not ideal at present but will be in the near future as it becomes independent of the port of Rotterdam. Again, issues on rail connectivity do not offer a convincing piece of evidence in the failure of the Ceres Paragon Terminal.

6.4.4.3   Barge

Of all goods transport to and from Amsterdam, over one-third takes place through inland barge shipping. The port of Amsterdam connects to the Rhine River through the Amsterdam–Rhine canal, thereby connecting the city to the European hinterlands. Both industrial and consumer markets in the Netherlands, Germany, Austria, and Switzerland can be served rather quickly (no obvious congestions) and efficiently (economies of scale versus speed). The big advantage the Amsterdam–Rhine canal has over the waterways connecting Rotterdam and Antwerp to the river systems is the absence of current. This leads to considerable time and fuel gains. When the situations for all three modalities are added up, it does not form a very convincing piece of evidence in the case of hinterland connectivity being an important reason for the failure of the Ceres Paragon Terminal.

6.4.5   HANDLING COSTS OF CONTAINERS

The Ceres Paragon Terminal, up to 2005, was not able to attract carriers on a contractual basis. Fixed cost, therefore, added up in the absence of a frequent and considerable income. As a final means to attract customers and to cover a fraction of the costs, the terminal decided to offer its services for relatively low handling costs. This port choice criterion was identified by main stakeholders in the sector as most important. The tariffs charged by the terminal formed only a marginal part of those of competing terminals. It seemed to have no effect on carriers, however.

6.4.6   REVIEW

When we review the port choice criteria for the port of Amsterdam, a number of flaws can be identified: the lock complex and the North Sea canal having its effect on the port's nautical access, and the port's intermodal linkage influence on hinterland connectivity. None of them seem to lead to such a fatal judgment for the Ceres Paragon Terminal. Even adding up the flaws can not fully explain the failure of the terminal, as most of the flaws are not yet reality and can be overcome in the future, especially when these are contrasted with the terminal's capabilities and the low handling tariffs charged. In the next section, we study if some political issues offer an explanation.

6.5   OPPORTUNITIES AND THREATS

In this section, we discuss the opportunities and threats in relation to other competitive ports from a political perspective. Opportunities arise from capacity shortages and delayed expansion of competitive terminals and impact of alliances. Currently, massive TEU volumes need to be transported. Terminal capacity in Northwest Europe is insufficient to guarantee an optimized transport chain. Therefore, plans for capacity expansion of established terminals have been made and are being realized, but will take a relatively long time before being operable. The port of Rotterdam can extend its capacity by realizing Maasvlakte 2. However, the first ships will not berth before 2012, and this will only happen following an optimistic scenario. Numerous political decisions need to be made before laying the foundation stone. Any postponement of political decisions can be advantageous for the port of Amsterdam. In these days of rapidly growing containerized cargo volumes versus marginal increases in transshipment capacity, the Ceres Paragon Terminal is being offered a unique chance. Its directly available transshipment capacity might be a decisive factor in establishing itself during this transition phase for the competition.

With the takeover of perhaps the most powerful member, P&O Nedlloyd, the future of the Grand Alliance was all but guaranteed. However, plans have been presented in which cooperative activities between the New World Alliance (American President Lines, Hyundai, and Mitsui O.S.K. Lines) and the Grand Alliance are initiated.[22] These plans enable the Grand Alliance to retain, and even increase, the number of departures on a selection of loops. The Grand Alliance will have a greater degree of freedom in the selection of these loops with the disappearance of P&O's veto against shifting activities from established ports to Amsterdam. The cooperative activities between both alliances can open up new opportunities for the Ceres Paragon Terminal.

Threats can be identified by discussing developments in neighboring ports and potential decisions made by the founder of the Ceres Paragon Terminal. As explained, the massive port of Antwerp needs to deal with tidal differences. However, recent agreements between the Dutch and Belgian governments will tackle this problem. The Westerschelde is about to be dredged and, as a result, will be made more accessible for the latest generation of vessels, in contrast to what is the case in the port of Amsterdam. Furthermore, the port of Antwerp signed a protocol agreement with the second Belgian port of Zeebrugge in which both ports entered into a partnership, joining forces to strengthen their competitive position regarding international container traffic. In the long term, the agreement might lead to a single massive Antwerp-Zeebrugge port complex. Strangely enough, contrary to the Belgian situation, on a national level Amsterdam seems to encounter competitive actions from the port of Rotterdam.

The Greek-American stevedore Kritikos, founder of the Ceres Paragon Terminal, decided to keep his 50% share in the terminal at least up to the end of 2006. In 2006, Kritikos prevented the Ceres Paragon Terminal being sold to the port of Rotterdam. The decision to whom Kritikos will sell his share might have a significant impact on the future of the port of Amsterdam.

6.6 CONCLUSIONS

With respect to important port choice criteria, the container ambitions of the port of Amsterdam seem justified, although a number of problems have to be overcome. In particular, infrastructural issues with regard to the lock complex, the access canal, and hinterland connectivity have to be solved but are not insuperable.

With a revolutionary and highly productive concept, an enormous growth market, a forthcoming capacity shortage with competitors, an advantageous geographical location, and low handling costs, a number of important (success) factors have presented themselves to make things work. The Ceres Paragon Terminal has not been able to position itself in the Northwest European container port arena so far. When strengths and opportunities are well managed and the odds regarding weaknesses and threats are favorable, the terminal has a chance to become well visited. Given the fact that Antwerp and other competitive European terminals are planning massive expansions and cooperative actions, a cooperation between the Dutch ports of Rotterdam and Amsterdam might be necessary to guarantee high national TEU throughput in the long term.

REFERENCES

1. Amsterdam Port Authority. 2005. Eerste Klant voor Ceres Paragon container terminal Amsterdam. Press release May 30, 2005.
2. Kroon, R. 2004. Seriously Ceres: The port of Amsterdam positioned in the competitive North-west European container port. MSc thesis, Vrije Universiteit Amsterdam, the Netherlands.
3. Veldman, S. J., and Bückmann, E. H. 2003. A model on container port competition: An application for the West European container hub-ports. *Maritime Economics & Logistics* 5:3–22.
4. Lirn, T. H., Thanopoulou, H. A., Beynon, M. J., Beresford, and A. K. C. 2004. An application of AHP on transhipment port selection: A global perspective. *Maritime Economics & Logistics* 6:70–91.
5. Song, D.-W., and Yeo, K.-T. 2004. A competitive analysis of Chinese container ports using the analytic hierarchy process. *Maritime Economics & Logistics* 6:34–52.
6. Notteboom, T. E. 2002. The interdependence between liner shipping networks and intermodal networks. Paper presented at Proceedings of the International Association of Maritime Economists (IAME), Panama City, Panama.
7. Wang, T.-F., Song, D.-W., and Cullinane, K. 2002. The applicability of data envelopment analysis to efficiency measurement of container ports. Paper presented at Proceedings of the International Association of Maritime Economists (IAME), Panama City, Panama.
8. Sowinski, L. 2002. King of the hill doesn't always mean prince of ports. *World Trade* 15:40–42.
9. Langen, P. W., van der Lugt, L., and Eenhuizen, J. H. A. 2002. A stylised container port hierarchy: A theoretical and empirical exploration. Paper presented at Proceedings of the International Association of Maritime Economists (IAME), Panama City, Panama.
10. Temmerman, W. C. J. 2002. Problematische scheepvaartpassages op het Noordzeekanaal. MSc thesis, Technic University, Delft, the Netherlands.
11. CPB (Centraal Plan Bureau). 2003. Beknopte analyse van de overslag in de Amsterdamse haven. CPB note to Ministry of Verkeer en Waterstaat. www.cpb.nl.
12. Amsterdam Port Authority. http://www.amsterdamport.nl (accessed June 2004).
13. Amsterdam Port Authority. http://www.amsterdamport.nl (accessed September 2006).
14. Ceres. http://www.ceresglobal.nl (accessed June 2004).
15. Bonnes, K. 2004. Interview with tug boat captain of Svitek Weismuller. December. Amsterdam, the Netherlands.
16. CPB (Centraal Plan Bureau). 2001. Analyse zeetong Noordzeekanaalgebied: een second opinion. CPB note to Rijkswaterstaat Noord-Holland. http://www.cpb.nl.
17. van Oord, G. 2004. Interview with logistics manager with the Amsterdam Port Authority. Amsterdam, the Netherlands.
18. Koopmans, C. C. 2003. *De baten van de sluis: nieuwe inzichten*, Stichting voor Economisch Onderzoek der Universiteit van Amsterdam (SEO). Publication 684, Amsterdam Port Authority.
19. Drewry Shipping Consultants. 2003. *The annual review of global container terminal operators*. Drewry: London.
20. Zeetoegang IJmuiden. http://www.zeetoegangijmuiden.nl (accessed September 2006).
21. Rotterdams Dagblad, het. 2004. *P&O Nedlloyd gelooft niet in Ceres*.
22. Verberckmoes, S. 2005. *Grand alliance en New World Alliance bundelen krachten*, pp. 1, 14. Nieuwsblad NT Transport.

7 Inland Ports
Current Operations and Future Trends

Edwin Savacool

CONTENTS

ABSTRACT

This chapter analyzes existing inland port operations and port planning and predicts future development trends. Inland port operations are classified by the type and method of cargo handling, to provide transportation and distribution planners sufficient information to begin the process. A review of concepts considered for future inland port development is also provided.

Inland ports are cargo handling sites remotely located from traditional air, land, and coastal transportation infrastructure. Like most traditional ports, the ideal inland port is able to handle multimodal and intermodal freight operations, has the ability to process international trade, and provides value-added services expected by shippers. As both the private sector and the Department of Defense (DoD) become more focused on rapid global distribution and efficient supply chains, inland ports will become more important. Transportation infrastructure and distribution planners are recognizing that inland ports may also promote more efficient multimodal corridors to support traditional terminals, especially in the case of ocean ports. The emerging efforts to develop an integrated inland port system within the United States are evaluated and summarized in this chapter.

7.1 INTRODUCTION

The growth in the volume of U.S. import containers has created both hope and concern for distribution process stakeholders in both the public and private sectors. There is hope in regions surrounding deep-water ocean ports that the increased cargo flow will create significant economic growth. In fact, this growth is seen in regions around inland ports as good-paying logistics jobs are created. However, significant concerns have emerged in areas that currently lack efficient port cargo-clearance capability. This problem is caused by inadequate transportation infrastructure. Traffic congestion encompassing major seaports and long queues at container terminal gates is a primary source of air pollution; increased energy consumption, loss of revenue for low-margin truck drayage operators, and damaged transportation infrastructure caused by the number of container-bearing trucks on the roadway are other problems created by increased volume and poor cargo-clearance capabilities. The incorporation of inland port processing facilities connected by adequate rail service to an international port facility with on-dock rail can mitigate or eliminate many, if not all, of these concerns.

Many factors have influenced the influx of import containerized freight, including changes in the U.S. economy, increased international sourcing, advances in transportation and information technology, regulatory structures that enable international transportation markets, and the reduction of trade barriers. Trade with Canada and Mexico is likely to continue its upward trend, affecting trucking and rail shipments especially in central states, while continued trade growth with Pacific Rim nations will impact port and rail containerized cargo throughput.[1] Without proper planning, this expected growth will negatively impact the performance of transportation infrastructure and will intensify freight security issues. While freight security once focused primarily on cargo shrinkage, principally cargo theft, in the post-9/11 era an equal or greater focus must be placed on preventing the use of our transportation system as a weapon by terrorist organizations.

There are numerous options available to international shippers to improve their inventory distribution from the purchase order vendor to the door of the customer. Shippers have the option to avoid certain congested areas. However, for public and private transportation planners in regions currently experiencing the negative impacts that can accompany increased cargo volume, the problem is very complex. The task for the public sector is to mitigate or eliminate the negative impacts of the ever-increasing volume of goods movement within the United States while ensuring the realization of the positive economic benefits of this increased flow. Solving the issues related to increased goods movement requires that the public sector work closely with the private transportation sector and shippers to establish and maintain a reliable, secure, and environmentally friendly transportation network.

Regional planners have a number of options, including the development of new infrastructure, expansion of current infrastructure, employment of advanced cargo handling and information management technologies, and improvement of operational practices at ports. As is often noted, there is limited land currently available at U.S. international ocean ports, making the option of port expansion either infeasible or too costly. In some cases the only viable options are those that are based on improved use of existing port land masses through better integration of an ocean terminal as a single node in an integrated end-to-end supply chain distribution system. These improvements are often enabled by relocating port processing functions to an inland port.

This chapter provides a broad overview of how the increased use of inland ports can directly support the improved flow of containers through congested international ocean terminals. The chapter will address how inland ports can be integrated as nodes within the end-to-end distribution process to relieve ocean port congestion while adding value to the distribution process, supporting economic growth, and providing the Department of Defense ensured access to strategic seaports.

7.2 INLAND PORTS DEFINED

The definition of inland ports has been greatly expanded since the concept of inland ports was first established in the United States by the Army Corps of Engineers. The first use of the term *inland ports* by the Corps of Engineers was to distinguish riverside ports from deep-draft ocean harbors.[2] While inland ports have a broader definition today, riverside ports are still considered to be a valuable part of the inland port network. Supporting navigation by maintaining and improving channels was the Corps of Engineers' earliest Civil Works mission, dating to federal laws in 1824 authorizing the Corps to improve safety on the Ohio and Mississippi rivers and several ports. Currently, of the 25,000 miles of inland, intracoastal, and coastal waterways and channels in the United States, approximately 12,000 miles constitute the commercially active inland and intracoastal waterway system currently maintained by the Army Corps of Engineers.[3]

As containerized imports have continued to increase, the definition of inland ports has been expanded. Today inland ports are defined more commonly as shipping, receiving, and distribution centers designed to relieve the congestion at increasingly busy seaports.[4] Inland ports now include cargo handling facilities removed from traditional air, land, and coastal gateways. Inland ports are being integrated with traditional ports by transportation corridors such as the Mid-Continent International Trade and Transportation Corridor (MCITTC) discussed in more detail later in this chapter.[2]

In the United States, ports of entry were historically located along land and coastal borders between countries. However, the increased volume of international air transportation, both passengers and freight, has resulted in the development of large inland airport hubs, which perform the traditional governmental port-of-entry functions. As with airport hubs, inland ports provide the means to move the processing of international containerized cargo to locations away from traditional ports of entry at land and coastal borders. Therefore, the definition for selected inland ports has been expanded to include "a location where international trade is processed."[5]

7.3   INLAND PORTS: AS AN INTEGRATED NODE
IN AN END-TO-END DISTRIBUTION NETWORK

Given the need to improve flow through traditional ports, the ability of inland ports to process international trade remote from congested land border and traditional maritime ports will have a positive impact on the end-to-end distribution network. An inland port directly linked with an international ocean terminal can support more efficient port clearance and cargo inspection and provide additional value-added services. At an inland port containerized freight can be transferred to a different mode and transloaded for store-to-door delivery, freight can be stored for distribution as vendor management inventory, or components can be assembled into consumer products. This consolidation of value-added services away from congested ports has significant potential to improve distribution flow and add significant capacity to ocean terminals and traditional border ports of entry.[5]

Depending on the land mass available, an inland port can be designed to provide a wide range of distribution services, including multimodal transportation (truck, rail, air, water, and pipeline transportation), warehousing, freight consolidation, and manufacturing services. This "one-stop" distribution service has the potential to attract shippers concerned with efficient supply chains. It is important for shippers to maintain a distribution pattern that provides the lowest transportation costs possible while receiving the most efficient and reliable movement of inventory to the point of use. This shipper viewpoint is necessary in order to achieve the lowest cost of inventory. Integrating inland ports into a regional and possibly a national distribution network, along with the consolidation of services at the inland facilities, will assist port, regional, state, and national transportation planners in improving the efficiency and effectiveness of the transportation system within the United States. An integrated regional inland port has the potential to increase capacity of international ocean ports, reduce traffic congestion and pollution, and create good-paying jobs. Inland ports will not only provide jobs associated with the direct handling of cargo, but also create opportunities for distribution and manufacturing positions that are attracted to the area because of the inland port services.[5]

7.4   INLAND PORTS AND NATIONAL SECURITY

While the commercial sector and regional population are the primary beneficiaries of an inland port, the Department of Defense would also benefit from an inland port network integrated with the strategic seaport, rail, and highway networks. Inland ports present a unique and timely opportunity for DoD to address concerns associated with ensured access to strategic ports short of presidential declaration of a national emergency. Further, an inland port presents a unique opportunity for DoD to collaborate with commercial and other governmental stakeholders to develop technology and business process solutions. The solutions can address future commercial and DoD throughput requirements, support homeland defense, and provide assistance to natural disaster relief efforts. The objective is to minimize the disruptive impact of military force deployments and natural disaster relief support efforts. As will be described below, selected ports, rail lines, and highway networks are currently designated as strategic assets and are monitored by several federal agencies. However, designated strategic ports, rail lines, and highways are not currently managed as an integrated network incorporating inland ports as connecting or interface nodes in a national system of distribution corridors.

The oversight of the strategic port, rail, and road network is an ongoing process by two primary federal agencies: the Department of Transportation (DOT) and the DoD. The DOT Maritime Administration (MARAD) continuously monitors strategic commercial ports through scheduled port readiness assessments, site inspections, and survey reports that are provided by the commercial ports. Annual port planning orders (PPOs) are issued to designated commercial ports and necessary revisions are made according to existing port conditions. MARAD also chairs the steering and working groups of the National Port Readiness Network (NPRN), which has been comprised of up to nine federal agencies that have responsibilities for supporting the movement of military forces through U.S. ports. Continual efforts are made to improve coordination and NPRN initiatives at both the national and local levels.[6]

Under its Railroads and Highways for National Defense program, DoD, with the support of the DOT, ensures the nation's rail and highway infrastructure can support defense emergencies. Rail infrastructure is managed under the Strategic Rail Corridor Network (STRACNET), which consists of approximately 39,000 miles of rail lines considered important to the national defense. Approximately 5,000 miles of track is designated as essential to allow sufficient connection of military facilities. The military places heavy and direct reliance on railroads to integrate bases and connect installations to maritime ports for force deployment requirements. Therefore, it is important to maintain an integrated network of mainlines, connectors, and clearance lines to support the movement of heavy and oversized equipment. To ensure that military needs are properly considered by the class I and short-line railroad industry, the DoD Surface Distribution and Deployment Command (SDDC), a subordinate command of the U.S. Transportation Command (USTRANSCOM), identifies facilities of the railroad infrastructure important to national defense.[7]

The Strategic Highway Network (STRAHNET) is a system of public highways that provide access, continuity, and emergency transportation of personnel and equipment in support of military force deployments, disaster relief, and homeland defense. The STRAHNET consists of approximately 61,000 miles of roadway designated by the Federal Highway Administration in partnership with DoD as being of strategic importance. The network is comprised of about 45,400 miles of interstate and defense highways and 15,600 miles of other public highways. The STRAHNET is complemented by about 1,700 miles of connectors, which are additional highway routes linking more than 200 military installations and ports to the network.[8]

To better integrate the national transportation network for defense and commercial distribution, several programs have been sponsored by the Department of Defense. Both programs are considered to have significant potential as national systems. The programs currently are being developed regionally. Regional Agile Port Intermodal Distribution (RAPID) has an East Coast orientation and Strategic Mobility 21 has a West Coast focus. Strategic Mobility 21 is an Office of Naval Research (ONR)–sponsored multiyear advanced logistics technology demonstration program designed to investigate and design the development of an integrated, dual-use (commercial and military) distribution network within Southern California. The program's early years have an initial regional focus, with future years planned to concentrate on the end-to-end military and commercial distribution networks. Strategic Mobility 21 is working in concert with the Center for the Commercial Deployment of Transportation Technologies (CCDoTT), a California State University Foundation activity, and the Southern California Logistics Airport Authority on the development of the former George Air Force Base as an inland port. Strategic Mobility 21 plans include future work with other regional inland port initiatives to develop a dual-use national distribution network.[9]

On the East Coast, the Delaware River Maritime Enterprise Council (DRMEC) and the Howland Group have developed the Regional Agile Port Intermodal Distribution (RAPID) System using collaborative business processes and technologies. The effort is sponsored by DoD, MARAD, and the Commonwealth of Pennsylvania. The RAPID System is an operating network of infrastructures, people, processes, and technologies with an end-to-end focus on efficient and secure cargo movement. RAPID was created in response to the military requirement to move cargo quickly and securely through a strategic port by sharing critical information on cargo movements internally and with law enforcement, emergency management, and trusted commercial stakeholders. The RAPID System has evolved from an operating proof of concept to the DoD beta site for advanced global distribution solutions in 5 years. RAPID provides a collaboration center that directly supports DoD port operations. RAPID serves as a hub for gathering and sharing of information required within a common operating picture (COP) for military personnel and other key stakeholders. The RAPID Center's capabilities have evolved to the point where they are exportable to any DoD terminal operation.[10] Since January 2004 DRMEC has successfully handled twenty-three military vessels at the strategic Port of Philadelphia and 2,789,742 square feet of cargo.

Both Strategic Mobility 21 and RAPID System are creating a positive climate as well as establishing the capabilities required for the future development of a dual-use national distribution network.

7.5   INLAND PORT CLASSIFICATION

In considering the use and integration of inland ports, it is constructive to review their classification. As previously noted, inland ports and related concepts have been implemented or proposed to solve a number of issues with the focus on freight movement. While there are many inland port models, the basic form of the evolving inland port concept is embodied in the Virginia Inland Port (VIP) located inland at Front Royal, Virginia. The VIP employs a rail shuttle linking the ocean terminals located in Newport News and Norfolk, Virginia, with the Front Royal inland terminal. The VIP functions as a satellite to the ocean terminals.[5–11]

The basic VIP inland port concept has been expanded in planning other inland port facilities. Inland port facilities currently planned and implemented within the United States include container and railcar storage depots, air cargo consolidation facilities, transloading services, Free Trade Zone designation, truck and longer-combination-vehicle (LCV) parking, and agile port concept container sorting (container block formation and block swapping).[11]

The CCDoTT has conducted research on the addition of Agile Port Systems (APS) to operations incorporating satellite inland port terminals. Experiments and demonstrations are being conducted by CCDoTT to determine the value of incorporating APS or components with a port operation. An initial CCDoTT demonstration indicated that the flow of commercial and military force deployments through international ocean terminals would be improved with an APS and that the average U.S. port throughput capacity could be increased by up to 300%.[12] Additional information on the agile port concept is provided later within this chapter.

While the VIP model of development has gained interest, not all inland ports, existing or proposed, follow the same model of development. To fully understand inland ports, it is important to recognize that facilities can be categorized in different ways, yet still focus on the processing of international trade. The categories of inland ports focus on inland waterway ports, air logistics parks, trade and transportation center inland ports, and maritime feeder networks or corridors.[5]

While inland ports can be categorized by the functions they perform, it is important to note that inland ports are constantly changing as they mature through the life cycle. The life cycle begins with the initial research and planning phases through physical implementation, expansion, long-term operation, and refinement. The constantly changing distribution patterns in the world demand this flexibility to maintain the proper balance between cargo movement, congestion control, environmental management, and infrastructure development.

7.5.1 INLAND WATERWAY PORTS

As defined by the Army Corps of Engineers, inland waterway ports are the original inland port in the United States. Current inland waterway ports handle bulk commodities, including grain, coal, petroleum, and chemicals in addition to general break bulk and containerized cargo. Inland waterways provide the ability to efficiently transport large volumes of bulk commodities over long distances. Individual barges are lashed together to form a "tow." On smaller waterways tows normally consist of four to six barges. On the larger rivers with locks, such as the Ohio, Upper Mississippi, Illinois, and Tennessee rivers, 15-barge tows are common. Such tows move approximately 22,500 tons of cargo as a single unit. A single 15-barge tow is equivalent to about 225 rail cars or 870 tractor-trailer trucks. Tows can consist of over forty barges on the Mississippi River below its confluence with the Ohio River, well over 50,000 tons of cargo. As a result of this efficiency, inland waterways can enable fuel consumption savings, reduce air pollution, reduce traffic congestion, better transport safety, and lessen noise pollution.[3]

The integration of inland waterway ports with the international transportation network can help the overall distribution network in several innovative ways. As an example, cargo moved through the Ports of New York and New Jersey could be connected with the Pittsburgh waterway ports via the Port Authority of New York and New Jersey Port Inland Distribution Network (PANYNJ-PIDN) to more efficiently move freight to the final inland destination. Currently the PIDN is designed to move containerized freight from the ocean terminals to Pittsburgh by rail with local cargo distribution by truck. Creating an intermodal interface with an inland waterway port could add additional options for the Port Authority of New York and New Jersey and the local Pittsburgh community.[11] Additional information is provided on the current PIDN concept in the section 7.5.4.

7.5.2 Logistics Airports

Several dedicated air cargo ports or logistics airports have been established in the United States in recent years. Logistics airports have been primarily established on former Air Force bases closed as a part of the DoD Base Realignment and Closure (BRAC) process. Air cargo ports can exist in conjunction with passenger airports or as dedicated cargo facilities. Customs, distribution facilities, and in some cases manufacturing centers are all components of logistics airports that classify them as inland ports.

An example of this type of inland port is the Southern California Logistics Airport (SCLA) being developed at the former George Air Force Base located near Victorville, California. This facility is being developed by Stirling International into a 5,000-acre master-planned multimodal business and industrial airport complex. The facility offers large industrial sites of 20 to 100 acres. The development plan includes a fully dedicated multimodal logistics and e-business fulfillment center served by truck, rail, and air transportation systems. The airfield has a distinct advantage in being able to operate 24 hours a day, 7 days a week (24/7) without curfew. The airfield has become a significant link in the international supply chain with daily cargo flights from Asia.[5-11]

Development of the facility is aided by several incentive programs. Government incentive programs include a 1,940-acre designated foreign trade zone. The facility also provides on-site 24/7 secured U.S. Customs port-of-entry support. A container freight station and bonded warehouse with refrigeration service is available. The facility is included in the California Local Agency Military Base Recovery Area (LAMBRA) program. The LAMBRA program was developed to attract reinvestment and reemploy displaced government and contract workers and has tax incentives similar to those for enterprise zones that are binding for a period of 8 years. The SCLA is also the largest redevelopment district in California (60,000 acres).[13] Several major studies are under way to determine how the capabilities of the SCLA and other inland ports fit into the overall Southern California cargo movement network and how they should be supported for future development.

7.5.3 Trade and Transportation Center Inland Ports

Inland ports can be placed in a general class of locations where border processing of trade is shifted inland and multiple modes of transportation and a variety of value-added services are offered. A significant number of these sites are located at former Air Force bases that have been closed due to the BRAC legislation. These sites are often referred to as brown fields.* However, some of these sites have been devel-

---

* Abandoned, idled, or underused industrial or commercial facilities where expansion or redevelopment
  is complicated by real or perceived environmental contamination.

oped at green-field locations. One example of a green-field development is located in Alliance, Texas, at the inland port facility called the Alliance Texas Logistics Park (Alliance).

Green-field and brown-field development to facilitate international trade require different approaches. In the Air Force base situation, the air transportation assets already exist but other modes of transportation may be either inadequate or not available. At green-field sites airfield assets most likely need to be built, but they can be located near a class I mainline. Each location provides differing benefits. A closed base already has significant infrastructure investment like runways, buildings, and telecommunications, which can be converted from military use to distribution or manufacturing functions. At a green-field location, the site can be developed to better fit the current flow of international trade and meet the stakeholders' requirements. As a result of the basis for development, the services provided for this class of inland port are extremely variable. This type of inland port can range from a single site where intermodal connections and manufacturing centers are located, to an entire city or region, such as Alliance, that facilitates international trade through the encouragement of inland port activities.

The Alliance Texas Logistics Park is located 15 miles north of downtown Fort Worth and 15 miles west of the Dallas–Fort Worth International Airport. Alliance is one of the largest green-field master-planned inland port developments in the country. Alliance is a development that integrated and expanded existing air, rail, and highway to connect the inland port with both domestic and international markets. Alliance has been designated a foreign trade zone and an enterprise zone.[11-14]

Hillwood, which is a Perot company, refers to the development as a "transportation-driven community." The Hillwood complex is an example of an inland port that was originally planned as a "reliever" industrial airport, but gradually evolved during development into a multimodal facility offering stakeholders many transportation and value-added services. The facility business park includes over 140 companies with approximately half from Fortune 500, Global 500, and the Forbes List of Top Private Companies. Economic growth has been impacted by the investment of more than $5 billion by the companies, creating 24,000 full-time jobs.[11]

Alliance includes a major Burlington Northern and Santa Fe Railway (BNSF) operated 735-acre intermodal yard. The BNSF is one of the four remaining class I transcontinental railroads in North America. To support the facility, Alliance designated 1,500 acres immediately east of the intermodal yard for rail stakeholders to develop freight distribution centers. The BNSF intermodal terminal services have been provided at Alliance since 1994.

The Alliance inland port facility is located on both the main line of the BNSF and the North American Inland Ports Network (NAIPN). The NAIPN is a working group in North America's Super Corridor Coalition (NASCO), which advocates the interests of inland ports along the Mid-Continent International Trade and Transportation Corridor (MCITTC). The NASCO corridor is depicted in Figure 7.1. The intent of the NAIPN is to network inland ports and improve the efficiency of international commerce throughout the mid-continent of North America. In addition to Alliance, current inland port members of the NASCO include Kansas City SmartPort, Inland Port San Antonio, Winnipeg, and the Bajio/Central Inland Ports initiative.[2-15]

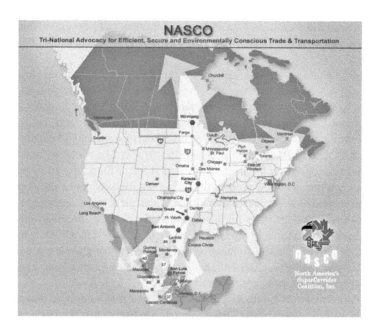

FIGURE 7.1 North America's Super Corridor Coalition.

7.5.4 MARITIME FEEDER INLAND PORTS (NETWORKS AND CORRIDORS)

These projects link together inland ports, seaports, and related developments into operating networks or corridors. These networks and corridors can be used as network links to other inland port nodes, ocean terminals, and distribution centers.

The Port Authority of New York and New Jersey Port Inland Distribution Network (PANYNJ-PIDN) is an example of a maritime feeder inland port network. Working with public and private entities, the PANYNJ is developing a network of inland distribution hubs at key customer locations in the northeastern United States. The goals of PIDN are to improve the efficiency of cargo moving through the port to inland points, improve air quality in the region, and reduce truck congestion on highways. The Port of Albany was the first northeast port to participate in the PIDN. The PANYNJ estimates that the Port of Albany service alone has eliminated 20 million vehicle miles traveled (VMT), accounting for approximately 130,000 gallons of fuel saved and the elimination of 88 tons of toxic emissions.[11–16]

7.6 INLAND PORTS: AS A COMPONENT OF AN AGILE PORT SYSTEM

Several Agile Port System (APS) concepts have been studied for implementation at satellite inland ports. As an example, over the past 8 years, CCDoTT and ONR have assessed and developed analysis tools such as dynamic simulation models to investigate and demonstrate APS concepts that have inland ports integrated as an integral component of the system. From the start of the program, the focus has been

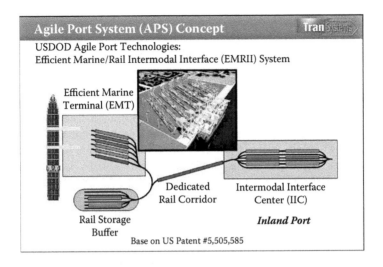

FIGURE 7.2 Efficient Marine/Rail Intermodal Interface (EMRII) System.

on developing a dual-use APS. The term *dual use* refers to the intended design utility of the APS for both the commercial and military sectors. The CCDoTT dual-use APS concept, named the Efficient Marine/Rail Intermodal Interface (EMRII), was demonstrated at Fort Lewis, Washington, and the Port of Tacoma. A schematic diagram of the complete EMRII System concept and its components is provided in figure 7.2.

In June 2003, a full-scale demonstration of the commercial Efficient Marine Terminal (EMT) component of the EMRII System was performed. The demonstration focused on the marine terminal component of the concept and was completed at the Port of Tacoma's Washington United Terminal (WUT) in Tacoma.

The limited EMT demonstration determined that the implementation of the commercial EMT concept and associated processes has the potential to achieve operating cost savings of up to approximately 40%. This equals an estimated cost savings of over $350,000 per single vessel while handling 12,000 20-foot equivalent units (TEUs). Additionally, it was estimated that the average U.S. port could realize up to a 300% increase in throughput capacity through the implementation of the EMT.[12] This increase in throughput capacity would have a very positive impact on our national economy and the ability of the DoD to deploy forces on an as-needed basis.

After the successful completion of the commercial EMT demonstration and identification of significant cost savings and capacity increases, it became clear to the stakeholders that a full-scale demonstration of the entire EMRII System, as depicted in figure 7.2, must be performed and analyzed to determine the maximum potential benefits of the APS concept.

7.7 FUTURE INLAND PORT TRENDS

Efforts to reduce congestion and improve goods movement in the United States will require a more integrated and cooperative approach to capacity expansion and improved productivity of existing transportation assets. DOT is developing a framework for government officials, the private sector, and private citizen groups to take steps to relieve current congestion issues through public-private partnerships.[17] With the growing congestion in the United States and deteriorating and inadequate infrastructure, without a concerted effort we could experience a significant slow-down in goods movement by 2011.[18]

Port partnerships that enable the inland port concept and relieve local congestion appear to be increasing in number and depth of cooperation. The Ports of Oakland and Sacramento have formed an alliance with a goal of shipping containers to Sacramento by barge instead of trucking them within the region. In a similar fashion, the Ports of Tacoma and Olympia, Washington, have formed a partnership to create a joint intermodal and logistics center on approximately 750 acres purchased by the Port of Tacoma.[19] The vision for the land, south of Olympia near Maytown, Washington, includes facilities designed for handling the growing volume of international and domestic cargo, economic growth, and reduced reliance on truck transportation.[20] In addition to the development of inland ports, the development of connecting corridors is continuing to generate interest at both the local and national level.

7.8   CONCLUSION

Inland ports have become important nodes in an integrated transportation network. When properly designed, developed, and integrated, they have significant impact on the overall efficiency of cargo distribution. Inland ports have proven that when integrated with an appropriate rail network, they can relieve traffic congestion and pollution, improve safety and cargo security, and enable more efficient shipper supply chain logistics. Just as important, inland ports are able to enable economic growth through the attraction of businesses that create good-paying positions for the region.

REFERENCES

1. U.S. Department of Transportation, Bureau of Transportation Statistics. 2003. *U.S. International Trade and Freight Transportation Trends*, BTS03-02, pp. 1–4. Washington, DC.
2. Schmitt, K. 2006. Inland ports, redesigning America's landscape to fit a multi-modal mission. *Defense Transportation Journal*, 62(1):18–19.
3. U.S. Army Corps of Engineers, Institute for Water Resources. 2000. *Inland Water Navigation: Value to the Nation*, 1–4. http://www.usace.mil/docs/InlandNavigation.pdf
4. Wikipedia. 2006. Inland port. http://en.wikipedia.org/wiki/Inland_port.
5. Leitner, S. J., and Harrison, R. 2001. *The identification and classification of inland ports*, 26–27, 41–42. Report 0-4083-1, Center for Transportation Research, University of Texas at Austin.
6. National Port Readiness Network. http://www.globalsecurity.org/military/agency/dot/nprn.htm.
7. Strategic Rail Corridor Network (STRACNET). http://www.globalsecurity.org/military/facility/stracnet.htm.
8. Strategic Highway Network (STRAHNET). http://www.globalsecurity.org/military/facility/strahnet.htm.

9. Mallon, L. G., and Hwang, J. 2001. *Strategic Mobility 21: Network centric joint force deployment and distribution process, Advanced Logistics Technology Demonstration (ALTD) joint operational concept*, 1–4. Center for the Commercial Deployment of Transportation Technologies. ftp://www.foundation.csulb.edu/CCDOTT/Deliverables2004/task%203,20/task%203,20.2.pdf.

10. McInnis, P. 2006. LTC, The 841st Transportation Battalion utilizes RAPID Center to operate Seaport Tactical Operations Center (STOC). http://www.drmec.org/index.php?subaction=showfull&id=1159191713&archive=&start_from=&ucat=&.

11. The Tioga Group, Railway Industries, and Myer, Mohaddes Associates. 2006. *Inland port feasibility study*, 1–89. Technical report, prepared for the Southern California Association of Governments.

12. TranSystems. 2003. *Efficient marine terminal, full scale demonstration*, 1.1–1.16. Technical report, prepared for the Center for the Commercial Deployment of Transportation Technologies.

13. Global Access Logistics Center at Southern California Logistics Airport (SCLA). http://www.logisticsairport.com/solutions.php.

14. Alliance. A world of commerce. http://www.alliancetexas.com/Alliance/.

15. North American Inland Ports Network (NAIPN). http://www.nascocorridor.com/pages/ports_network/ports_network.htm.

16. The Port of New York and New Jersey. Fact sheet, improving land access. http://www.panynj.gov/DoingBusinessWith/seaport/pdfs/02_01_01_03_LG.pdf.

17. U.S. Department of Transportation. 2006. *National strategy to reduce congestion on America's transportation network*, 1–6. http://isddc.dot.gov/OLPFiles/OST/012988.pdf.

18. Edmonson, R. G. 2006. Elusive goal. *Journal of Commerce* 7:31.

19. Rosynsky, P. T. 2006. Helping hand. *Journal of Commerce* 7:28–31.

20. Port of Tacoma and Port of Olympia. South Sound Logistics Center: Olympia and Tacoma Port Commissions approve strategic partnership. Joint news release. http://www.portoftacoma.com/newsreleases.cfm?sub=68&lsub=858.

8 Inland Terminal Concepts

Athanasios Ballis

CONTENTS

ABSTRACT

Inland terminals are valuable elements of the transportation system as they facilitate freight transport activities, support environmentally friendly modes, relieve cities and ports from congestion, and offer warehousing and distribution logistic services. Various inland terminal types exist, and even more (confusing or even conflicting) terms are used for their designation. The current chapter presents the basic operational concepts reflected by the existing terminal types, mainly in terms of logistic activities (attempting in parallel to tackle the fuzziness in terminology). Furthermore, a number of innovative technological concepts proposed for the enhancement of inland terminal performance are presented.

8.1 INTRODUCTION

Traditionally, the transport needs of national and international trade were facilitated by maritime ports, railway terminals, and airports that offer the space, equipment, and operating environment for the handling and storage of freight commodities. However, it is now recognized that a growing amount of trade is being processed at inland sites.[1] As a result of continuous evolution (following the technological and organizational advancements in the sectors of transport and logistics) and adaptation to the market conditions, today's face of inland terminals is polymorphic. In the current chapter, the main organizational concepts reflected by the existing inland port types (section 8.2) and a number of selected innovative technological concepts proposed for the advancement of inland ports (section 8.3) are described. Finally, conclusions are drawn for the usefulness and future potentialities of the inland terminal types (session 8.4).

8.2 INLAND TERMINAL TYPES AND ASSOCIATED ORGANIZATIONAL CONCEPTS

Various inland terminal types exist nowadays, and a large variety of terms are used for their designation. The terminology concerning inland terminal is characterized by a lot of fuzziness, as different terms are frequently used to describe the same terminal type, while the same term may be used to describe different facilities.

In the American literature the term *inland port* has a broad meaning. In a recent study investigating the various inland port schemes,[1] an inland port is defined as a site located away from traditional land, air, and coastal borders with the vision to facilitate and process international trade through strategic investments in multimodal transportation assets by promoting value-added services as goods move through the supply chain. In the same study, four basic terminal types were identified: trade and transportation center inland ports, inland waterway ports, maritime feeder inland ports, and air cargo ports. On the contrary, in Europe the term *inland port* is related (almost exclusively) to the inland waterways network. According to the official terminology,[2] inland ports form part of the trans-European network for the interconnection between the waterways and other modes of transport. Terminals serving rail–road and rail–rail transshipments and offering temporary storage of unitized units (containers, swap bodies, semitrailers) are noted as combined transport terminals. The term *intermodal terminal* is wider as it includes railway, inland waterway, and deep-/short-sea transshipment installations for unitized units. The terms *freight village, logistic center/platform/park, transport center, city logistic center, urban distribution center,* and so on are used to denote sites specially organized for carrying out logistics activities.[3] When such services are offered in sites outside, but organizationally linked to ports, the terms port *logistic activity zone, hinterland terminals,* and *dry port* are used.

Despite the diversity in terminology, there are common technical/organizational concepts for inland ports. In the following paragraphs, these concepts are analytically presented. In order to avoid confusion, the generic term *inland terminals* is used for all inland-located terminal types.

8.2.1 Trade and Transportation Center Inland Ports: Logistics Centers

The trade and transportation center inland ports can be looked at as locations where border processing of trade is shifted inland and multiple modes of transportation are offered. These inland terminal types are at locations where value is added to goods.

In the United States, many of these sites are located at Air Force bases that have recently been closed, while other sites have materialized at green-field locations.[1]

The logistics centers developed in Europe (the terms *freight villages, logistic platforms, logistic parks, transport centers, urban distribution centers*, and *city logistic centers* are also used) as well as in the United States (where the *term global freight village* is used)[4] offer similar functionalities. The first initiatives of sites specially organized for carrying out logistics activities in Europe started in the 1960s in response to the need for restructuring cities and expelling trucks and transport firms from city centers. The first such terminals in Europe, named Centres Routieres, created in the wider Paris region in France (notably Garonor and Sogaris), aimed to reduce the traffic of heavy trucks in the city through the establishment of distribution networks based on smaller trucks (relief from traffic congestion and city logistics are among the main reasons for the implementation of the freight village concept in the United States[5]). The next stage saw development toward the need to improve basic sectorial services and provide a functionally and economically satisfactory supply for the logistic operators and freight transport firms.[6] Furthermore, the enhancement of intermodal transport that required the transport of sea containers by rail to hinterland increased the demand for inland transshipment facilities. Later on, during the 1980s, two additional driving forces contributed to inland terminal development: (1) capacity restrictions in seaports that generated needs for external handling and storage facilities (either in the port environment or in the hinterland), and (2) initiatives of the national railways to stop the decrease of the market share of rail freight by promoting intermodal solutions.[7,8]

A logistics center is the hub of a specific area where all the activities relating to transport, logistics, and goods distribution, for both national and international transit, are carried out on a commercial basis, by various operators. The operators may be either owners or tenants of the buildings or facilities (warehouses, distribution centers, storage areas, offices, truck services, etc.). In addition, bank, postal, insurance services, and, in certain cases, Customs infrastructure may also be accommodated.[3] Such activities usually concern the transformation of the inbound shipments, from suppliers and manufacturers, to outbound shipments to customers (who may be other distributors, end users, or manufacturers). As inbound shipments often contain a significant number of stock, keeping units of one or a limited number of commodity types, while outbound shipments usually consist of few stock, keeping units of many commodity types by more than one supplier, blending (taking relatively homogeneous input streams and creating customer-specific "blends") and impedance matching (synchronization of relatively large and infrequent inbound flows of each stock, keeping units to relatively small and frequent outbound flows) are required. In addition, value-adding services like repackaging (e.g., from a pallet into smaller quantities), labeling for specific customers, and assembling (e.g., repackaging printers with destination-specific power cords and documentation) may be provided.[9,10]

The consolidation of warehousing, distribution, and logistics management services at one location increases the attractiveness of inland terminals. Furthermore, the freight villages evolve alliances among the entities responsible for the transport, storage, and distribution services, which can generate significant reduction in the number of trucks' vehicle kilometers. Nevertheless, the creation of such facilities is not always an easy task as conflicts may occur between actors involved,[11] and conditions for their financial viability should be provided.[12]

The European experience revealed that transport operators located inside logistics centers have reached higher productivity than firms located outside, in terms of intermodal activities turnover, total traffic volume, and total intermodal flows. Moreover, it has emerged that firms located inside logistics centers or in the immediate environment are more likely to use intermodal transport than firms located outside.[8,13] The latter outcome justifies the initiatives of many national railways that promote the concept of transportation center inland ports due to their positive impact to intermodality and especially to the share of rail market. Nevertheless, a technicality should be mentioned: the existence of railway lines parallel to a warehouse allows for direct train-to-truck cross-docking as well as for train-to-storage operations, but the required infrastructure (a platform at the level of wagon floor instead of truck shipping/receiving docks) makes the truck-to-truck cross-docking impractical (no dock levelers and dock seals exist). When the rail tracks are seldom used, the option for the rail access to warehouse may not balance the loss of truck-to-truck, cross-docking operation. Yet it should be noted that in the current European operating environment (where the utilization of rail in freight villages is usually very low), many new warehouse facilities provide rail interfaces, as the owners prefer designs having a long-term vision.[14]

8.2.2 INLAND WATERWAY PORTS

The inland port is an old terminal type but its role has been upgraded in recent decades due to the need for the promotion of environmentally friendly modes as well as due to its involvement in certain logistic chains. Inland waterways are offering an alternative transport way that allows for the reduction of road traffic congestion and the associated environmental externalities providing a cost- and energy-efficient mean (1 ton of cargo travels 59 miles by truck, 202 miles by rail, or 514 miles by barge per gallon of fuel[15]). In the United States, 25,000 miles of navigable waterways exist on rivers, the Great Lakes, and intercoastal waterways.[16] Today's major inland waterway ports provide one of the most efficient means for transport of bulk commodities like grain, coal, petroleum, and chemicals in addition to general cargo.[1]

Similarly, in Europe, inland ports are important elements of the transport system either as urban/regional logistic centers or as industry-supporting facilities.[17] The inland waterways network is considerably shorter and less dense than networks of other modes of transport, but the majority of highly developed European areas are located close to waterways.[18]

Although there are many common features between seaports and inland ports, certain differences should be mentioned. First, seaports have a clear national and even supranational functionality (the hinterland of most seaports stretches beyond national/state borders) while inland ports are merely considered of local importance (having a local or regional hinterland). Second, seaports aim to (and many of them have achieved) withdraw most of their cargo handling activities from the highly dense populated regions or cities, while inland ports are in most cases still located in the center of urban regions in order to facilitate logistics and urban goods distribution. Third, local community groups are entirely aware of the economic benefits from the existence of a seaport in terms of employment and economy. On the contrary, inland ports possess less economic scale. In cases of scarcity of land, inland ports may have to defend their existence/expansion against the pressure for redevelopment to other land uses that are more attractive to local community.[19]

8.2.3 MARITIME FEEDER INLAND PORTS

A maritime feeder inland port (the terms *dry port, hinterland terminal,* and *port logistic activity zone* are also used for such terminal types) is most closely related to traditional maritime ports. The increasing problem with transporting goods to and from the traditional port through the city has, together with the expensive costs of establishing new docks, and so on, created favorable conditions for the establishment of maritime feeder inland ports, which can handle the port-related activities (including Customs clearance and registration) and potentially attract value-added services.[6]

Furthermore, the need for the establishment of a maritime feeder inland port may arise from the logistic development of a traditional port. In the preliminary stages of the logistic development process, a port is limited to fundamentally technical activities (e.g., warehousing, container consolidation and deconsolidation in the container freight station, silo storage, dry bulk blending and bagging, etc.). However, in an advanced logistic development stage, the logistic functions are dominant, covering specific segments of the distribution chain and channeling of flows. In the container field, these functions are grouped under a maritime feeder inland port[7] that acts as a trade and transportation center in the hinterland of the port.

In the United States, maritime feeder inland ports are typically located 50 to 250 miles away from the port.[20] This distance will allow for fast delivery to the maritime port and potentially enough shift away from the highways serving the maritime port to be effective at relieving congestion.[21]

8.2.4 AIR CARGO PORTS

Air cargo represents a small, in terms of volume, but very significant in terms of value share of the transport market, which grows at a significant pace since in addition to mail, express, and emergency cargoes, new market niches of perishable commodities and freight for just-in-time deliveries are nowadays transported by air.

Three options exist for the air transport phase: by passenger airplanes using a part of their belly hold capacity, by combi airplanes sharing main deck space between passenger and cargo, and by all-cargo/freighter airplanes.[22] In the ground/airport the transport and processing of these cargoes is performed by a specialized system that includes dedicated loading units, transport modes, equipment, airport installations, and commercial organization. Customs, distribution facilities, and in some cases manufacturing centers are all components located in the wide area of airport (the terms air cargo port,[1] air freight, or air cargo terminal/center are used for the designation of these installations).

8.3 TECHNOLOGICAL CONCEPTS

Transport systems continuously face the challenge to reduce the cost entailed in the transshipment of cargoes and improve the quality of service offered by implementing more effective modes, equipment, and management/control strategies for the terminals and the network. Boosted by globalization and global economy growth, the transportation needs have already shown considerable growth, and these trends are likely to continue in the future, increasing the pressure to terminals and networks. Therefore, focus should be given to the proposed innovations in all relevant fields of transport, logistics, terminal planning and operation, network organization, and so forth. Within this framework, the development of innovative technological concepts, especially for unitized cargoes, is an issue high on the agenda of the transport industry, and a number of new promising systems are already in mature design or in demonstration phase. Many of these concepts have been evaluated,[23–26] and in some cases they have been further developed, through systematic work carried out within research projects.[27] The vast majority of these new transshipment systems/concepts are tailored to one (or more) of the four basic network operating concepts: direct trains running between two terminals without handling on the way (block trains or shuttle trains); group trains or feeder systems (to link terminals of the same region through short, feeder links and to perform the long-distance transport in one complete train); liner trains (which are partly loaded/unloaded at each terminal stop, offering regular service and enabling the integration of terminals with lower demand in the network of intermodal transport); and the hub-and-spoke system (which increases the connectivity of medium and small terminals).

Each of the above operating concepts imposes different organizational requirements as well as equipment configuration. For example, the hub-and-spoke system requires the parallel processing of many transport modes in order to achieve direct mode-to-mode transshipments and reduces the intermediate storage of loading units, while liner trains require the handling of a relatively smaller number of modes but at a very short time window.

A subset of these technological concepts is applicable to inland ports. More specifically, technological concepts suitable for small- to medium-size rail–road facilities, automatic guided vehicles, and barge terminals have the potentiality to be included in a typical inland port terminal. The following paragraphs provide examples of such technological concepts. It must be noted that concepts for large-scale rail–rail and rail–road terminals also exist[25,29,31] but have been excluded from the current presentation, as they concern a very limited number of terminals.

8.3.1   INNOVATIVE RAIL–ROAD TRANSSHIPMENT CONCEPTS

Terminals serving rail–road transshipments are usually equipped with reach stackers and rail-borne gantry cranes. In order to overcome the limitations of the above conventional equipment and provide the additional transshipment capacity necessary for future growth,[28] a number of innovative new systems/concepts have been developed.[29–31] Two of these concepts are suitable for inland ports, as they may contribute quite positively to the reduction of cargo handling time, which is a typical request in logistics. These concepts are:

- The handling of trains on electrified tracks. This concept is applied to networks where freight trains are circulated on electrified tracks (the majority of the main European railway lines are electrified and are used by both passenger and freight trains). As the loading tracks of the rail–road terminals are not electrified (to allow the handling of the loading units by gantry cranes or reach stackers), a diesel locomotive is required to move the train from the electrified lines to the transshipment tracks (which is interpreted in terms of cost, complexity, and delays). In order to eliminate this problem, three methods have been proposed:

1. The use of a slewing catenary on the loading track. Following the arrival of the train, the catenary withdrawal device is moved to one side (over the entire length of the train) and allows portal cranes or reach stacker operations to be carried out.
2. By allowing the train to coast with momentum. The railway line is electrified up to the transshipment area (see figure 8.1) but not in the transshipment lanes (so there is no catenary over the length of the train). The train enters the terminal with the pantograph lowered and stops when the electric locomotive is positioned under the overhead on the far side, so that it is able to move off by raising up its pantograph.
3. Special handling equipment that can work under the catenary. One of the proposed technologies makes use of a rotary telescopic arm with a spreader (see figure 8.1).

- Each of the above systems has a main weakness (cost, reliability/applicability, and slow operation, respectively) that imposes friction to its implementation. Nevertheless, in certain cases the time savings are so important that they can overcome the above disadvantages.
- The use of fast transshipment devices. Various innovative designs have been proposed based on multiple spreaders, overhead grapple arms, or high-rack constructions, but until today they remain on paper or in the demonstration phase.[30,31] Another innovative concept (also in the demonstration phase) that may have future market potentiality is the "moving train" (see figure 8.1). A special shunting locomotive moves the wagons slowly in front of the handling equipment (while in the conventional system the handling equipment is moved along the stationary train). A high-speed automatic crane serves the moving trains and a semiautomatic crane performs the storage operations. A cross-conveyor is included for transferring the loading units between the two cranes. A significant advantage of the system (that may be very useful for rail-to-warehouse operations) is that storage is concentrated in a given area while in the conventional systems the storage is done parallel to the tracks.[32]

8.3.2 THE CONCEPT OF AUTOMATED GUIDED VEHICLES

An automated guided vehicle (AGV) is a vehicle driven by an automatic control system that serves as a driver. A U.S. company installed the first AGV in 1954. Since then, this technology has evolved as a result of research performed by European and American companies. AGVs are making initial inroads for port applications in many parts of the world. In the port of Rotterdam, AGVs are also used for the transport of containers between the port quay and the nearby railway terminal. The use of AGVs is anticipated in certain future barge terminal designs (see section 8.3.3) as well as for the transport of containers between maritime and inland ports (see chapter 10).

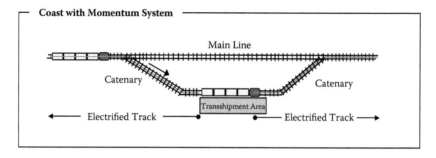

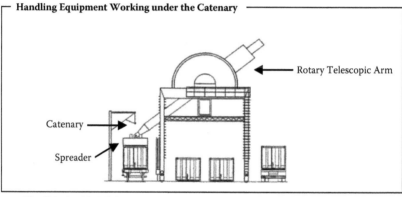

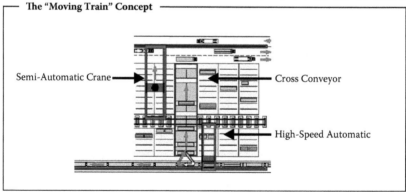

FIGURE 8.1 Proposed innovative road–rail transshipment technologies. (Information from European Commission/DG-TREN.[25])

8.3.3 Innovative Barge Concepts

Barge transport is a relatively low-cost, environmentally friendly transport mode that should be further promoted as an alternative to road transport. To this aim, various innovative designs have been proposed[33] (see figure 8.2).

One cluster of systems/concepts is based on the perception that the cost effectiveness and the realization of a large share of the growing hinterland transport can only be achieved by bundling (consolidating) the transport flows, increasing the scale of

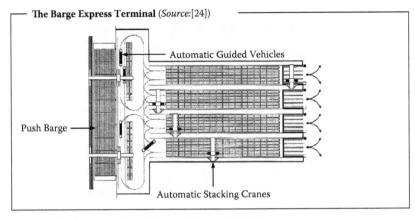

The Barge Express Terminal (*Source:*[24])

Automatic Guided Vehicles

Push Barge

Automatic Stacking Cranes

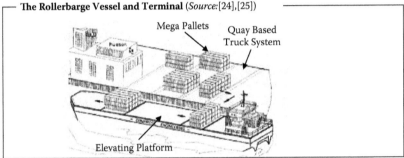

The Rollerbarge Vessel and Terminal (*Source:*[24],[25])

Mega Pallets   Quay Based
               Truck System

Elevating Platform

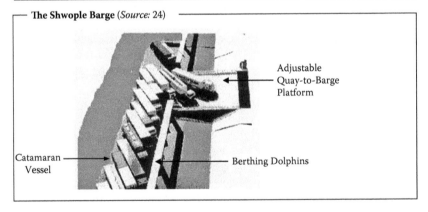

The Shwople Barge (*Source:* 24)

Adjustable
Quay-to-Barge
Platform

Catamaran
Vessel   Berthing Dolphins

FIGURE 8.2 Proposed innovative transshipment technologies in inland waterways.

operation, and introducing automated handling at barge terminals. The following innovative concepts have been proposed:[25]

1. Systems that integrate advanced technological elements in order to attract large-scale barge container transport (like the Barge Express and the FAMAS Barge Service Centre). In the Barge Express system the quay cranes can be operated automatically or manually. For the automated

transshipment, the push barge must remain almost still and horizontal; therefore, adaptations to the berth (i.e., a jetty) and the crane are necessary. The transport between the quay cranes and the stacking cranes is performed by AGVs. Automated stacking cranes (ASCs) perform the loading/unloading of AGVs or storage operations and semiautomatic truck loading/unloading.[34] The FAMAS Barge Service Centre is a dedicated barge terminal type that is also equipped with ASCs for the yard operations, AGVs for the transport of loading units, and an overhead bridge crane that is moved perpendicularly to the quay (on jetties) and straddles the barge.[25]

2. Systems that make use of grouped loading units in order to reduce the loading/unloading time. The Rollerbarge system consists of megapallets (that can accommodate many containers or other loading unit types), special barges (equipped with an elevating platform that allows the horizontal transfer of megapallets on board), a quay-based track system, as well as a yard storage and truck/train service subsystem.[35] The roll-flats (or roll-trailers) system is a similar technology (megapallet, quay-based track system) where the lifting platform is not on the barge but in the river, parallel to the quay. The system includes flat-top barges and the lifting platform elevates the barge to align the level of deck and quay. Both systems are in a conceptual stage.

3. Systems based on the roll-on/roll-off concept. The Shwople Barge system consists of an adjustable quay-to-barge platform, a series of berthing "dolphins," and a wide catamaran vessel that allows semitrailers or trucks up to 18 m in length and up to 44 tons to be stowed diagonally on board. The berthing dolphins allow the vessel movement fore and aft to align with the adjustable quay-to-barge platform (the gap is bridged by link spans). The vehicles can enter/exit the vessel in any order and therefore a sequence of port calls can be implemented.

The realization of the barge systems/concepts that require large container flows has to wait for better market conditions. In Europe, the market for container barge transport is expected to be more and more segmented into a carrier haulage segment and merchant haulage segment.[36] The growing importance of carrier haulage transport contributes to bundled transport flows and will probably support the realization of the associated barge systems. Nevertheless, additional requirements should be satisfied. The Shwople Barge for example, provides a technically efficient "bus for trucks" system that suffers from a significant disadvantage: in parallel to all major rivers (at least in Europe), a road exists, and the speed of the truck is normally greater than the speed of a barge. On the other hand, the barge can travel for 24 hours and under specific operating conditions (congested roads, implementation of regulations that limit the truck-driving hours) can provide a viable alternative.

Another cluster of concepts is based on the idea of self-unloading barges and includes four variants: In the Port-hopper variant the tow of the vessel is equipped with an elevating platform that allows a vehicle to be driven inside the vessel where a shipboard traveling crane unloads the vehicle and stacks the loading unit on board.

The other variants (one-container call line vessel, bow transshipment, and sideways transshipment) perform the loading/unloading by various on-board crane types.[25]

The initial concept of the self-unloading barges was to provide services with limited infrastructure requirements and in many places along a river in order to reduce the road transport trip (and the associated cost); yet these claims were not fully realized in practice: the loading units cannot be left unattended, and therefore a terminal facility is required.[33]

8.4   CONCLUSIONS

Inland terminals are valuable elements of the transportation system as they facilitate freight transport activities, support environmentally friendly modes, relieve cities and ports from congestion, and offer warehousing and distribution logistic services. The modern inland terminal may have many faces: trade and transportation center inland port, logistics center, inland waterway port, maritime feeder inland port, and air cargo port.

As the growth of global economy has led to higher transport needs, the role of inland terminals, especially in the case that advanced logistic services are offered, is expected to be enhanced. Consequently, innovative organizational and technological concepts should be used in order to improve the current performance level. The vast majority of the innovative technical concepts proposed for the inland terminals remain on paper or in the pilot demonstration phase, pending for higher transport flows and market maturity. On the other hand, a limited number of innovations (automatic guided vehicles, automatic stacking cranes) have successfully passed the pilot phase and fight for their market penetration. This is definitely a sign of hope that may encourage the research efforts toward more efficient inland terminals.

REFERENCES

1. Leitner, S., and Harrison, R. 2001. *The identification and classification of inland ports*. Research report 4083-1, Centre of Transportation Research, Texas Department of Transportation, Austin.
2. Decisions 1692/96/EC and 1346/2001/EC. http://eur-lex.europa.eu (accessed September 5, 2006).
3. EUROPLATFORMS EEIG. 2004. *Logistics Centres: Directions for use*, http://www. gvzregensburg.de/downloads/What_is_a_Freight_Village.pdf. (Accessed December 20, 2002).
4. Weisbrod, E. R., Swiger, E., Muller, G., Rugg, F. M., and Murphy, M. K. 2002. Global freight villages: A solution to the urban freight dilemma. CD-ROM. In *Proceedings of the 81st TRB Annual Meeting*, Washington, D.C.
5. Zografos, K., and Regan, A. 2002. Current challenges for intermodal freight transport and logistics in Europe and the US. CD-ROM. In *Proceedings of the 81st TRB Annual Meeting*, Washington, D.C.
6. NeLoc. 2004. Service concept report for Logistic Centres.
7. Ministerio De Fomento, Puertos del Estado, Spanish State Ports Agency. *Guide for developing logistic activity zones in ports*, Spanish State Ports Agency IAPH, Madrid, ISBN 84-88975-36-8.

8. NeLoc. 2003. *Best practice handbook for logistics centres in the Baltic Sea region*, http://www.neloc.net/Reports.html (Accessed December 20, 2007).

9. Govindaraj, T., Blanco, E. E., Bodner, D. A., Goetschalckx, M., McGinnis, L. F., and Sharp, G. P. 2000. Design of warehousing and distribution systems: An object model of facilities, functions and information, 1099–1104 *Proceedings of the 2000 IEEE International Conference on Systems, Man & Cybernetics*, Nashville, TN (Institute of Electrical and Electronic Engineers, Inc. ISBN 0-7803-6583-6).

10. Rouwenhorst, B., Reuter, B., Stockrahm, V., Van Houtum, G. J., Mantel, R. J., and Zijm, W. H. M. 2000. Warehouse design and control: Framework and literature review. Invited review. *European Journal of Operational Research* 122:515–33.

11. Tsamboulas, D., Kapros, S., and Ballis, A. 1997. Conflicts and co-operation in the freight villages creation process. Paper presented at Proceedings of the 2nd Convention of the European Association for the Advancement of Social Sciences, Conflicts and Co-Operation, Nicosia, Cyprus.

12. Tsamboulas, D., and Kapros, S. 2003. Freight village evaluation under uncertainty with public and private financing. *Transport Policy* 10:141–56.

13. EC-DG TRANSPORT. 1999. *Quality of freight villages structure and operations*, research project. Summary report FV-2000. ftp://ftp.cordis.europa.eu/pub/transport/docs/summaries/integrated_fv-2000_report.pdf (Accessed December 20, 2007).

14. Ballis, A. 2006. Freight villages: Warehouse design and rail link aspects. CD-ROM. In *Proceedings of the 85th TRB Annual Meeting*, Washington, DC.

15. Nachtmann, H. 2002. *Economic evaluation of the impact of waterways on the state of Arkansas*.

16. Muller, G. 1999. Intermodal freight transportation, 4th ed. Washington, D.C.: Eno Transportation Foundation and the Intermodal Association of North America, Washington, DC.

17. Dooms, M., and Haezendonck, El. 2004. An extension of "Green Port Portfolio Analysis" to inland ports: An analysis of a range of eight inland ports in Western Europe, 1324–1341. In *Proceedings of the Annual Conference of the International Association of Maritime Economists* (IAME), Izmir.

18. Rydzkowski, W., and Wojewodzka-Krol, K. 2003. The role of inland waterways in the process of the enlargement of the EU. Paper presented at Seminar on Transport Infrastructure Development for a Wider Europe, Paris.

19. Dooms, M., and Macharis, C. 2003. A framework for sustainable port planning in inland ports: A multistakeholder approach. Paper presented at the 43rd European Congress of the Regional Science Association (ERSA), University of Jyvaskyla, Finland.

20. Harrington, L. 1991. Landlocked shippers use inland port. *Transportation and Distribution*, 82–86.

21. Leitner, S., and Harrison, R. 2001. *The identification and classification of inland ports*. Research report 4083-1, Centre of Transportation Research, Texas Department of Transportation, Austin http://www.utexas.edu/research/ctr/pdf_reports/4083_1.pdf (Accessed December 20, 2007).

22. Ashford, N., Stanton, H. P., and Moore, C. A. 1997. Airport technical services. In *Airport operations*, 2nd ed., 281–310. New York: McGraw-Hill.

23. European Commission/DG-Transport. 1995. *Future optimum terminals for intermodal transport*, SIMET research project. Doc Euret/411/95, Brussels.

24. European Commission/DG-TREN. 1997. *Deliverable D2: New-generation terminal and terminal-node concepts in Europe*, TERMINET research project, European Commission, Brussels, 85–88 (Project Reference: IN-96-SC.1204, Coordinator: Technische Universiteit Delft, The Netherlands).

25. European Commission/DG-TREN. *Innovative technologies for intermodal transfer points*, ITIP research project. Contract 2000-AM.10005, http://www.eutp.org/download/itip/D1/annex3.pdf (Accessed December 20, 2007).

26. European Commission/DG-TREN. 1997. *Deliverable 1: Maritime terminals and harbor economic at interfaces with intermodal transports*, IQ research project, European Commission, Brussels (Project Reference: IN-95-SC.0313 Coordinators: ME & P (UK), Institut für Wirstschaftspolitik und Wirtschaftsforschung (DE), Insitut National de Researche sur les Transport et Leur Sécurit Arcueil, (FR). Netherlands Organisation for Applied Scientific Research, TNO (NL), LT Consultants Ltd (FI), TRT Trasporti e Territorio Srl (IT), Közlekedéstudományi Intézet RT (HU), and NEA Transport Research and Training (NL).

27. European Commission/DG-TREN. 1998. *Deliverable D8: Advanced handling (transshipment, internal transport and storage)*, IMPULSE research project. European Commission, Brussels (Project Reference: IN-95-Sc.0255 Coordinator: Krupp Fördertechnik Gmbh, Essen, Germany).

28. Ruesch, M. 2002. Standardization related to transfer points, EUTP clustering meeting. RAPP Ltd. Engineers and Planners, Zurich.

29. Woxenius, J. 1998. Development of small-scale intermodal freight transportation in a systems context. Doctoral thesis, Report 34, Department of Transportation and Logistics, Chalmers University of Technology, Göteborg, Sweden.

30. Woxenius, J. 1998. *Inventory of transshipment technologies in intermodal transport.* International Road Transport Union (IRU), Geneva.

31. Ballis, A., and Golias, J. 2002. Innovative transshipment in the combined transport sector in Europe: Inventory and expert system approach. CD-ROM. In *Proceedings of the 81st TRB Annual Meeting*, Washington, D.C.

32. Ballis, A. 2004. Advanced rail and maritime system demonstrations in Europe. *Transportation Research Record* 1873:89–98.

33. Ballis, A., and Stathopoulos, A. 2002. Innovative transshipment technologies: Implementation in seaports and barge terminals. *Transportation Research Record* 1782: 40–48.

34. TRAIL Onderzoekschool. 1996. *Barge EXpress; innovatief perspectief voor de internationale binnenvaart.* Delft University.

35. Huijsman, H. 1995. Rollerbarge. Paper presented at Intermodal '95, Section Innovative Intermodal Technology, Amsterdam.

36. Wendling, K. 1994. *Logistieke dienstverlening via de binnenvaart in de vervoersrelatie.* Nederland-Duitsland: Den Haag.

9 Maglev Freight Conveyor Systems

Kenneth A. James

CONTENTS

ABSTRACT

Existing sea-based ports are typically surrounded by major metropolitan areas, requiring movement of shipping containers through those areas, which places unwelcome strain on the existing infrastructure. While a number of the nation's ports—New York, New Orleans, Oakland, Houston, and so forth—experience this container movement problem, the paradigm example is supplied by the Port of Los Angeles/Long Beach (hereby referred to as the "Port"), the nation's largest and most important port. Almost one-half of the nation's port-related traffic passes through the Los Angeles metropolitan area on its way into the interior, as shown in figure 9.1.

The Alameda rail corridor was developed to help accommodate the unprecedented growth of container traffic generated by the Port's intensifying customer base. However, this project's anticipated capacity has not been fully realized due to constraints on train building and rail throughput at the corridor terminus and the unplanned transition to transshipping approaches. The impact of freight movement on the Los Angeles community has not significantly been reduced. A number of terminals at the Port must truck containers to the terminus of the Alameda Corridor, 4 and 20 miles from the Port, imposing significant congestion and concomitant diesel pollution to the surrounding community. Costly proposals to expand the area's existing highways in conjunction with a growing recognition of the dangers of diesel particulate

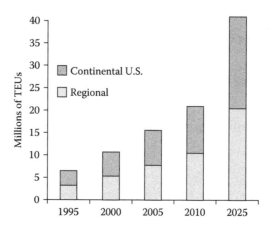

FIGURE 9.1 Container growth at the ports of LA/LB.

emission (DPE) have prompted a novel approach to the container movement challenge. This approach utilizes a proven Maglev conveyor-belt technology that shows promise for both short-haul urban freight movement and inter-state-bound containers. The application of this technology to container freight movement inside the Port and beyond its confines will reduce both highway congestion and pollution throughout the Los Angeles area.

9.1 INTRODUCTION

A research arm of California State University at Long Beach (CSULB), the Center for the Commercial Deployment of Transportation Technologies (CCDoTT), has studied this issue for several years. CCDoTT developed the *agile port concept*,[1] which involves moving containers from the Port, where storage space is scarce, to inland ports or intermodal container transfer facilities (ICTFs), where containers can be redirected to local transshippers or organized into transcontinental trains. The weakest link in this scenario is the dedicated link between the Port and remote facilities. Thus, for the last 3 years CCDoTT has prioritized the definition of a new paradigm in moving containers out of the Port. The approach selected by CCDoTT uses freight-optimized Maglev technology for a variety of supply-chain applications. This is very different from the concept of passenger Maglev because freight-optimized Maglev has a known and predictable ridership—freight containers, which all have the same destinations: ICTF facilities.

This container movement paradigm has several requirements. The first is to accommodate projected Port growth so the economic base of not only the Southern California region, but also the entire county, can continue to grow. The second is to relieve pressure on the existing highway infrastructure, which cannot well handle even its current load. The third is to improve the quality of life, not just at the Port but throughout the region. An adage appropriate to the CCDoTT paradigm shift, attributed to Einstein, is: "You cannot solve problems with the same technologies that caused them." These three requirements define the parameters of a new container transport approach.

Growth of the Port is essential because of the potential for new jobs and increased wealth within the region. A recent study shows that newly created logistics jobs have in fact more than made up for the loss of jobs in the manufacturing sector due to industry moving from Southern California. Logistics-related jobs are typically higher paying than manufacturing jobs that require similar skills.[2] Acreage at the Port rents for upwards of $250,000 per acre per year, producing income for cities and state agencies.[3] Supplies for military sustainment have historically passed through the Port, and military planners need to continue to be able to count on the Port as a means of shipping supplies to military depots overseas. Most importantly for the nation, almost one-half of all the imports to this country come through the Port.[4] To meet the projected future container influx, Port throughput must be increased and yet the size of the Port cannot be increased; clearly, there is not room for the needed expansion. A container movement approach should have the capability of moving an additional 5 million or more 40-foot containers per year from the Port.

Traffic problems on Southern California freeways are legendary. The estimate of $11 billion per year in productivity losses in Los Angeles and Orange counties due to freeway congestion is not surprising.[5,6] Adding more containers from the Port year after year will likely bring the region to a standstill. Even if local governments, political action groups, and community leaders could agree on how and where rail or highway could be expanded, these means of container transport still have a wide footprint (land surface area utilized) and cannot easily be elevated. To reduce the stress on the existing Southern California infrastructure, a container movement approach should be capable of high throughput but have a smaller footprint than road or rail expansion projections.

Stationary sources of pollution such as electrical power plants have made great strides in reducing air pollution with massive "scrubbers," and automobile technology has also improved over the years, resulting in a marked improvement of air quality for the Southern California region. One pollutant, however, remains problematic: diesel particulate emissions (DPEs). This pollutant is different from gaseous pollutants in that it is localized to areas where diesel engines operate such as the Port, truck/train intermodals, and along freeway and rail corridors. The effects of DPEs are devastating. More than thirty human epidemiological studies have found that diesel exhaust increases risk for several types of cancer, and a 1999 California study found that diesel exhaust is responsible for 70% of cancer risk from air pollution.[7] Only recently has the danger of having homes and schools close to sources of DPEs been recognized. Figure 9.2 shows an Air Quality Management District (AQMD)[8] study of how DPE is concentrated around the Port and transport paths. To alleviate the severity of the DPE problem for the entire community, a container movement approach should exploit fixed power sources that produce minimal pollution.

The aforementioned economic, congestion, and pollution issues facing urban freight movement involving the Port produce conflicting constraints to balancing Southern California's economic future with the region's quality of life. As the international trade industry (ships, trucks, and trains) has been identified as a major source of pollution due to the heavy use of diesel power, Port expansion plans have

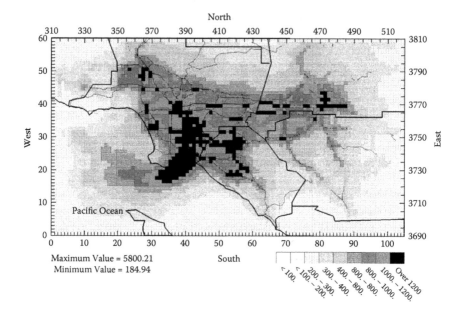

FIGURE 9.2 Model estimated cancer risk per million population for the Los Angeles basin.

run into a community environmental road block; more rigorous Environmental Impact Reports (EIRs) are being demanded. Responding to community pressures, some elected officials are discussing placing limits on Port emissions; these would also serve to limit Port growth. California state legislators have fired off a barrage of bills aimed at regulating and changing the way goods are handled, workers are compensated, and pollution is curbed at California ports and transportation hubs. Several of these bills add constraints to operations while others add costs to the movement of containers both within and beyond the Port region. From an economic perspective, these bills impact the economies of the Port and cargo movement, and therefore affect the cost of doing business. The Maglev approach represents a win-win solution: moving containers in sufficient number and speed to allow continued economic growth, while alleviating congestion and pollution concerns throughout the Southern California basin.

9.2 NEW PARADIGM FOR CONTAINER MOVEMENT

Maglev is a solution than can help solve the problems created by the technology responsible for the congestion and pollution Southern California is faced with today. It provides the needed balance between more and better jobs of an expanding economy and a quiet, clean, and safe environment for the people who have those jobs. It cannot be too often reiterated: "One cannot solve problems with the same technologies that caused them."

Maglev is not a new concept. After several years of development, the world's first commercial urban Maglev and high-speed Maglev passenger lines have gone

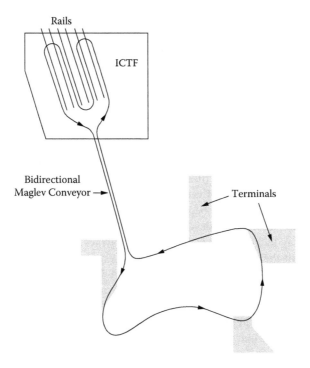

FIGURE 9.3 Maglev freight conveyor concept schematic.

into service in Japan and China, respectively. Application of Maglev technology to a freight-only system is an innovative alternative to conventional road or rail infrastructure. The environmental and community constraints on expanding conventional means of container transport through the Los Angeles basin indicated as early as 2002 that a Maglev freight system had costs comparable to highway and rail for moving containers through urban areas.[9] The referenced study involved an electromagnetic suspension (EMS) design by Transrapid, the German developer of the world's first commercial Maglev system. Recent work at Lawrence Livermore Laboratory and General Atomics (GA) has shown that an electrodynamic suspension (EDS) Maglev also has significant potential benefits for transporting containers.

Since Maglev guideways are typically envisioned as moving in one direction (the conveyor-belt metaphor applies here), a continuous-loop configuration for Port to ICTF container movement as shown in figure 9.3 is considered the most practical embodiment of a Maglev container transport system for the Port.

9.3   MAGLEV URBAN FREIGHT TECHNOLOGIES

Maglev technology is a way of floating container carriages utilizing a magnetic field to move them along a guideway that does not contain any moving parts. Maglev technology was conceived decades ago; however, only in the last 50 years has it been applied to real-world situations. There are two types of Maglev technologies:

the first, EMS Maglev uses electronic feedback control to lift the carriage with *attractive* magnetic force. This system was developed by the German firms Siemens and Krupp in a joint venture named Transrapid. The Transrapid carriage is pulled forward by a linear synchronous motor (LSM) that is similar to a typical electric motor, but unwound and laid lengthwise along the guideway.

The second Maglev type is EDS, which was conceived in the United States during the 1960s, and later developed in Japan. EDS employs a magnet moving above a conductive plane producing an opposite image of the magnet and generating magnetic *repulsion* that causes carriages to lift away from the guideway. Forty years ago the only magnets powerful enough to be used for this form of Maglev were superconducting magnets, which at that time were laboratory oddities. It was not until the late 1980s with the development of rare-earth magnets such as neodymium-iron-boron (NbFeB) that EDS technology became realizable without cryogenics. EDS still had to wait until the 1990s for the development in the Lawrence Livermore Laboratories of Halbach array technology.[10] Halbach magnet arrays are geometric arrangements of magnets with their fields oriented in such a way that the total magnetic field is concentrated on one side of the array. Today GA has licensed the technology and built a full-scale 400-foot EDS Maglev test track at its headquarters in San Diego. This system also uses a linear synchronous motor for propulsion. The advantage of the EDS is its passive nature: there are no on-board power supplies to generate the lift forces as in EMS (all that is needed is forward motion, generated by the LSM windings in the guideway). In addition, an EDS suspension leads to significantly greater air gaps, resulting in more lenient guideway construction tolerances, with resultant cost savings potential.

9.4   EVOLUTION OF THE MAGLEV FREIGHT CONCEPT

Three years ago, CCDoTT approached Transrapid with the concept of freight Maglev. At that time Transrapid was the only provider of commercial Maglev systems. Transrapid systems have carried over 6 million passengers for 2.2 million miles with a 99.9% on-time service record. Transrapid recognized the economic advantages of CCDoTT's "container conveyor" approach and began working with CCDoTT on a first-order model of a port to inland intermodal system. Figure 9.4 shows the Transrapid freight-optimized design with attracting magnets lifting the carriage.

Transrapid engineers performed a propulsion-power, system architecture analysis, also shown in figure 9.4.[9] Since that first design, Transrapid has determined that single stacking of containers on a freight carriage allows use of the guideway structures presently used by their commercial passenger system and now promotes both freight and passenger service with their systems.

The electrodynamic suspension approach, developed by GA, is passive in that once the carriage, which initially rests on wheels, is propelled by the linear synchronous motor to a velocity of around 5 mph, lift is achieved. Figure 9.5 shows a schematic of the passive magnet Halbach array configuration relative to the transposed conductor guideway. GA has built a full-scale prototype of a passenger EDS Maglev system at its facility in San Diego consisting of a carriage, guideway, and power distribution system. The system is presently the only full-scale Maglev in the

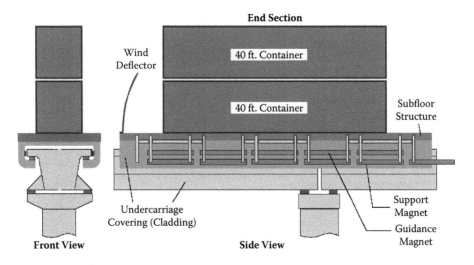

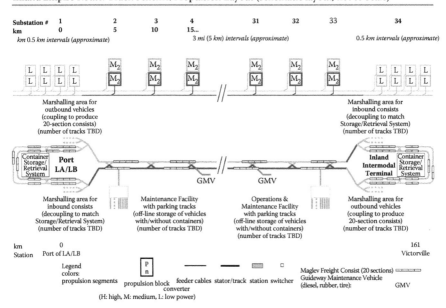

FIGURE 9.4 Transrapid freight-optimized Maglev carriage and system architecture.

United States. Figure 9.5 also shows experimental results from system tests demonstrating the magnitude of the required velocity for "lift-off" as well as the measured drag as a function of velocity.

An area where significant Maglev system cost optimization can be realized involves the guideway and associated components. The EMS system with its

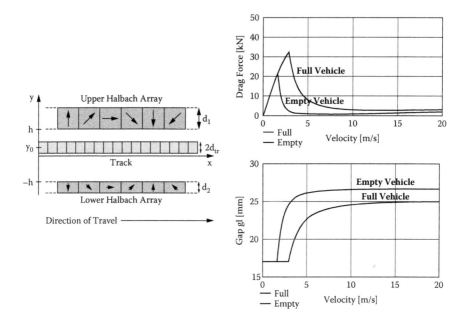

FIGURE 9.5 General Atomic's EDS schematic with results for full-scale freight prototype.

electronic feedback control, operating with a nominal air gap on the order of 10 mm, has inherently tight tolerances, between the lifting magnets and the guideway. Camber and support spacing of an elevated guideway for an EMS system are critical design factors. An EDS system lifts away from the guideway—on the order of

FIGURE 9.6 World's first container mover on a Maglev system.

"Hybrid Girder" Guideway

FIGURE 9.7 Urban Maglev guideway steel-fiber-reinforced concrete (SFRC) hybrid girder.

20 to 30 mm—allowing more versatility in guideway design, with more lenient tolerances in component fabrication and assembly.

Like Transrapid, GA proposes enhancing its passenger Maglev carriage to carry shipping containers.

On June 8, 2006, a 20-foot loaded container was placed on the test carriage to become the world's first container moved by a Maglev system (figure 9.6). GA has also considered various forms of prefabricated guideway sections as shown in figure 9.7. These sections utilize a new form of concrete structure designed specifically for Maglev systems.[11]

9.5 MAGLEV ADVANTAGES BEYOND ELIMINATING DIESEL PARTICULATES

While utilization of stationary power sources eliminates diesel particulates, the fact that Maglev does not use wheels and derives traction from a linear motor adds significant advantages to Maglev container transport.

When containers are placed on a railcar, the entire weight of the container is distributed onto the wheels, each of which has a contact area of a few square centimeters. The same container loaded onto a truck chassis puts all its weight on the tires, with contact area on the order of tens of square centimeters. In both cases the container's weight is focused onto small areas of rail or road and generates a moving pressure wave on the rail or road. This pressure wave can account for misalignment of the rails on railway tracks and unevenness of the cement plates on highways and washboards on asphalt surface streets. Maglev distributes the weight of a container over a large bank of magnets on the order of tens or even hundreds of thousands of square centimeters, which effectively eliminates the severe pressure gradients of wheeled vehicles. The result is a transport technology that produces minimum stress on the guideway infrastructure and thus is the most reliable and economical method to *elevate* freight transport. The readily elevated Maglev has a very small footprint, which gives it accessibility to a large variety of rights-of-way and minimizes both community and environmental impact. Also, the carriage itself does not have to support the container weight over axels, and thus its structure can be uniformly distributed over its length. The benefit of this architecture is that Maglev has the largest payload-to-carriage ratio of any land transport.

As explained earlier, the freight Maglev system uses a linear synchronous motor where the guideway is the active portion of the motor and magnets on the carriage are the passive portion. Maglev technology has unique enabling features, which make it ideal for carrying cargo. The linear synchronous motor and friction-free magnetic suspension result in a system that can accelerate much faster than conventional wheeled systems (0.15 g acceleration is typical); this leads to high throughput (short headways). In addition, the magnetic propulsion system can handle much greater grades (Maglev design typically involves grades of 10%; this can be compared to 6% grades needed for cargo). Maglev carriages powered by linear synchronous motors require no exposed overhead electrical wires or electrified "third rail," which contributes positively to overall safety evaluations. Since the carriage makes no contact with the guideway during operation, traction is produced by the motor rather than friction between wheel and road or rail, enabling Maglev to function safely in all climate conditions.

Linear induction motors can also be used for Maglev propulsion; the difference here is that the guideway is passive and the active portion of the motor is on the carriage, which does require an electrical power pickup. Linear induction motors are inherently less efficient than linear synchronous motors and require strict alignment of the active and passive components to maintain their efficiency. There are applications, however, many of which involve long distances and a small number of carriages, where linear-induction-propelled Maglev makes economic sense. Port container operations require large numbers of carriages on short-haul conveyor loops, making linear synchronous motors the obvious choice for this application. Yet another advantage of in-guideway linear synchronous motors is carriage control. Computer switching of power to short lengths of guideway where the carriage is located ensures cost-effective operation by restricting the application of power to particular sections of the guideway. The guideway sections adjacent to activated guideways produce a natural safety buffer between carriages.

The combination of unmanned operation and high-speed, elevated operation provides enhanced security benefits. Unmanned operation allows for automated, nonstop x-ray and gamma-ray scanning at inspection portals. Constantly moving containers on the conveyor are not readily tampered with, and image processing with closed-circuit cameras is simplified by the relatively static view offered by an elevated guideway.

9.6   PROPOSED MAGLEV CONVEYOR SYSTEMS

With the recent development of the easily elevated, nonpolluting Maglev container Electric Cargo Conveyor (ECCO) concept by the CSULB College of Engineering through CCDoTT funding, two commercial embodiments for container transport—both with military application—have become apparent: (1) an immediate ECCO application for moving 5,000+ containers per day between the Port and the I-710 Corridor ICTFs, and (2) a longer-term ECCO conveyor system capable of moving containers between the Port and both the inland warehouse concentrations and transcontinental rail terminals at Victorville and Beaumont.

The obvious immediate application for Maglev technology is a feeder system from terminals to ICTFs, which has the potential to reduce or eliminate short-haul trucking from terminal to Alameda Corridor ICTFs and railheads. Such a feeder system would get containers out of the Port and would eventually be part of a larger and more comprehensive regional Maglev freight system. A number of terminals at the Port do not have any capacity for moving containers directly by rail, so all such containers are destined to pass through the Los Angeles basin on their way to the rest of the country. As a conservative estimate, upwards of 2 million containers are drayed from those terminals to railheads between the Port and downtown Los Angeles as demonstrated in the photograph in figure 9.8.

The first application of the ECCO was defined in a 2006 contract from the port of LA to GA (with CSULB's ECCO concepts supporting the system architecture)[12] to do a preliminary cost estimate for an ECCO system between the Port and ICTFs near the Port moving more than 5,000 containers per day. A possible route is shown in figure 9.9.

FIGURE 9.8 Container traffic on I-710.

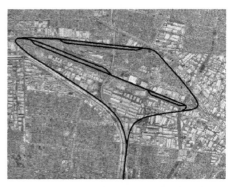

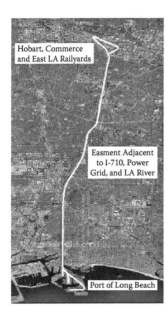

FIGURE 9.9 I-710 Corridor ECCO alignment.

The team of CSULB, GA, and nationally recognized civil engineering and railroad signaling safety companies determined the cost of a totally elevated port system to be $90 million per mile. This is considerably less than a trenched rail corridor ($125 million per mile) or new, at-grade freeway additions ($150 million per mile) that can carry the same container volume but slash though the community and impact the environment. By way of contrast, the operating cost of the proposed Port ECCO system is $2.20 per container per mile, of which $1.00 is the cost of electricity. The present cost of trucking containers along the I-710 from the Port to the major ICTFs of around $125[3] indicates that the ECCO system's "fare box" has the potential to pay not only for its day-to-day operation but also for its amortized capital costs. Conservative estimates based on EMFAC 2001 pollution show that the replacement of drayage truck trips with an ECCO system along I-710 would reduce diesel particulates emission by 77 tons per year and oxides of nitrogen by 1,532 tons per year.[13]

As shown in figure 9.10a, the ECCO is completely compatible with current Port operations. As it is elevated—except when configured for at-grade spurs or sidings—the ECCO would not impact existing rail or truck-gate operation. The system is planned to utilize the same equipment and labor procedures that are presently used at terminals to load and unload dock rail. However, to fully utilize the capacity of the

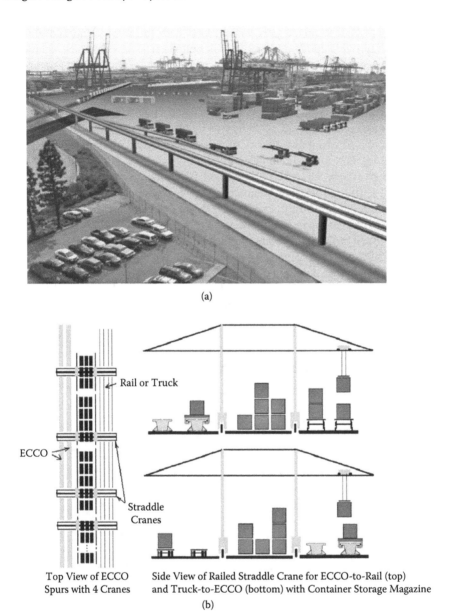

(a)

Top View of ECCO   Side View of Railed Straddle Crane for ECCO-to-Rail (top)
Spurs with 4 Cranes   and Truck-to-ECCO (bottom) with Container Storage Magazine

(b)

FIGURE 9.10 (a) ECCO compatibility with road and rail at the Port. (b) Automated ECCO load and unload.

ECCO system, automated equipment with a computer-managed container storage magazine is required (figure 9.10b).

Application of Maglev to cargo movement can bring the same nonpolluting, quiet, efficient, and reliable service as it has brought to existing Maglev passenger systems. The recently developed ECCO technology can move containers out of the Port to ICTFs, inland port, and beyond with the same clean and efficient attributes, improving upon both road and rail options. The best long-term solution to Port growth, congestion, and pollution is to complement existing as well as proposed road and rail expansion in Southern California with the ECCO system, minimizing movement by truck of containers from the Port to inland warehousing complexes, and cumbersome rail movement of containers through the Los Angeles basin to the origins of transcontinental rail at Victorville and Beaumont.

The increased speed and density of a dedicated express container transporter connecting the Port to the Inland Empire as well as Victorville, a railhead for the Burlington Northern Santa Fe (BNSF), and Beaumont, a railhead for the Union Pacific (UP), showed Maglev technology could accommodate Port growth and carry an additional 5 million containers per year.[9] The elevated Maglev with its narrow footprint can carry more containers than a much wider freeway because of the consistent speed (averaging 90 mph) of the containers on the conveyor system. The benefits of such a Port-to-inland corridor approach are numerous. Container traffic bound for the continental United States would be separated from commuter traffic and trucks servicing distributors and manufacturers within the region, making freeways more useful. Reduced congestion would lessen the need to expand freeways. Less congestion would allow for more reliable military movement to the Port. An unintended benefit from the commercial Maglev regards plans for a military staging area at the Southern California Logistics Center, the former George Air Force Base. The cost for land in inland areas is cheaper for warehouse transshippers: $250,000 per acre per year at the Port versus $250 per acre per year in Hesperia.[3]

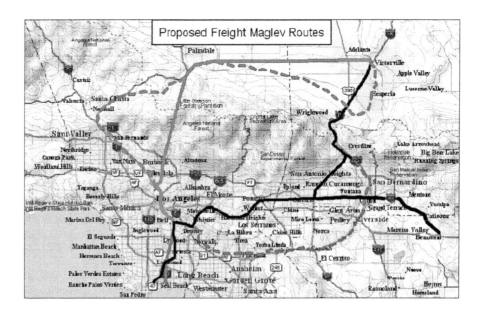

FIGURE 9.11 I-710 Corridor ECCO alignment.

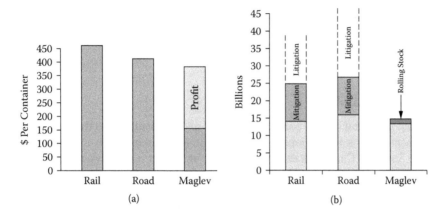

FIGURE 9.12 (a) Comparison of transportation costs to inland intermodals and rail yards. (b) Comparison of rail, highway, and Maglev capital costs to carry an additional 5+ million containers per year.

CCDoTT considered a number of rights-of-way as shown on the map in figure 9.11. Perhaps the most promising route is the one that follows I-15 through the Cajon pass, as shown in figure 9.11. Such a route is possible because the Maglev freight-optimized systems have an ability to climb steep grades. As previously mentioned, both the EMS and EDS freight-optimized Maglev systems are projected to be able to carry containers up a 6% grade, while rail can only handle 3%. This is why trains must take a circuitous route through the pass and require expensive tunneling. The 6% maximum grade for freight Maglev matches the maximum grade allowed on the interstate highway system, suggesting Maglev rights-of-way along interstate medians.

The chart in figure 9.12a shows the projected operational costs of sending a container from the Port to an inland intermodal terminal via the ECCO system.[12] The energy required per trip for Maglev was estimated to be approximately 500 KWh per 40-foot container.[12] Conventional lift-on/lift-off handling costs were added to this energy expense to arrive at the Maglev freight operational cost. The one-way cost projections for the road and rail container transport along the same route were determined by quotes from trucking firms[14] and the trial Alameda Corridor "shuttle train,"[15] respectively. Maglev costs fare very well when compared to conventional truck portage or shuttle trains presently under evaluation and include the inestimable added benefit of negligible air and noise pollution. Note that the significant profit in Maglev transport for amortization of capital costs still allows for fare savings over road and rail.

Any thorough examination of capital costs as in figure 9.12b must reflect the small footprint and ease of elevation of the Maglev freight system, which makes its construction cost less than the cost of expanding highway and rail in the crowded Los Angeles basin. Highway costs are based on construction of a four-lane elevated truck expressway with on and off ramps[16] transitioning to widened freeways to allow for dedicated truck lanes. Rail costs are based on having to "trench" (drop below road level) several miles to eliminate grade crossings through East Los Angeles.[17] These are all very expensive propositions; Maglev technology can not only lead to very significant system capital cost savings, but also possesses operating cost margins that would encourage private investment.

9.7 CONCLUSIONS

Moving large numbers of containers quickly and efficiently from the Port to transcontinental trains, transshippers, and Inland Empire warehouses is vital to the health of the Southern California economy. Equally important is the physical health of the region's citizens. A technology that moves containers without emission of noise or air pollutants and can contribute to the economy while reducing traffic congestion is desperately needed.[18] The possibility of economic and environmental harmony is demonstrated with the ECCO system. This chapter presents such a technology—Maglev—and describes how that technology can be projected onto the region's goods movement infrastructure. including possible routes, container throughput volume, capital expenditures, and operational costs.

REFERENCES

1. CCDoTT Pacific Northwest. 2004. Agile port system demonstration plan. CCDoTT.
2. Husing, John E. 2005. Southern California Association of Governments logistics & distribution: An answer to regional upward social mobility economics & politics. http://www.scag.ca.gov/goodsmove/pdf/HusingLogisticsReport.pdf (accessed April 29, 2005).
3. CCDoTT. 2005. Interview with Long Beach Terminal port operator. December 10.
4. Manny, A. 2005. California Marine transportation system infrastructure needs. http://www.mxsocal.org/MARITIME-INFORMATION/SoCal-MTS-Advisory-Council/California—MTS-Infrastructure-Needs-Report.aspx (accessed August 15, 2005).
5. Schrank, D., and Lomax, T. 2005. *The 2005 Urban Mobility Report.* http://www.campo texas.org/pdfs/2005urbanmobilityreportfinal.pdf (accessed November 13, 2005).
6. Texas Transportation Institute. The Texas A&M University System. http://mobility. tamu.edull (accessed May, 2005).
7. Bailey, D., et al. 2004. Harboring pollution: The dirty truth about U.S. ports. Natural Resources Defense Council. http://www.nrdc.org/air/pollution/Port/Port2.pdf (accessed June 13, 2005).
8. Multiple Air Toxics Exposure Study in the South Coast Air Basin (MATES-II). 2000. South Coast Air Quality Management District Governing Board. http://www. aqmd.gov/matesiidf/matestoc.htm (accessed June 2006).
9. Transrapid International Container Conveyor System Preliminary Study. 2003. Orangeline Development Authority and Transrapid International-USA, Inc.
10. Heller, A. A new approach for magnetically levitating trains and rockets. http://www. llnl.gov/str/Post.html (accessed September 20, 2006).
11. Venkatesh, M., and Jeter, P. 2006. Guideway steel fiber reinforced concrete hybrid design. Paper presented at the International Maglev Conference, Dresden, Germany, August 2006.
12. Conceptual Design Study for the Electric Cargo Conveyor (ECCO) System. 2006. Final report, General Atomics, June 2006.

13. Model simulation of EMFAC 2002 program. http://arbis.arb.ca.gov/msei/onroad/latest_version.htm (accessed October 5, 2005).
14. R & C Trucking. 2005. Phone survey, October 5, 2005.
15. CCDoTT meeting with Alameda Corridor's chief executive officer John Doherty and chief engineer Art Goodwin, April 2005.
16. MTA. 2005. MTA board adopts report on proposed $5.5 billion overhaul of congestion-plagued I-710 freeway from port to Pomona freeway. http://www.mta.net/press/2005/01_january/metro_012.htm (accessed January 24, 2005).
17. Southern California Regional Strategy for Goods Movement. A plan for action. http://www.scag.ca.gov/goodsmove/rePortmove.htm (accessed June 23, 2005).
18. Anon. Let us have Maglev. Editorial. *Long Beach Press Telegram*, May 8, 2005 (accessed January 3, 2007).

10 Container Movements with Time Windows

Hossein Jula

CONTENTS

ABSTRACT

Along with all the benefits of containerization come increased operational complexities. The growth in the number of containers has already introduced congestion and threatened the accessibility to many terminals at port facilities. Traffic congestion at ports magnifies traffic congestion in the adjacent traffic networks, which in turn affects the trucking industries on three major service dimensions: travel time, reliability, and cost.

The increasing traffic congestion at and around the container terminals necessitates the investigation of efficient, reliable, and systematic ways of handling containers. In this chapter, the time window appointment system, which has recently been introduced as a way to reduce congestion, is investigated. Furthermore, the container movement by trucks in metropolitan areas with time constraints at origins and destinations is modeled as an asymmetric multi-traveling salesman problem with time windows (m-TSPTW) with social constraints. Different variations of the m-TSPTW are studied, and solution methods are reviewed and evaluated.

10.1 INTRODUCTION

In the trucking industry, time is money. The ability of a trucking company to succeed economically rests on its ability to move goods reliably and efficiently, with minimal delay. In many traffic networks, especially in major cities, traffic congestion has already reduced mobility and system reliability, and has increased transportation costs. In addition to contributing to truck drivers' inefficiency, traffic congestion is a major source of air pollution (especially diesel toxins), wasted energy, increased maintenance cost caused by the volume of trucks on roadways, and so no. With the expected substantial increase in the volume of international and national containers, together with the anticipated growth in the number of personal vehicles in use, it is expected that the condition of traffic congestion will only get worse, unless careful planning is initiated.

There are numerous ways to improve traffic congestion, and therefore reduce transport times associated with goods movements. Options include developing new and expanding current facilities, deploying advanced technologies, and improving operational characteristics and system management practices. It should be noted that the scarcity of land in major cities has made the option of developing new facilities, if not infeasible, significantly costly.

The focus of this chapter is to investigate methods to improve the operational characteristics of container movements by developing techniques that can be easily implemented. More specifically, in this chapter, we investigate methods for improving the scheduling of trucks that are used to transfer containers among marine terminals, intermodal facilities, warehouses, and end customers. Each of these customers/facilities may have imposed time window constraints on pickup/drop-off containers.

The time window system arises naturally in the trucking industry due to the commitments made to the end customers/facilities and due to the limited availability of certain resources at their locations. In addition, in order to reduce congestion, and thus pollution, at and around marine terminals, local authorities may enforce time appointment systems at container terminals. For instance, the so-called Lowenthal Bill limits the allowable time trucks can idle in and around seaports in California to 30 minutes. Hence, many container terminals in Southern California have started implementing time appointment systems at their gates.

In the sequel, we investigate the route planning for a fleet of trucks in metropolitan areas where containers should be delivered to and picked up from customers, intermodal terminals, and container ports within specific time windows.

10.2 CONTAINER MOVEMENT PROBLEM

The container movement problem in metropolitan areas can be stated as follows. A set of containers needs to be moved in a metropolitan area by a local trucking company. A set of trucks, initially located at the company's depot (which hereafter will be called depot), is deployed to move these containers among the depot, end customers, and service stations (STs). Service stations include marine terminals and intermodal facilities. Associated with each container is a time window imposed by customers and STs for pickup or delivery at origin and destination points.

We assume that each truck can only serve a single load at a time (e.g., one 40-foot equivalent unit [FEU] container), and that each driver may not be at the wheel for more than a certain number of hours during a day (working shift). In other words, each driver has to drive his truck back to the depot before his shift ends. This time limit is a social constraint enforced by trucking companies to conform to local or federal laws. Considering a relatively small transportation network (such as a metropolitan network), we assume that a driver can serve even the farthest customer in the network during a single working day.

Assuming that the trucking company knows all its daily loads to be delivered/picked up a priori, the objective is to minimize the total cost of providing service to the customers and STs within their specified time constraints. In other words, the objective is to select some or all of the trucks and assign a set of containers to each such that collectively all containers are processed within their time constraints and the total cost (distance) is minimized.

10.2.1 PROBLEM MODELING

Let V be the set of M vehicles labeled v_m, $m = 1, 2, ..., M$, working for the trucking company, that is, $V = \{v_1, v_2, ..., v_M\}$. We assume that, at any time, a vehicle $v_m \in V$ can transfer at most a single container l from its origin to destination. We denote by $O(l)$ and $D(l)$ the origin and destination of container l, respectively. In addition, we also assume that container l must be picked up from its origin during a specific period of time known as the pickup time window and denoted by $[a_{O(l)}, b_{O(l)}]$. Likewise, container l must be delivered at its corresponding destination during a delivery time window denoted by $[a_{D(l)}, b_{D(l)}]$. With a pickup/delivery time window $[a_k, b_k]$, we assume that if a vehicle arrives at the pickup/delivery location at any time $t < a_k$, it waits till time $t = a_k$ to be served (i.e., getting loaded/unloaded). If it arrives at that location at any time $a_k \leq t \leq b_k$, it will be immediately served. The vehicle, however, cannot be served if it arrives at that location at any time $t > b_k$.

We also define task l as the sequence of jobs associated with picking up container l from its origin, $O(l)$, transferring it from its origin to destination, $D(l)$, and delivering it at the destination.

To some vehicles, sets of tasks may be assigned. Vehicles with assigned tasks are called *active* vehicles, while those without any tasks are *inactive*. Let v_m be an active vehicle and $|v_m|$ be the total number of tasks assigned to that vehicle. Let $l_{im} \in v_m$ be the ith task assigned to vehicle v_m, that is, $v_m = \{l_{1m}, l_{2m}, ..., l_{|v_m|m}\}$. The sequence of the tasks performed (visited) by vehicle v_m is called route m. Route m, denoted by $r_m = [p, l_{1m}, l_{2m}, ..., l_{|v_m|m}, p]$, starts from depot p, visits all tasks $l_{im} \in v_m$, and ends at the same depot.

Figure 10.1 shows a typical route r starting from the depot of the trucking company and ending at the same depot. Solid lines in the figure illustrate the travel between the origin and destination of a container, while dashed lines indicate empty traveling between the destination of the last drop-off and the origin of the next pickup. The ovals represent tasks to be performed on route r.

Let $l_{im}, l_{jm} \in v_m$ be the ith and jth scheduled tasks assigned to vehicle v_m, respectively, and let the jth task be performed immediately after the ith task. Let $c_{O(l_{im})D(l_{im})}$

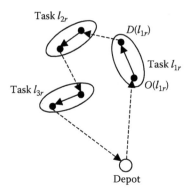

FIGURE 10.1 A route r with three tasks. The route starts from the depot and ends at the same depot. The large empty circle denotes the depot location. Each small black circle denotes the origin (O) or the destination (D) of a container.

denote the cost of performing task l_{im}. This consists of the cost of picking up container l_{im} at its origin $O(l_{im})$, moving the container from its origin to destination $D(l_{im})$, and delivering the container at its destination. Also, let $c_{D(l_{im})O(l_{jm})}$ be the cost of empty vehicle traveling between the destination of l_{im} to the origin of l_{jm}. In addition, let $c_{P,O(l_{1m})}$ and $c_{D(l_{|v_m|m}),P}$ represent the costs of empty traveling from the depot to the first task (origin of the first container) and from the last task (destination of the last task) to the depot, respectively.

Furthermore, let $F(v_m)$ signify the cost associated with performing all tasks assigned to vehicle v_m on route m in the specified order. The cost $F(v_m)$ can be expressed as

$$F(v_m) = \sum_{i=1}^{|v_m|} c_{O(l_{im})D(l_{im})} + \sum_{i=1}^{|v_m|-1} c_{D(l_{im})O(l_{i+1,m})} + c_{P,O(l_{1m})} + c_{D(l_{|v_m|m}),P} + f_{act} \quad (10.1)$$

where $l_{i+1,m}$ is the immediate task on route m after task l_{im}, and f_{act} is the vehicle activation cost indicating the cost of bringing a vehicle into service.

The trucking company's objective is to determine the optimal routes for M vehicles working for the company such that the completion of handling all containers results in minimum total cost:

$$J = Min \sum_{m=1}^{M} F(v_m). \quad (10.2)$$

Let us assume that the fleet of vehicles at the company is homogeneous. Consequently, the cost $c_{O(l_{im})D(l_{im})}$ in (10.2) is fixed and is independent of the assigned vehicle and the order of tasks performed on the route. That is, the total cost in (10.2) is only affected by the cost of empty vehicle traveling between the destinations and

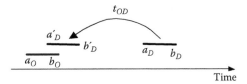

FIGURE 10.2 A typical relation between the time window at the origin $[a_O, b_O]$, destination $[a_D, b_D]$, and the time window at the destination shifted back in time $[a'_D, b'_D]$.

origins of tasks as well as vehicles' activation costs. Thus, the problem of interest is reduced to finding the best feasible assignment and sequencing of N tasks to M vehicles such that the total empty travel cost of vehicles is minimum.

In figure 10.1, route m is said to be feasible if it satisfies the time window constraints at the origins and destinations of all assigned tasks and the total time needed for traveling on the route is less than a certain amount of time called the working shift (time) and denoted by T. In other words, it is required that all tasks assigned to a vehicle be done within the time horizon T.

We define the *time window* of task l_{im} as the period of time in which performing task l_{im} should be started. This time window can be expressed in terms of (1) the time window at the origin, $[a_{O(l_{im})}, b_{O(l_{im})}]$, (2) the time window at the destination, $[a_{D(l_{im})}, b_{D(l_{im})}]$, and (3) the travel time between the origin and destination, $t_{O(l_{im})D(l_{im})}$. Figure 10.2 demonstrates a typical relationship between these three elements. For the sake of simplicity, we eliminate all subscripts l_{im} in figure 10.2. Thus, the time window denoted by $[a'_D, b'_D]$ in figure 10.2 is the time window at the destination shifted back in time by $t_{O(l_{im})D(l_{im})}$.

Figure 10.3 presents all possible relationships between time windows $[a'_D, b'_D]$ and $[a_O, b_O]$. Dashed areas in figure 10.3 indicate the time windows at origin node O

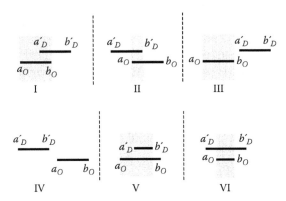

FIGURE 10.3 All possible relations between the time window at the origin and the time window at the destination shifted back in time. The dashed area presents the portion of the time window at origin O during which a vehicle can be loaded and still meet the time window constraint at destination D.

during which vehicle v_m can be loaded, and yet it meets the time window constraint at destination D. For instance, case I in figure 10.3 indicates that if v_m starts task l_{im} at any time within the interval $[a_O, b_O]$, it would be served at both the origin and destination nodes. Case I also indicates that if the loaded vehicle leaves the origin O at any time $\tau \in [a_O, a'_D]$, it will reach destination D prior to time a_D. Therefore, the vehicle has to wait at destination D for a period of time equal to $a'_D - \tau = a_D - t_{OD} - \tau$. If the vehicle leaves origin O at any time $\tau \in [a'_D, b_O]$., it will be unloaded at D without any delay. Moreover, if the vehicle reaches the origin O at time $\tau \in (b_O, b'_D)$, it cannot be loaded at the origin even though the vehicle can meet the time constraint (i.e., can be unloaded) at the destination.

According to figure 10.3, the time window of task l_{im}, $[a_{l_{im}}, b_{l_{im}}]$, can be calculated by

$$a_{l_{im}} = a_{O(l_{im})}$$

$$b_{l_{im}} = \min \ (b_{O(l_{im})}, b_{D(l_{im})} - t_{O(l_{im})D(l_{im})}). \tag{10.3}$$

Note that the possibility of $b_{l_{im}} < a_{l_{im}}$ in (10.3) corresponds to case IV in the same figure, which is an infeasible case and may not occur in a real situation.

We also define the *service time* of task l_{im} as the time needed for carrying container l_{im} from its origin to its destination (i.e, between $O(l_{im})$ and $D(l_{im})$), together with the pickup time at $O(l_{im})$ and drop-off time at $D(l_{im})$. The service time includes the waiting time at the destination, if any.

In summary, the container movement problem with time windows can be stated as follows: M vehicles are located at the depot and have to perform N tasks. The objective is to select some or all of the vehicles and assign routes (i.e., a set of tasks in a specific order) to them such that in the collection of all routes together, the cost is minimized and each task is performed exactly once within a specified time window. The problem as described falls in the class of the multi-traveling salesman problem with time windows (m-TSPTW).

Note that since the travel cost between each of two tasks, say i and j, depends on the direction of movement, that is, $c_{D(l_j)O(l_i)} \neq c_{D(l_i)O(l_j)}$, the container movement problem is, in fact, asymmetric. Recall also that it is required that the total time needed for a vehicle to visit all assigned tasks on its route and return to the depot be less than a certain amount. This time limit is an additional constraint to the m-TSPTW and will be called the social constraint. Therefore, the container movement by trucks in metropolitan areas with time constraints at origins and destinations can be modeled as an asymmetric m-TSPTW with social constraints.

10.3   TRAVELING SALESMAN PROBLEM WITH TIME WINDOWS

In this section, the mathematical formulation for the m-TSPTW problem is presented, the variants of the problem are studied, and existing solutions are briefly reviewed and evaluated.

10.3.1 PROBLEM FORMULATION

Let $G = (ND, A)$ be a graph with node set $ND = \{P_o, P_d, N\}$ and arc set $A = \{(i, j)$ $| i, j \in ND\}$. The nodes P_o and P_d represent the single depot (origin–depot and destination–depot), and the set $N = \{1, 2, ..., n\}$ is the set of tasks to be performed. To each arc $(i, j) \in A$ a cost c_{ij} and a duration of time t_{ij} are associated, representing the cost and the time of traveling between nodes i and j, respectively. In addition, to each node $i \in ND$, a service time s_i and a time window $[a_i, b_i]$ are associated. As described earlier, the service time s_i is the duration of time for a vehicle to perform task i, and $[a_i, b_i]$ is the time window of task i, respectively. An arc $(i, j) \in A$ is said to be feasible if and only if $a_i + s_i + t_{ij} \leq b_j$. Let V be the set of vehicles v. A route r^v in the graph G is defined as the set of tasks (nodes) assigned to vehicle v such that the time when the service begins at node $j \in r^v$ is within the time window of that node. Let

$x_{ij}^v = 1$ if arc $(i, j) \in A$ is traveled by vehicle v, and 0 otherwise, and

T_i^v be the time when service begins at node i by vehicle v.

The m-TSPTW can be formulated as follows:
Minimize:

$$\sum_{v \in V} \sum_{(i,j) \in A} c_{ij} x_{ij}^v \tag{10.4}$$

subject to:

$$\sum_{v \in V} \sum_{j \in N \cup \{P_d\}} x_{ij}^v = 1 \qquad \forall i \in N \tag{10.5}$$

$$\sum_{v \in V} \sum_{j \in N} x_{P_o j}^v \leq |V| \tag{10.6}$$

$$\sum_{j \in N \cup \{P_d\}} x_{ij}^v - \sum_{j \in N \cup \{P_o\}} x_{ji}^v = 0 \qquad \forall i, j \in ND, \forall v \in V \tag{10.7}$$

$$x_{ij}^v \left(T_i^v + s_i + t_{ij} - T_j^v \right) \leq 0 \qquad \forall i, j \in ND, \forall v \in V \tag{10.8}$$

$$a_i \leq T_i^v \leq b_i \qquad \forall i \in ND, \forall v \in V \tag{10.9}$$

$$x_{ij}^v \in \{0, 1\} \qquad \forall (i, j) \in A, \forall v \in V \tag{10.10}$$

The objective function (10.4) expresses the total cost. Constraints (10.5) require that only one vehicle visit each node in N. Constraints (10.6) ensure that at most $|V|$ number of vehicles are used. To use exactly $|V|$ vehicles, the inequality should be replaced by equality. Constraints (10.7) guarantee that the number of vehicles leaving node j is the same as the number of vehicles entering that node. Therefore, constraints (10.5) to (10.7) together enforce that at most $|V|$ number of vehicles visit all nodes in N only once. Constraints (10.8) enforce the time feasibility condition on consecutive nodes. Constraints (10.9) specify the time window constraints at each node. Finally, constraints (10.10) are the binary constraints.

The m-TSPTW formulation in (10.4) through (10.10) is a nonlinear model since constraints (10.8) are nonlinear constraints. Considering the binary constraints in (10.10), one can easily linearize constraints (10.8), resulting in the m-TSPTW model, as follows:

$$T_i^v + s_i + t_{ij} - T_j^v \leq \left(1 - x_{ij}^v\right) M \qquad \forall i, j \in ND, \forall v \in V \qquad (10.11)$$

where M is a large constant.

In addition to constraints (10.4) to (10.10), we also require that the following constraint be met in order to implement the social constraints imposed on truck drivers:

$$T_{P_d}^v - T_{P_o}^v \leq T \qquad \forall v \in V \qquad (10.12)$$

Constraints (10.12) require that each driver shall not be on the wheel more than a certain number of hours, T, in each single day. Note that by adding (10.12) to the m-TSPTW model, we assume that truck drivers have to return their trucks back to the depot before the end of their shifts.

In the following, we will describe different variants of the m-TSPTW and briefly review existing solutions.

10.3.2 DETERMINISTIC TSPTW

When the number of vehicles involved is exactly one, the problem is reduced to the single-vehicle TSPTW. More precisely, in the TSPTW, constraints (10.6) are an equality with $|V| = 1$. In the TSPTW, a single vehicle, initially located at the depot, must serve a number of geographically dispersed customers such that each customer is served within a specified time window. The objective is to find the optimum route with minimum total cost of travel.

In the literature, the TSPTW has not received as much attention as its special case, the traveling salesman problem (TSP),[1-5] in which the time window constraints are relaxed. Although any problem instance of the TSP is feasible, the existence of the time window constraints may lead to an infeasible instance of the TSPTW. That is why even finding a feasible solution to the TSPTW is an NP-complete problem.[6,7] In general, as the widths of the time windows increase, it is easier to find a feasible solution to the TSPTW as the TSPTW approaches the TSP.[8] However, as the widths narrow, it is computationally much easier to find the optimal solution, if feasibility sustains.

Due to the complexity of the TSPTW, very few research efforts have been focused on finding the exact solution to the TSPTW. Dumas et al.[9] used dynamic programming (DP) enhanced by a variety of elimination tests to optimally solve the TSPTW. These tests took advantage of the time window constraints to significantly reduce the number of arcs in the graph in order to eliminate the states. The authors managed to solve problems with up to 200 nodes with small window size, and 80 nodes with larger time windows.

Focacci et al.[10] proposed a hybrid exact algorithm for solving the TSPTW based on a constraint programming framework, which finds feasible paths in conjunction with a propagation algorithm to find the optimum solution. The proposed algorithm managed to optimally solve asymmetric TSPTW problems with up to seventy nodes. Pesant et al.[11] also developed a constraint logic programming model for solving problems for up to eighty nodes with relatively tight time windows.

Branch-and-bound-based algorithms have also been considered and used for solving the TSPTW. Baker[12] developed a branch-and-bound algorithm based on an integer programming formulation. The proposed algorithm was able to solve problem instances for up to fifty nodes. Christofides et al.[13] presented a branch-and-bound algorithm on the state space relaxation of a dynamic programming formulation. The algorithm was able to solve problem instances for up to fifty nodes with moderately tight time windows. Ascheuer et al.[7] developed formulations for asymmetric TSPTW and solved various problem instances for up to fifty to seventy nodes to optimality using branch-and-cut methodology. Ascheuer et al.[7] concluded that "dynamic programming and implicit enumeration techniques may outperform [their] cutting plane method, in particular when the time windows are rather tight."

Finally, Langevin et al.[15] (504) reported a methodology for solving problems with up to forty nodes using a two-commodity flow formulation.

Given the inherent computational difficulty of the TSPTW, a variety of heuristics have been reported in the literature. Lin and Kernighan[16] proposed a heuristic algorithm based on a k-interchange concept for the TSPTW. The method involves the replacement of k arcs currently in the solution with k other arcs.

Calvo[17] proposed the use of a new heuristic method for the TSPTW based on solving an auxiliary assignment problem. To find better solutions, the algorithm uses two objective functions. When the algorithm gets trapped in a local minimum, it uses the second objective function to widen its neighborhood region.

Gendreau et al.[18] developed a generalized insertion heuristic algorithm for TSPTW. The algorithm gradually builds a route by inserting at each step a vertex onto the route and performing a local optimization. Once a feasible route has been determined, a postoptimization algorithm is used to improve the objective function.

Metaheuristic methods have also been reported in the literature to solve the TSPTW. Unlike local search heuristics that terminate once a local optimum has been reached, metaheuristic methods explore a larger subset of the solution space in the hope of finding the near-optimal solution. Ohlmann and Thomas[19] presented a variant of the simulated annealing method called compressed annealing approach to solve the TSPTW. Carlton and Barnes[8] solved the TSPTW using a tabu search approach.

10.3.3 Deterministic m-TSPTW

The multi-traveling salesman problem with time windows (m-TSPTW) can be stated as follows: m salesmen are located in a city (i.e., node $n + 1$) and have to visit n cities (nodes: 1, ..., n). The task is to select some or all of the salesmen and assign tours to them such that in the collection of all tours together, the cost (distance) is minimized and each city is visited exactly once within a specified time window.

A special case of the m-TSPTW is the multi-traveling salesman problem (m-TSP) in which the time window constraints are relaxed. More precisely, referring to formulations (10.4) to (10.10), the m-TSP is the relaxation of the m-TSPTW when constraints (10.8) and (10.9) are relaxed. An instance of the m-TSP can be easily transformed to an equivalent instance of the TSP by creating $m - 1$ (where m is the number of salesmen) additional nodes at the location of the node $n + 1$,[1] and determining the incident edges for additional nodes. Hence, the TSP solution methods can be used to find the solution to the m-TSP.

In contrast to the relationship between the m-TSP and the TSP, the m-TSPTW cannot be converted into an equivalent TSPTW. The transformation of an instance of the m-TSP to the TSP problem is based on the fact that the TSP and the m-TSP are both problems in one dimension: space. The TSPTW and m-TSPTW, however, are problems in two dimensions: space and time. An instance of the m-TSPTW may be feasible in time, while any polynomial time transformation of that instance to an instance of the single-vehicle TSPTW would be infeasible. In other words, in general, such a transformation does not exist.

On the other hand, the TSPTW problem is a special case of the vehicle routing problem with time windows (VRPTW) in which the capacity constraints are relaxed (for extensive surveys and bibliography regarding the vehicle routing problem and its variants, see references 20–25). Consequently, one may think of applying the VRPTW solution methods to the m-TSPTW by relaxing the capacity constraints in the VRPTW. Although the idea looks very rational, the experimental results show otherwise. In their work, Dumas et al.[9] (367) state:

> Even though the TSPTW is a special case of the VRPTW, the best known approach to the latter problem[26] is not well suited to solve the TSPTW. This column generation approach would experience extreme degeneracy difficulties in this case.

In response to these difficulties, researchers have sought methods tailored for the m-TSPTW.

Lee[27] developed two heuristics based on the vehicle scheduling problem (VSP) for the m-TSPTW. The VSP algorithms are exact in that they can find the optimal solution to the VSP in polynomial time. However, solutions found by VSP algorithms may be infeasible for the m-TSPTW. Two construction heuristics are developed to assign each customer to a route. Improvement heuristics are then developed to combine the initial routes.

Wang and Regan[28] described an iterative solution method for the m-TSPTW using time windows discretization. At each iteration, they generate an overconstrained and an underconstrained version of the problem. The former provides a feasible solution and an upper bound on the cost, whereas the latter gives a lower bound on the cost of the optimal solution. The method was implemented on small-size problems with twenty vehicles and seventy-five nodes.

Jula et al.[29] developed the following three methodologies to solve the m-TSPTW with additional social constraints: (1) an exact two-phase dynamic programming,

TABLE 10.1

Comparing Exact, Hybrid GA, and Insertion Methods (w = 2 hours)

No. of Nodes	Dynamic Programming		Genetic Algorithm[a]		Insertion Method	
	Cost	CPU Time[b]	Cost	CPU Time	Cost	CPU Time
7	13.69	0.38	13.69	0.22	17.69	0.11
10	23.88	1.70	23.88	0.50	26.96	0.11
15	35.38	326.4	35.38	0.76	42.25	0.16
20	NA[c]	NA	52.44	15.86	59.91	0.27
30	NA	NA	98.97	61.61	107.0	0.49
50	NA	NA	179.4	191.91	194.45	1.45
100	NA	NA	338.9	1876	359.08	4.67

[a] In average (based on the results of 10 trials; each trail runs for 1,000 iterations).

[b] In seconds, on an Intel Pentium 4, 1.6 GHZ, coded in Matlab 5.3 developed by MathWorks.

[c] NA: The result could not be obtained.

(2) a hybrid methodology consisting of dynamic programming in conjunction with genetic algorithms (GAs), and (3) an insertion heuristic method. They compared these methodologies on a Euclidean plane in which the coordinates of nodes were uniformly distributed. For each node, the earliest time to start service, that is, a_i, was generated randomly using a uniform random generator between 9:00 A.M. and 5:00 P.M., the time window length was set to be 2 hours, and the service time was generated randomly between 30 minutes and 2 hours. Table 10.1 summarizes and compares the results obtained from the exact, hybrid genetic algorithm (GA), and insertion methods.

As indicated by table 10.1, the exact method is efficient for relatively small-size problems consisting of a few nodes. The hybrid GA is capable of finding the optimal solutions for small-size problems and near-optimal solutions for medium- to large-size problems. The insertion heuristic method is computationally very fast and is able to find relatively good solutions for large-size problems.

It should be mentioned that many practical applications can be modeled as an instance of the m-TSPTW, and thus can be solved using optimization techniques developed for the m-TSPTW. For example, in this chapter, we have shown that the container movement problem with time appointment system can be modeled as a variant of the m-TSPTW. Other applications include the ship scheduling problem,[30,31] the bus scheduling problem,[32–34] the daily aircraft scheduling problem,[35] the lot sizing and scheduling problem,[36] and the unmanned aerial vehicle (UAV) routing.[37]

10.3.4 STOCHASTIC TSPTW

Most existing methods for vehicle route planning assume known static data in an environment that is time varying and uncertain by nature, which limits their widespread applicability. For instance, in the deterministic TSPTW the customer demands and locations, travel costs, and travel times are known a priori. In the real world, however, operations in any traffic network contain a fairly high degree of uncertainties, including arrival of new orders, cancellation of existing orders, variable waiting times, and variable travel times due to traffic congestion.[38–40]

The stochastic traveling salesman problem with time windows (STSPTW) arises whenever some or all elements of the TSPTW are random. Similar to its deterministic counterpart, the STSPTW is a special case of the stochastic vehicle routing with time windows (SVRPTW) in which the capacity constraints are relaxed. Although the research efforts on the STSPTW have been very scant, during the last two decades, the stochastic vehicle routing problem (SVRP) and the SVRPTW have been the subject of a relatively wide body of research.

The most studied area in SVRP has been the VRP with stochastic demands (VRPSD) and with stochastic customers (VRPSC).[41–46] Despite its importance and practical application, especially in major cities with traffic congestion, research efforts on the stochastic VRP with stochastic travel times (VRPST) have been limited.[47–49] The objective in the VRPST problem is to find the optimal vehicle routes in the presence of random travel and service times. Most developed VRPST solution methods require the knowledge of the distribution of the sum of the travel and service times along the routes.[45,47]

Jula et al.[50] investigated the STSPTW in which traveling times along the roads and service times at the customer locations are stochastic processes. They developed a methodology to estimate the vehicle arrival time at customer locations and proposed an approximate solution method based on dynamic programming to find the least-cost route in the STSPTW.

The results from various experiments reported in Jula et al.[50] indicated that the developed methodology was able to estimate the mean and standard deviation of the arrival times at the nodes of stochastic networks with very small errors. The results also showed that the approximate algorithm was successful in solving both stationary and nonstationary STSPTW problems with up to eight nodes with fairly wide time windows.

10.3.5 Dynamic TSPTW

The TSPTW is said to be dynamic if the information (input) on the problem is made known to the decision maker or is updated concurrently with the determination of the route.[51] Hence, the solution methods for the dynamic TSPTW should incorporate real-time traffic information and real-time changes in the customer demands and locations. In other words, the solution algorithm should be dynamic in the sense that, as the vehicle proceeds to the next destination, the algorithm should update the route assigned to the vehicle in response to the new information.[52]

As a new request is placed by a customer in real time, the dynamic algorithm may examine the feasibility and cost of including the new request into the current existing routes. Hence, the algorithm may recommend one of the following actions: accept the request, reject the request, or postpone the decision until further information (may be about other real-time requests) becomes available.

To add the accepted request to the existing routes, two different algorithms may be adopted:

1. An insertion method in which a fair amount of requests are known beforehand and are scheduled statically. Then, real-time requests are incorporated (inserted) in the initial solution framework.[53,54]
2. A fast static algorithm in which the dynamic problem is solved as a sequence of fast static problems. The algorithm is applied repetitively when new requests are considered.[55,56]

It should be noted that, in the presence of a high degree of dynamisms, it is widely expected that optimal solutions for routing (i.e., a sequence of *all* customers) will be outperformed, over time, by algorithms that are more local (i.e., considering the next node) in nature.[57]

When the travel time along the route is uncertain and time dependent, the arrival time at the customer locations will be uncertain too. Whereas in the past, once the drivers left their depots, it was difficult for them to adjust their routes according to traffic congestion, new advanced technologies make accurate real-time routing a possible reality. Jula et al.[58] studied methods of improving routing in uncertain and dynamic environments. They developed a technique to predict the travel times on the arcs of a dynamic network based on a predictor corrector form of the Kalman filter. In this technique, available historical data are used for predicting the travel times, and real-time measurements are used to correct and update the prediction at each instant of time. Other dynamic travel time predictor techniques can be found in references 59–61. Jula et al.[58] also used the predicted travel times on the arcs to estimate the arrival times at the nodes of the network, a step that can be used for dynamic route planning.[62]

10.3.6 OTHER EXTENSIONS AND VARIATIONS OF THE M-TSPTW

- **Multiple depots.** In the m-TSPTW, all vehicles are initially located at one depot. An extension of the m-TSPTW is the one in which vehicles may be positioned at multiple depots with different locations or may start their routes at different points. The problem is to simultaneously assign each customer to a depot and to construct routes for each vehicle according to the m-TSPTW rules (also see Cordeau et al.[63]).
- **Multiple time windows.** The traditional TSPTW, in which a single time window is assigned to each customer, can be extended to include multiple time windows. The TSP with multiple time windows may occur when the delivery of cargos cannot be done continuously within a single time window, or the customers are not open for delivery every hour of the day or every day of the week.[64,65] Pesant et al.[66] studied the TSPTW problem with multiple time windows based on constraint programming models.
- **Soft time windows.** Two types of time windows can be considered at customer locations: hard and soft. The hard time window, which has been considered in this chapter, is the constraint that cannot be violated. However, in the soft time window, the constraint can be violated at a cost. The cost of violating the soft time window usually appears as a penalty cost in the objective function.[67,68]

- **Precedence constraints.** In the TSPTW with precedence constraints, given a customer i, it is required that a given set of customers be visited before visiting customer i. The problem is investigated by Mingozzi et al.,[69] in which the authors described an exact algorithm based on dynamic programming to solve the problem.

10.4 CONCLUSION

In this chapter, the container movement by trucks in metropolitan areas with time constraints at origins and destinations was investigated. We showed that the problem can be modeled as an asymmetric multi-traveling salesman problem with time windows (m-TSPTW) with social constraints. The m-TSPTW and its variants were studied and solution methods were reviewed and evaluated. In particular, we investigated the deterministic, stochastic, and dynamic variants of the problem.

Since the problem is NP-hard, as the number of customers increases, finding the optimal solution becomes a daunting task. The current exact algorithms can solve problems with up to 200 nodes with tight window size, and 70–80 nodes with wide time windows. For larger problems, a variety of heuristic and metaheuristic solutions have been reported and used.

It should be noted that traffic networks, in particular in major cities, are time varying and uncertain by nature, which limits the applicability of deterministic m-TSPTW. Whereas in the past, once the drivers left their depots, it was difficult for them to adjust their routes according to traffic congestion, new advanced technologies make accurate real-time routing a possible reality. Hence, it is widely expected that the dynamic TSPTW, which incorporates real-time traffic information and real-time changes in the customer demands and locations, will be the solution of tomorrow.

REFERENCES

1. Reinelt, G. 1994. *The traveling salesman: Computational solutions for TSP applications*. Lecture Notes in Computer Science, vol. 840. New York: Springer-Verlag.
2. Johnson, D. S., and McGeoch, L. A. 1997. The traveling salesman problem: A case study in local optimization. In *Local search in combinatorial optimization*, ed. E. H. L. Aarts and J. K. Lenstra, 215–310. New York: John Wiley & Sons.
3. Lawler, E. L., Lenstra, J. K. A., Rinnooy Kan, H. G., and Shmoys, D. B., eds. 1991. *The traveling salesman problem: A guided tour of combinatorial optimization*. New York: John Wiley & Sons.
4. Jünger, M., Reinelt, G., and Rinaldi, G. 1997. The traveling salesman problem. In *Annotated bibliography in combinatorial optimization*, ed. M. Dell'Amico, F. Maffioli, and S. Martello, 199–221. New York: John Wiley & Sons.
5. Gutin, G., and Punnen, A. P., eds. 2002. *The traveling salesman problem and its variations*. Combinatorial Optimization Series, vol. 12. Boston: Kluwer Academic Publishers.
6. Savelsbergh, M. W. P. 1985. Local search in routing problems with time windows. *Annals of Operations Research* 4:285–305.
7. Ascheuer, N., Fischetti, M., and Grötschel, M. 2001. Solving the asymmetric traveling salesman problem with time windows by branch-and-cut. *Mathematical Programming* 90:475–506.

8. Carlton, W. B., and Barnes, J. W. 1996. Solving the traveling-salesman problem with time windows using tabu search. *IIE Transactions* 28:617–29.

9. Dumas, Y., Desrosiers, J., Gelinas, E., and Solomon, M. M. 1995. An optimal algorithm for the traveling salesman problem with time windows. *Operations Research* 43:367–71.

10. Focacci, F., Lodi, A., and Milano, M. 2002. A hybrid exact algorithm for the TSPTW. *INFORMS Journal on Computing* 14:403–17.

11. Pesant, G., Gendreau, M., Potvin, J. -Y., and Rousseau, J.-M. 1998. An exact constraint logic programming algorithm for the traveling salesman problem with time windows. *Transportation Science* 32:12–29.

12. Baker, E. K. 1983. An exact algorithm for the time-constraint traveling salesman problem. *Operations Research* 31:938–45.

13. Christofides, N., Mingozzi, A., and Toth, P. 1981. State-space relaxation procedures for the computation of bounds to routing problems. *Networks* 11:145–64.

14. Ascheuer, N., Fischetti, M., and Grötschel, M. 2000. A polyhedral study of the asymmetric traveling salesman problem with time windows. *Networks* 36:69–79.

15. Langevin, A., Desrochers, M., Desrosiers, J., Gelinas, S., and Soumis, F. 1993. A 2-commodity flow formulation for the traveling salesman and the makespan problems with time windows. *Networks* 23:631–40.

16. Lin, S., and Kernighan, B. 1973. An effective heuristic algorithm for the traveling salesman problem. *Operations Research* 1:498–516.

17. Calvo, R. W. 2000. A new heuristic for the traveling salesman problem with time windows. *Transportation Science* 34:113–24.

18. Gendreau, M., Hertz, A., Laporte, G., and Stan, M. 1998. A generalized insertion heuristic for the traveling salesman problem with time windows. *Operations Research* 43:330–35.

19. Ohlmann, J. W., and Thomas, B. W. 2007. A compressed annealing heuristic for the traveling-salesman problem with time windows. *INFORMS Journal on Computing* 19:80–90.

20. Golden, B. L., and Assad, A. A., eds. 1988. *Vehicle routing: Methods and studies.* Amsterdam: North Holland Publication.

21. Laporte, G. 1992. The vehicle routing problem: An overview of exact and approximate algorithms. *European Journal of Operational Research* 59:345–58.

22. Fisher, M. 1995. Vehicle routing. In *Network routing*, vol. 8, *Handbooks in operations research and management science*, ed. M. O. Ball, T. L. Magnati, C. L. Monma, and G. L. Nemhauser, 1–33. Amsterdam: Elsevier Science.

23. Toth, P., and Vigo, D. 2002. *The vehicle routing problem.* Monographs on Discrete Mathematics and Applications. Philadelphia: SIAM.

24. Bräysy, O., and Gendreau, M. 2005. Vehicle routing problem with time windows. Part I. Route construction and local search algorithms. *Transportation Science* 39: 104–18.

25. Bräysy, O., and Gendreau, M. 2005. Vehicle routing problem with time windows. Part II. Metaheuristics. *Transportation Science* 39:119–39.

26. Desrochers, M., Desrosiers, J., and Solomon, M. 1992. A new optimization algorithm for the vehicle routing problem with time windows. *Operations Research* 40:342–54.

27. Lee, M. S. 1992. New algorithms for the m-TSPTW. Ph.D. dissertation, University of Maryland, College Park.

28. Wang, X., and Regan, A. C. 2002. Local truckload pickup and delivery with hard time window constraints. *Transportation Research* 36B:97–112.

29. Jula, H., Dessouky, M., Ioannou, P., and Chassiakos, A. 2005. Container movement by trucks in metropolitan networks: Modeling and optimization. *Transportation Research* 41E:235–59.

30. Appelgren, L. H. 1969. A column generation algorithm for a ship scheduling problem. *Transportation Science* 3:53–68.

31. Appelgren, L. H. 1971. Integer programming methods for a vessel scheduling problem. *Transportation Science* 5:64–78.

32. Swersey, A., and Ballard, W. 1983. Scheduling school buses. *Management Science* 30:844–53.

33. Bianco, L., Mingozzi, A., and Ricciardelli, S. 1995. An exact algorithm for combining vehicle trips. In *Computer-aided transit scheduling*, ed. J. R. Daduna, I. Branco, and J. Paixao, 145–72. Lecture Notes in Economics and Mathematical Systems, vol. 43. Berlin: Springer-Verlag.

34. Desaulniers, G., Lavigne, J., and Soumis, F. 1998. Multi-depot vehicle scheduling with time windows and waiting costs. *European Journal of Operational Research* 111:479–94.

35. Desaulniers, G., Desrosiers, J., Solomon, M. M., and Soumis, F. 1997. Daily aircraft routing and scheduling. *Management Science* 43:841–55.

36. Drexl, A., and Kimms, A. 1997. Lot sizing and scheduling: Survey and extensions. *European Journal of Operational Research* 99:221–35.

37. O'Rourke, K. P., Bailey, T. G., Hill, R., and Carlton, W. B. 2000. Dynamic routing of unmanned aerial vehicles using reactive tabu search. *Military Operations Research Journal* 6.

38. Powell, W. B., Jaillet, P., and Odoni, A. 1995. Stochastic and dynamic networks and routing. In *Network routing*, Vol. 8, *Handbooks in operations research and management science*, ed. M. O. Ball, T. L. Magnati, C. L. Monma, and G. L. Nemhauser, 1–33. Amsterdam: Elsevier Science.

39. Powell, W. B. 1996. A stochastic formulation of the dynamic assignment problem with an application to truckload motor carriers. *Transportation Science* 30:195–219.

40. Psaraftis, H. N. 1995. Dynamic vehicle routing: Status and prospects. *Annals of Operations Research* 61:143–64.

41. Bertsimas, D. J. 1992. A vehicle routing problem with stochastic demand. *Operations Research* 40:574–85.

42. Bertsimas, D. J., and Simchi-Levi, D. 1996. A new generation of vehicle routing research: Robust algorithms, addressing uncertainty. *Operations Research* 44:286–304.

43. Branke, J., Middendorf, M., Noeth, G., and Dessouky, M. M. 2005. Waiting strategies for dynamic vehicle control. *Transportation Science* 39:298–312.

44. Gendreau, M., Laporte, G., and Seguin, R. 1995. An exact algorithm for the vehicle routing problem with stochastic customers and demands. *Transportation Science* 29:143–55.

45. Gendreau, M., Laporte, G., and Seguin, R. 1996. Stochastic vehicle routing. *European Journal of Operational Research* 88:3–12.

46. Secomandi, N. 2001. A rollout policy for the vehicle routing problem with stochastic demands. *Operations Research* 49:796–802.

47. Laporte, G., Louveaux, F., and Mercure, H. 1992. The vehicle routing problem with stochastic travel times. *Transportation Science* 26:161–70.

48. Lambert, V., Laporte, G., and Louveaux, F. 1993. Designing collection routes through bank branches. *Computers & Operations Research* 20:783–91.

49. Kim, S., Lewis, M. E., and White III, C. C. 2005. Optimal vehicle routing with real-time traffic information. *IEEE Transactions on Intelligent Transportation Systems* 6:178–88.

50. Jula, H., Dessouky, M., and Ioannou, P. A. 2006. Truck route planning in nonstationary stochastic networks with time windows at customer locations. *IEEE Transactions on Intelligent Transportation Systems* 7:51–62.

51. Psaraftis, H. N. 1995. Dynamic vehicle routing: Status and prospects. *Annals of Operations Research* 61:143–64.
52. Schmitt, E. J., and Jula, H. 2006. Vehicle route guidance systems: Classification and comparison. Paper presented at Proceedings of the IEEE Intelligent Transportation Conference, Toronto.
53. Wilson, N. H. M., and Colvin, N. H. 1977. Computer dial-a-ride algorithms research project. Technical report R-76-20, Department of Civil Engineering, Massachusetts Institute of Technology.
54. Madsen, O. B. G., Ravn, H. F., and Rygaard, J. M. 1995. A heuristic algorithm for a dial-a-ride problem with time windows, multiple capacities and multiple objectives. *Annals of Operations Research* 60:193–208.
55. Psaraftis, H. N. 1988. Dynamic vehicle routing problems. In *Vehicle routing: Methods and studies*, ed. B. L. Golden and A. A. Assad, 223–48. Amsterdam: North Holland Publication.
56. Larsen, A., Madsen, O. B. G., and Solomon, M. M. 2004. The a priori dynamic traveling salesman problem with time windows. *Transportation Science* 38:459–72.
57. Powell, W. B., Towns, M. T., and Marar, A. 2000. On the value of optimal myopic solutions for dynamic routing and scheduling problems in the presence of user noncompliance. *Transportation Science* 34:67–85.
58. Jula, H., Dessouky, M., and Ioannou, P. (In press). Real-time estimation of travel times along the arcs and arrival times at the nodes of dynamic stochastic networks. *IEEE Transactions on Intelligent Transportation Systems*.
59. Rice, J., and van Zwet, E. 2004. A simple and effective method for predicting travel times on freeways. *IEEE Transactions on Intelligent Transportation Systems* 5:200–7.
60. Chien, S. I. J., and Kuchipudi, C. M. 2003. Dynamic travel time prediction with real-time and historical data. *Journal of Transportation Engineering* 129:608–16.
61. Wu, C. H., Ho, J. M., and Lee, D. T. 2004. Travel-time prediction with support vector regression. *IEEE Transactions on Intelligent Transportation Systems* 5:276–81.
62. Dessouky, M., Ioannou, P., and Jula, H. 2004. *A novel approach to routing and dispatching trucks based on partial information in a dynamic environment.* Metrans Technical Report 03-01.
63. Cordeau, J. F., Laporte, G., and Mercier, A. 2001. A unified tabu search heuristic for vehicle routing problems with time windows. *Journal of the Operational Research Society* 52:928–36.
64. Bell, W., Dalberto, L. M., Fisher, M. L., Greenfield, A. J., et al. 1983. Improving the distribution of industrial gases with an on-line computerized routing and scheduling optimizer. *Interfaces* 13:4–23.
65. Christiansen, M., and Fagerholt, K. 2002. Robust ship scheduling with multiple time windows. *Naval Research Logistics* 49:611–25.
66. Pesant, G., Gendreau, M., Potvin, J.-Y., and Rousseau, J. M. 1999. On the flexibility of constraint programming models: From single to multiple time windows for the traveling salesman problem. *European Journal of Operational Research* 117:253–63.
67. Balakrishnan, N. 1993. Simple heuristics for the vehicle routing problem with soft time windows. *Journal of the Operational Research Society* 44:279–87.
68. Taillard, E., Badeau, P., Gendreau, M., Guertin, F., and Potvin, J. -Y. 1997. A tabu search heuristic for the vehicle routing problem with soft time windows. *Transportation Science* 31:170–86.
69. Mingozzi, A., Bianco, L., and Ricciardelli, S. 1997. Dynamic programming strategies for the traveling salesman problem with time window and precedence constraints. *Operations Research* 45:365–77.

11 Intermodal Drayage Routing and Scheduling

Alan L. Erera and Karen R. Smilowitz

CONTENTS

ABSTRACT

Determining optimal and near-optimal daily operational plans for trucks used for local delivery tasks has been an important application area of operations research for nearly 50 years. The mathematical optimization problems that result in this context are typically called *vehicle routing problems* and are generally difficult to solve since they include both combinatorial sequencing and discrete-item packing as subproblems. The majority of the developed quantitative models and solution approaches in this area consider general routing problems in which vehicles consolidate the loads of many customers on tours that begin and end at a common depot, and thus individual vehicle tours are usually constrained by the physical capacity of the vehicle to stow goods as well as time constraints given by customer requirements and operational rules.

Intermodal drayage truck routing and scheduling problems represent a special class of vehicle routing problems called *full-truckload pickup and delivery problems*. Feasible routes in such problems are primarily constrained by time restrictions. In this chapter, we describe methods to improve container drayage truck routing and scheduling practice through the use of advanced systematic scheduling approaches based on information technology.

11.1   INTRODUCTION

Intermodal drayage refers to the local movement of intermodal freight transportation equipment (trailers and containers), both loaded and empty. Drayage service provides the "last-mile" link for the vast majority of freight moving via the international intermodal freight chain. For example, a large international import shipper may receive loaded 40-foot ocean containers at its regional distribution center via drayage service from a local container seaport. Export shippers moving manufactured product overseas may request empty containers and then load them for delivery to the port for export. While ocean carriers may arrange these container transfers, they are executed by drayage trucking services. Transfer of international containers by drayage, however, is not limited to last-mile scenarios. Drayage trucking services are also used to transfer containers between seaports and nearby off-port rail intermodal yards for connection to and from inland customer facilities. In another scenario, drayage services are used to move international ocean containers to transshipment warehouses where containers may be unloaded and their contents repacked into domestic rail containers or truck trailers.

Domestic intermodal equipment is also moved by drayage services. For example, railroads utilize drayage trucking to pick up and deliver loaded and empty intermodal equipment, such as 53-foot domestic containers. Railroads also often must organize "cross-town" drayage transportation of containers or trailers between rail yards that are not physically connected by rail infrastructure. Cross-town moves are prevalent in the United States in the major transportation hubs, such as Chicago, Kansas City, and Memphis, that provide connections between railroads that serve the East Coast and railroads that serve the West Coast. Finally, large long-haul trucking companies (both truckload and less-than-truckload [LTL] carriers) more frequently rely on rail intermodal transportation of trailers for long-distance moves between terminals or regions. In the United States, this is particularly true for moves between population centers on the Eastern seaboard or in the Midwest and those on the West Coast or in the Southwest. Usually, such trucking companies will coordinate drayage services with their own equipment. In the case of truckload trucking, drayage provides connections between customers and rail intermodal yards, while in the case of LTL trucking, drayage provides connections between break bulk or end-of-line terminals and the rail yards.

Although drayage moves are usually quite short in duration (anywhere from 15 minutes to a few hours of travel time), drayage transport cost is a substantial contributor to the total cost of an intermodal movement. An often-cited study from the mid-1990s estimates that drayage trucking cost accounts for about 40% of the total cost of a 900-mile rail–truck intermodal move.[15] A variety of factors are likely to keep the relative cost of the drayage component of intermodal transportation high, including

1. Growth in drayage service demand, due to continuing growth in global trade volumes. Government and industry sources predict that U.S. international trade tonnage may double between 2001 and 2010, along with a resultant doubling of total container trade from 16 million 20-foot equivalent units (TEUs) to 32 million TEUs.[10]
2. High labor costs, even though drayage truck drivers are the worst compensated segment of the trucking labor force.
3. Growing urban congestion, reducing the productivity of drayage service providers in most major markets.
4. Improving scale economies in ocean container transportation, primarily due to the introduction of larger vessels, that will continue to suppress costs for long-haul container movements.

Indeed, it seems highly likely that drayage will remain a "cost bottleneck" for intermodal transportation for the foreseeable future.

With this motivation, it is therefore critically important to consider opportunities for improving the operational efficiency of drayage service providers. This chapter will focus on methodology for drayage routing and scheduling decision support. When drayage firms (or other parties organizing drayage operations such as railroads, trucking companies, or ocean shipping lines) manage large fleets of trucks serving large sets of daily tasks within a local region, there exists significant opportunity for cost reduction through better routing and scheduling. In drayage, cost reduction is achieved primarily by reducing the amount of *empty mileage* traveled by the vehicle fleet. For a fleet of fixed size, reducing empty mileage may also lead to an increase in the number of revenue-generating tasks that can be performed each operating period per vehicle.

The remainder of this chapter is organized as follows. Section 11.2 introduces the drayage routing and scheduling problems to be considered. Section 11.3 presents a formal mathematical formulation of the routing and scheduling optimization problem in its most general form. Since drayage problems are special cases of difficult-to-solve optimization problems, it is not immediately clear that they are also difficult; however, it is established in section 11.4 that drayage problems are hard to solve for even the simplest problem variants. Finally, section 11.5 discusses methods for solving drayage problems, focusing on both exact and heuristic methods for solving the integer programming problem formulation.

11.2 DRAYAGE ROUTING AND SCHEDULING PROBLEMS

Consider now a single firm coordinating truck drayage operations, again noting that this firm may be a drayage firm or an external party like a trucking company, railroad, or ocean shipping line. The firm either owns or manages a fleet of trucks and drivers. Assume that at the beginning of each operating day, each truck and its associated driver will be dispatched from one of potentially many *depot locations* and will return to that same location by day's end. The truck fleet will be used each day to serve a set of equipment move requests, hereafter referred to as *tasks*. Each task will represent the movement of a loaded or empty trailer or container from one location to another location within the service region.

In general, drayage operations visit three types of locations in addition to truck depots. *Customer* locations will refer to shipper or consignee facilities. *Intermodal facility* locations will refer to rail intermodal yards where containers and trailers are loaded and unloaded from trains, and seaports where containers are loaded and unloaded from vessels. Finally, *equipment yard* locations will refer to storage facilities for empty containers and trailers that are not located on site at an intermodal facility. Each equipment movement will be loaded or empty. Loaded movements usually originate or terminate at a customer location; one exception is the so-called *cross-town* moves between rail intermodal facilities in gateway cities. *Inbound* loaded movements will refer to those originating from an intermodal facility and terminating at a customer, while *outbound* loaded movements originate at a customer and terminate at an intermodal facility. Figure 11.1 illustrates example movements of equipment in intermodal drayage operations.

Sometimes, drayage drivers must perform two tasks in sequence; that is, if a driver performs one task, then that same driver must perform a specific next task. A customer requesting an outbound loaded movement, for example, may request that the drayage truck bring in an empty trailer, and that the driver wait while the trailer is loaded. Such an operation is called *live loading*. There is a parallel operation

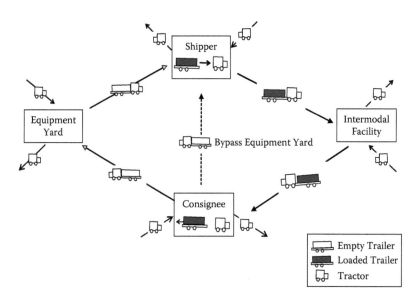

FIGURE 11.1 Typical truck movements found in intermodal trailer or container drayage operations. The intermodal facility might represent a seaport or a rail intermodal yard. (From K. Smilowitz. 2006. "Multi-Resource Routing with Flexible Tasks: An Application in Drayage Operations http://users.items.northwestern.edu/%7Esmilo/Smilowitz2006.pdf," IIETransactions, Volume 38, Number 7, June 2006, pp. 577–590 with kind permission of Springer Science and Business Media.)

known as *live unloading*, where the driver again waits with an inbound trailer until it is unloaded, and then moves away the empty. More prevalent (and usually more efficient) are so-called *drop-and-hook* movement combinations. In a drop-and-hook operation, a driver will arrive at a facility with one piece of equipment, drop it, and then pick up (or hook) another piece of equipment to move away from the facility.

While loaded movement tasks always specify a distinct origin location and destination location, this need not be the case with tasks specifying the movement of empty containers or trailers. Sometimes, within a service region, there may be multiple equipment yards or intermodal facilities from which a specific customer may source empty equipment or to which such equipment may be returned. It is therefore convenient to differentiate between *well-defined* tasks with specified origin and destination locations and *flexible* tasks where either the origin or the destination of the movement request may not be fixed. Smilowitz[19] presents this useful terminology.

Given a set of tasks requiring execution on a given operating day, the drayage coordinator seeks to utilize its fleet most effectively to serve the task set (assuming all tasks can be completed). If the coordinator has a fixed fleet available, one possible objective is to plan operations such that the total unnecessary travel distance required by the fleet is minimized. Note that unnecessary travel distance includes miles traveled without any equipment in tow (bobtailing) as well as miles traveled conducting flexible tasks with empty equipment in tow. If the coordinator is contracting individual vehicles to perform operations, or if driver costs include a fixed component (like benefit costs) that accrues on a daily basis regardless of miles traveled, an alternative objective is to minimize the total number of vehicles required to perform the set of tasks. These two objectives, of course, might also be combined with appropriate weights or used hierarchically (e.g., minimize vehicles, and break ties by minimizing unnecessary distance traveled).

Given an appropriate objective, the coordinator develops *scheduled routes* for a set of vehicles. The scheduled route for a vehicle specifies the order in which that vehicle should serve tasks (and the resultant order in which locations are visited), as well as the times at which each activity should begin. The primary constraints on drayage routes are related to timing considerations. At a minimum, each utilized driver is limited in the United States by so-called *hours-of-service* constraints that place upper bounds on both the amount of time spent driving and the amount of time spent on duty (including waiting, loading, and unloading) between rest periods. Currently, these limits are no more than 11 hours of drive time and 14 hours of duty time between rests no shorter than 10 hours. Other time constraints that affect drayage operations are operating time windows at the different facilities. Finally, individual tasks may have time constraints such as earliest available pickup times or latest available drop-off times; often, such deadlines are induced by so-called *cut times* for the loading of outbound trains or ocean vessels.

11.2.1 ADDITIONAL COMPLEXITIES

11.2.1.1 Dynamic and Stochastic Inputs

Decision support models for drayage operational problems typically assume that the set of all tasks (both well defined and flexible) for the upcoming period are known with certainty prior to the period so that plans can be made. This may not be the case in all applications. When previously unknown tasks may arise *during* the operating period, the operational setting is said to include *dynamic tasks. Stochastic and dynamic tasks* refer to those that may arise unexpectedly during the operating period, but for which the drayage coordinator has some prior information about the likelihood (or *probability*) that certain tasks may arise.

Another important input to drayage planning that may be both dynamic and stochastic is the travel time between pairs of customer locations. In the urban geographies where drayage operations typically occur, travel times can vary substantially by time of day. Furthermore, unexpected events like traffic incidents also lead to variability in travel times for one day to the next.

Finally, growing freight volumes at U.S. seaport terminals have led to increasing congestion in and around seaports. Often, long lines of drayage trucks wait at the terminal gates to gain access and then may wait again for substantial periods of time within the facility to complete their required tasks. The duration of such delays may vary significantly by the time of day and also may not be completely predictable.

11.2.1.2 Facility Access Restrictions

In response to the aforementioned problems of delay, many seaport terminal operators are now deploying access control systems. Under such systems, drayage operators can make reservations that allow truck access during specific time windows. The terminal operator then limits the total number of appointments available within each time window in order to manage congestion. Predictable workloads enable better resource scheduling and gate productivity, and reduced waiting time allows drayage firms to better utilize tractors and drivers.

Under typical scheduled access systems such as those deployed in California in response to regulation,[14] drayage operators can book appointment requests for hour-long time windows up to 2 weeks in advance of the access day. The only information that must be provided when booking is an identification code for the trucking firm. Therefore, operators can later decide which specific movements will utilize each appointment. Of course, such appointments complicate the planning problem for an operating day since they restrict when the firm is able to access the port. For more on this topic, the reader is referred to Namboothiri and Erera.[18]

11.3 DRAYAGE ROUTING OPTIMIZATION FORMULATION

In this chapter, we will focus on a static version of the drayage operations planning problem where all tasks are known prior to the operating period. Smilowitz[19] models this problem as a multiresource routing problem with flexible tasks (MRRP-FT). In the drayage context, the resources to be routed are drivers, tractors, and tow equipment (trailers or containers on chassis). Drayage drivers typically stay with their tractors during the course of an operating day, so we will treat the driver-tractor combination as a single resource that we will hereafter refer to as a vehicle. Furthermore, since we will also assume that sufficient pools of empty tow equipment are always available at yards or intermodal facilities, we will not track tow equipment explicitly in the modeling approach. Thus, we have a single-resource routing problem. During the operating period, each vehicle is first dispatched from one of a set $\mathcal{D}$ of depots where it is based. That vehicle must return to that same depot when its route is completed and can be away for no longer than Δ hours. Let $t_{k\ell}$ represent the travel time between any two locations k and ℓ, where travel times are assumed to be symmetric ($t_{k\ell} = t_{\ell k}$) and satisfy the triangle inequality ($t_{\ell k} \leq t_{\ell a} + t_{ak}$).

Tasks are classified as either well defined (meaning their origin and destination are both known) or flexible (either the origin or destination is unspecified). Let $\mathcal{T}$ be the set of all tasks, partitioned into two independent subsets $\mathcal{T}_w$ and $\mathcal{T}_f$ representing the well-defined and flexible tasks, respectively. For each flexible task, there may be multiple ways to execute it; these will be denoted feasible executions. Let ε_i be the set of all possible executions of flexible task $i \in \mathcal{T}_f$. Figure 11.2 depicts a well-defined task in part (a), and a flexible task with two possible executions in part (b). Lastly, let set $\mathcal{C}$ contain all prespecified *connections* between tasks that must be performed in sequence by the same driver; each element $(i, j) \in \mathcal{C}$ denotes that task $j \in \mathcal{T}$ must immediately follow task $i \in \mathcal{T}$. Note that i, j, or both may be flexible tasks, but the origin of task j must be the destination of task j.

Let $\mathcal{M}$ be the set of all *movements* that a vehicle resource may perform. Set $\mathcal{M}$ includes each task $j \in \mathcal{T}_w$ as well as all executions in ε_i for each $i \in \mathcal{T}_f$. Each movement $m \in \mathcal{M}$ then has a unique origin location o and destination location d. Furthermore, suppose each movement m has a *time window* $[a_m, b_m]$ specifying an interval between the earliest time and latest time a vehicle must be ready at the movement

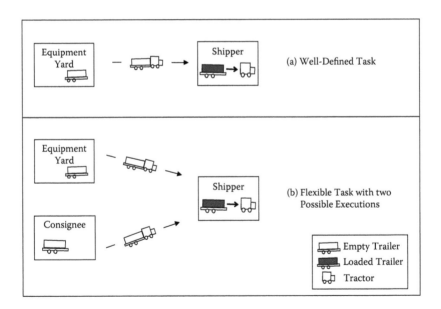

FIGURE 11.2 Well-defined (a) and flexible (b) drayage tasks. Note that the flexible task in part (b) is an empty move to the shipper with a flexible origin of either the consignee or the equipment yard. (From P. Francis, K. Smilowitz, and G. Zhang. "Improved Modeling and Solution Methods for the Multi-Resource Routing with Flexible Tasks: http://users.iems. northwestern.edu/%7Esmilo/FrancisSmilowitzZhang2006.pdfs," European Journal of Oprational Research, Volume 180 (3) 2007, pp. 1045–1059. Copyright Elsevier 2007.)

origin. Let $\mathcal{T}_m^O$ be the time required at the origin of movement m before dispatch, and let τ_m^D be the time required at the destination before task completion. It should be clear, then, that a common time window $[a_m, b_m]$ can always be constructed from a combination of time windows at the origin and destination under the assumption that no delay is allowed in transit.

The drayage routing problem (DP) in its most general form, then, is to determine a set of feasible vehicle routes serving tasks by performing movements such that some objective function is minimized. Each drayage vehicle can perform only a single task at a time, and thus a route can be represented as a sequence of movements $m \in \mathcal{M}$ to be executed between visits to its home depot $d : \{d, m_1, m_2, m_3, ..., m_p, d\}$. If the destination of movement m_k is not the origin of movement of m_{k+1}, such a sequence also implies bobtail vehicle travel between these locations. For a sequence to be feasible, its total duration must not exceed Δ and each of its component movements must be scheduled within its associated time window.

We now provide a mathematical programming formulation. The model is an extended form of a so-called *set partitioning* binary integer program. Suppose for each vehicle depot $d \in \mathcal{D}$ one could enumerate *each* feasible movement sequence (route) r for a single vehicle; denote the set of all such sequences $\mathcal{R}_d$, and let $\mathcal{T} = \cup_{d \in D} \mathcal{R}_c$. For each sequence $r \in \mathcal{R}$, we have the following information:

c_r: Cost of route r

$$a_{ri} : \text{Covering parameter:} = \begin{cases} 1 & \text{if movement } i \in \mathcal{M} \text{ is contained in route } r \\ 0 & \text{otherwise.} \end{cases}$$

Using a route indicates that a distinct resource is used. An appropriate cost c_r is thus a weighted combination of the per vehicle cost and the travel duration δ_r of r, which can be expressed as $c_r = F + \alpha \delta_r$ for some $\alpha \geq 0$.

The decision variables are then x_r and y_i, where x_r is used to select a subset of routes for execution and y_i is used to account for unserved tasks:

$$x_r = \begin{cases} 1 & \text{if route } r \in \mathcal{R} \text{ is selected} \\ 0 & \text{otherwise;} \end{cases}$$

$$y_i = \begin{cases} 1 & \text{if task } i \in \mathcal{T} \text{ is left unserved} \\ 0 & \text{0 otherwise.} \end{cases}$$

A general formulation of the drayage problem (DP) is then

$$\min \sum_{r \in \mathcal{R}} c_r x_r + \sum_{i \in \mathcal{T}} p_i y_i \tag{11.1a}$$

subject to

$$\sum_{r \in \mathcal{R}} a_{ri} x_r + y_i = 1 \qquad \forall i \in \mathcal{T}_w \tag{11.1b}$$

$$\sum_{r \in \mathcal{R}} \sum_{e \in \mathcal{E}_i} a_{re} x_r + y_i = 1 \qquad \forall i \in \mathcal{T}_f \qquad (11.1c)$$

$$\sum_{r \in \mathcal{R}_d} x_r \leq v_d \qquad \forall d \in \mathcal{D} \qquad (11.1d)$$

$$x_r \in \{0,1\} \qquad \forall r \in \mathcal{R} \qquad (11.1e)$$

$$y_i \in \{0,1\} \qquad \forall i \in \mathcal{T}, \qquad (11.1f)$$

where p_i is a penalty cost for not serving task i, and v_d is the size of the vehicle fleet available at depot d.

The objective function (11.1a) minimizes the total routing cost plus the penalty costs for not serving tasks. Equations (11.1b) are partitioning constraints that ensure that all well-defined tasks are executed exactly once. Equations (11.1c) are partitioning constraints for all flexible tasks. Equations (11.1d) restrict the number of routes that can be dispatched from a particular depot to the number of vehicles available there. Finally, equations (11.1e) and (11.1f) define the binary decision variables.

11.4 DRAYAGE ROUTING IS A HARD OPTIMIZATION PROBLEM

The simplest forms of the drayage routing problem defined in section 11.3 are still difficult-to-solve optimization problems for large task sets $\mathcal{T}$. In this section, we prove this statement formally. Consider a drainage problem of the form given by equation (11.1a–11.1f) where $F = 1$, $\alpha = 0$, $p_i = +\infty$, and $v_d = +\infty$, denote this problem UDP(V), the uncongested drayage problem with a vehicle minimization objective. Similiarly, let UDP(T) be the uncongested drayage problem with a travel time minimization objective, given by the problem (11.1a–11.1f) where $F = 0$, $\alpha = 1$, $p_i = +\infty$, and $v_d = +\infty$. In both problem settings, assume that time windows for all movements m are $[0, \Delta]$ and $\tau_m^O = \tau_m^D = 0$, and that no tasks are connected so that $\mathcal{C}$ is empty. Following the work in Namboothiri,[16] we show in this section that decision versions of these restricted drayage problems belong to the class NP-complete (see Garey and Johnson[8]).

11.4.1 COMPLEXITY OF UDP(V)

It is not too difficult to develop a polynomial transformation from the *bin packing problem* (BPP) to a restricted version of the drayage problem UDP(V), since minimizing the total number of required vehicles when duration is constrained by Δ is similar to minimizing the total number of bins with the same finite capacity.

Bin packing problem (BPP):
   Instance: n objects with integer weights $d_1, d_2, \ldots, d_n$ to be placed in bins, each with integer capacity Δ.
   Decision question: Is it possible to pack all n objects in $\leq v$ bins?

Restricted drayage problem R-UDP(V):
   Instance: An instance of UDP(V) with n well-defined tasks requesting inbound loaded movements from an intermodal facility to customer

locations $I_1, I_2, ..., I_n$. Vehicles are all based at a depot d colocated with the intermodal facility. Travel time between the facility and customer location I_i is $\frac{d_i}{2}$, and the vehicle duration limit is Δ.

Decision question: Is it possible to feasibly complete all n tasks with $\leq v$ vehicles?

Note that given any instance of BPP, an instance of R-UDP(V) can be constructed efficiently.

Theorem 11.4.1. UDP(V) is NP-hard.

Proof. It is easy to see that the decision version of UDP(V) is in *NP*, since verification of whether a solution is feasible can be performed in linear time in the number of tasks. Furthermore, it is easy to see that a "yes" instance of R-UDP(V) corresponds directly to a "yes" instance of BPP and vice versa. Each vehicle in a solution to R-UDP(V) travels from the facility to a customer and then back to the facility using total travel time d_i before it can complete another task. Since Δ total travel duration is available, every feasible solution to R-UDP(V) will also be a feasible packing for BPP and vice versa. Since BPP is *NP*-complete, UDP(V) is *NP*-complete and its optimization version is *NP*-hard.

11.4.2 COMPLEXITY OF UDP(T)

Problem UDP(T) can also be shown to be *NP*-hard by a polynomial transformation from BPP.

Restricted drayage problem R-UDP(T):
Instance: An instance of UDP(T) with n well-defined tasks requesting inbound loaded movements from an intermodal facility to customer locations $I_1, I_2, ..., I_n$. Vehicles are located at a single depot facility d. Travel time between the intermodal facility and customer location i is $\frac{d_i}{2}$, between the intermodal facility and the depot is $\frac{\epsilon}{2}$ (where $\epsilon > 0$), and between customer location i and the depot is $\frac{d_i + \epsilon}{2}$. The vehicle duration limit is $\epsilon + \Delta$.
Decision question: Is it possible to feasibly complete all n tasks with total duration $T \leq v \epsilon + \Sigma_{i \in N} d_i$?

Again, note that given any instance of BPP, an instance of R-UDP(V) can be constructed efficiently.

Theorem 11.4.2. UDP(T) is NP-hard.

Proof. Again, it is easy to see that a decision version of UDP(T) is in *NP* since feasibility can be verified in linear time in the number of tasks. It is also easy to verify that a "yes" instance for BPP is a "yes" instance for R-UDP(T). The tasks corresponding to the items in a bin j can be served in any order by a single vehicle, resulting in a route with duration equal to $\epsilon + \Sigma_{i \in N_j} d_i \leq \epsilon + \Delta$, where N_j is the set of items in bin j; since no more than v bins are used, the total duration of all tasks is no greater than $v \epsilon + \Sigma_{i \in N} d_i$. Similarly, a "yes" instance for R-UDP(T) is a "yes" instance for the BPP. Suppose the feasible solution to R-UDP(T) uses $k \leq v$ vehicles and a total travel duration of $T_k \leq v \in +\Sigma_{i \in N} d_i$. Consider vehicle j that performs tasks in set N_j. The minimum duration route for this vehicle is found by traveling from the depot to the intermodal facility (cost $\frac{\epsilon}{2}$), then performing all tasks (in any order) and returning to the depot from the last customer. Regardless of the order, this will require $\frac{\epsilon}{2} + \Sigma_{i \in N_j} d_i$. Since the total travel duration for any vehicle is no greater than $\epsilon +\Delta$, it is true that $\Sigma_{i \in N_j} d_i \leq \Delta$ for each vehicle j. Thus, a feasible solution to the bin packing problem would pack items into bins according to N_j for all $j = 1,...,k \leq v$. Since BPP is *NP*-complete, then *UDP(T)* is *NP*-complete and its optimization version is *NP*-hard.

11.5 SOLUTION TECHNIQUES

The general form of the drayage problem (DP) is quite similar to the well-studied class of problems known as *vehicle routing problems with time windows* (VRPTW); an excellent research overview for this class of problems is given by Cordeau et al.[4] Unlike the general VRPTW, feasible routes for the drayage problem are only constrained by time considerations, both customer time windows and total route duration, and not by the physical capacity of the vehicle to hold freight. Some research has also considered problems of this class, which are sometimes referred to as *pickup and delivery problems with full truckloads* or *vehicle routing problems with full truckloads*. Although this chapter is not meant to be a comprehensive literature review on this subject, the following subsection will provide the reader with some sources to consult for additional information.

11.5.1 RELATED LITERATURE

As mentioned, the book chapter by Cordeau et al.[4] provides a comprehensive overview of the VRPTW and surveys the literature for solving such problems, including advances in both exact approximation approaches that rely on integer programming, and suboptimal heuristic methods that can provide very high-quality solutions to large-scale problems with reasonable computation times. Most current work on heuristics focuses on so-called metaheuristics, such as tabu search, simulated annealing, and evolutionary algorithms, that seek to improve solutions using stochastic extensions of local search approaches. Another excellent set of surveys of heuristic solution approaches for VRPTW is given by Bräysy and Gendreau.[2,3]

Fewer papers focus specifically on routing problems with full truckloads. Arunapuram et al.[1] formulate the full-truckload routing problem similarly to the drayage formulation (DP) presented in section 11.3. The paper considers a multiple depot scenario, with a fixed fleet of vehicles allocated to each depot. An exact solution approach based on branch-and-bound with column generation is proposed to solve the problem. The computational results presented focus on test problems with characteristics somewhat consistent with drayage operations, in that all service request origins and destinations lie within a square with sides of length 100 miles. Problems with up to 100 lanes (roughly equivalent to tasks) are generally solved within a few minutes of computation time; however, the problems are tightly constrained by time windows with a duration of 30 or 60 minutes. When time window durations grow to 180 hours, average computation times grow to 15 minutes for fifty lane problems. Grönalt et al.[9] propose the use of savings-based heuristics for large-scale truckload routing problems. Comparisons to a lower bound indicate that methods perform with 1% of optimality for problems with loose time windows. Importantly, the test problems considered by the authors correspond more closely to truckload trucking operations where moves have durations of multiple days.

More specifically focused on drayage operations, Jula et al.[13] develop methods for addressing container movement problems using the *multiple traveling salesman problem with time windows* (*m*-TSPTW) model, appropriate for versions of the drayage problem with well-defined tasks in which a fixed fleet of vehicles is available. A two-phase exact solution approach is developed based on dynamic programming and is proposed to solve problems with up to 15–20 tasks. A metaheuristic based on genetic algorithms is proposed to solve large problem instances. Wang and Regan[21] also propose the use of the *m*-TSPTW model for drayage optimization problems with well-defined tasks. Importantly, this paper also considers the intraday dynamics of drayage operations. The paper proposes two heuristic methods that can be used to regenerate very high-quality vehicle routes at multiple decision times during the day as new information regarding the system state (such as new or cancelled customer requests, and current delays at intermodal facilities) is received.

11.5.2 Enumeration of $\mathcal{R}$

Given the power of today's personal computers, and advances in off-the-shelf software for solving linear and integer constrained optimization problems, many variations of the drayage problem formulation (DP) can be solved to optimality or near optimality for realistic instances using column-generation techniques. The primary difficulty in solving the problem is the potential size of the set $\mathcal{R}$, representing the set of all feasible routes for vehicles, with each $r \in \mathcal{R}$ covering some subset of the set of all possible vehicle movements $\mathcal{M}$. Note that each $r \in \mathcal{R}$ represents a route to be performed by a single vehicle, independent from the routes of all other vehicles.

Importantly, the task of enumerating each $r \in \mathcal{R}$ is fundamentally not difficult, even when each movement $m \in \mathcal{M}$ is constrained by a time window and the total duration of a vehicle route from the time it departs its origin depot d until the time it returns is constrained by Δ. In general, all feasible single-vehicle routes can be enumerated using a *tree-search* heuristic (such as depth-first search or breadth-first search) where each node in the tree represents a sequence of movements that can be completed by a vehicle. Building all routes, then, is simply the problem of determining all possible feasible *extensions* of each petal node in the search tree, where an extension refers to adding a single additional movement to the end of a route. Note that the primary problem in determining whether an extension is feasible is to determine the required duration of the resulting tour. Erera et al.[5] describe simple techniques for determining the minimum duration of a set of vehicle movements with time windows for a related problem found in less-than-truckload systems.

When drayage problems face tight time constraints or when relatively few tasks (and thus movements) need to be scheduled, enumerating all possible routes may yield route sets $\mathcal{R}$ with sufficiently small cardinality such that the problem (DP) can be solved directly with off-the-shelf integer optimization solver technology, such as ILog CPLEX. Ileri et al.[11] present numerical results for drayage problems faced by Schneider National, the largest truckload trucking carrier in North America. The authors consider drayage problems in which all tasks must be served ($p_i = +\circ$), and where drivers are partitioned into groups, with a limit on the number of available drivers in each group; a designated depot is one characteristic of a driver group, but the authors also group drivers by earliest and latest start times and differentiate between company and third-party drivers. Importantly, the model includes the concepts of flexible tasks and connected tasks. However, since time windows are assumed not to be restrictive at equipment pool locations, the authors can always choose the execution of the flexible task j performed between two well-defined tasks i and k such that its duration (and thus cost) is minimized.

Ileri et al.[11] develop a few important concepts for enumeration of $\mathcal{R}$. First, a pairwise *feasibility matrix* is maintained on the set of tasks for each driver group. Each entry (i,j) in this matrix indicates whether task j can possibly be served feasibly following task i using a driver from this group. This matrix is initialized to be true for all entries. Then, during enumeration, if it is determined that a route representing only task i (i.e., $\{d,i,d\}$) cannot be extended to a route serving i, then j (i.e., $\{d,i,j,d\}$), entry (i,j) is set to false. The feasibility matrix can speed subsequent search steps, since under the triangle inequality, j cannot follow i regardless of the tasks served prior to i. Also, at the completion of the breadth-first search route enumeration, *dominated* routes are discarded before solving the integer program. A route dominates another if it covers the same set of tasks using a driver from the same group and has lower cost.

For test problems representing daily drayage operations in the St. Louis region, the paper shows that exact solutions can be developed after route enumeration in no greater than 10 seconds of computation time using fairly standard hardware. The largest problem solved had 28 orders, where an order represents either a single well-defined task or two connected tasks where one is flexible and the other is well defined. This specific problem instance resulted in the enumeration of approximately 7,000 distinct vehicle routes; solving a {0,1} integer optimization model such as DP with 7,000 total variables and hundreds of constraints is not difficult in practice today.

11.5.3 Heuristic Column Generation

When route enumeration leads to an excessive number of feasible single-vehicle routes in $\mathcal{R}$, an alternative is to employ a *column-generation* decomposition. When solving an integer optimization problem like DP, a branch-and-bound search will typically be used to identify the optimal integer-valued solution; such searches are built into all commercial integer programming software. At each step of the branch-and-bound search, a linear relaxation of DP will be solved (perhaps with some additional variables that fix or bound specific variable quantities). Column-generation decomposition approaches to solving these linear programs do not require a priori enumeration of the complete column set $\mathcal{R}$. Instead, an iterative solution procedure is used that begins with some small subset $\mathcal{R}'$ of columns that contains a feasible solution. The linear program is then solved using column set $\mathcal{R}'$, yielding dual variable values. These duals are then used within a pricing subproblem that seeks to identify a column or columns in $\mathcal{R}\mathcal{R}'$ with negative reduced cost. If such columns are identified, they are added to $\mathcal{R}'$, and the process is repeated until no such improving columns are identified.

Heuristic column-generation procedures are used when it is computationally expensive to conduct column generation for each linear program solved during branch-and-bound. Heuristics typically utilize two approaches for improving computation time: (1) solving the column-generation pricing subproblem using a heuristic, and (2) limiting column generation to the root-node linear relaxation within the branch-and-bound. For drayage routing problems, many have applied approach 2, which we denote a *root column-generation heuristic*:

1. Let $\mathcal{R}'$ be a feasible set of routes covering all customer requests, assuming such a set exists.
2. Repeat.
3. Solve $DP^{LP}(\mathcal{R}')$, the linear relaxation of DP, using route subset $\mathcal{R}'$.
4. Solve pricing subproblem and add to $\mathcal{R}'$ all routes found with negative reduced cost $\bar{c}_j < 0$.
5. Until pricing subproblem identifies no routes to add.
6. Solve $DP(\mathcal{R}')$, the binary integer model DP using only route subset $\mathcal{R}'$.

Namboothiri[16] uses such a heuristic to solve loosely constrained versions of DP with $p_i = +\infty$ and a single depot d with $v_d = +\infty$. Furthermore, all tasks are well defined ($\mathcal{T}_f = \varnothing$). Given a solution to $DP^{LP}(\mathcal{R}')$ with some column set $\mathcal{R}'$, suppose $\pi = \{\pi_i\}$ represents the optimal dual variable values associated with constraints (11.1b). Then, the reduced cost $\bar{c}_r$ of any route $r \in \mathcal{R}$ is given by

$$\bar{c}_r = c_r - \sum_{j \in T} a_{rj} \pi_j. \qquad (11.2)$$

where T is the set of tasks covered by route r. The problem of determining one (or many) routes in $\mathcal{R}$ with negative reduced cost is an instance of the *elementary shortest-path problem with resource constraints* (ESPPRC), which is known to be a hard optimization problem.[6] To see that this pricing problem is an instance of ESPPRC, construct a network with nodes representing the depot (node 0) and each task. Let the travel time $\bar{t}_{ij}$ between nodes i and j be the total travel time from the origin of task i to the destination of task i, and then from the destination of task i to the origin of task j; a closed path from the depot (0) back to the depot (0) is feasible with respect to the resource constraints if it arrives at each node i within the time window for that task movement, and if the total path duration (the arrival time at the final task node k in the path added to $\bar{t}_{k0}$) is no greater than Δ. To determine the cost of a path (and thus a route), let the arc cost $\bar{c}_{ij} = \alpha \bar{t}_{ij} - \pi_j$. A minimum cost elementary path p^* from the depot to the depot will serve no task more than once. Furthermore, if such a path has total cost $\bar{c}(p^*)$ such that $F + \bar{c}(p^*) < 0$, then a negative reduced cost route has been identified to add to $\mathcal{R}'$.

For drayage problems with no time restrictions other than total route duration Δ, it may be computationally prohibitive to solve the ESPPRC exactly, even when only performing column generation at the root node. Therefore, Namboothiri[16] and Namboothiri and Erera[17] propose a heuristic approach for solving the pricing problem. The method, denoted a *layered-shortest-path* (LSP) heuristic, is a dynamic programming approach that ensures that routes are generated that satisfy the resource (time) constraints. However, the method does not use consumption of the time resource as a component of its state space; thus, it will not generate all nondominated paths from the depot to interim task nodes. As an alternative, the method uses the position (or layer) of a task within a route as a component of the state space. Elementary paths are built layer by layer by a labeling approach, where the best label for task i in position k on a route is found by considering all possible feasible extensions from tasks j in position $k-1$ on a route. Such an approach solves the ESPPRC heuristically in $O(n^3)$ time, where n is the number of tasks.

Results in Namboothiri[16] indicate that the LSP heuristic is an effective alternative to solving ESPPRC exactly, especially for problems that are fairly loosely constrained. When applied to the Solomon benchmark problems for the VRPTW and compared to an exact approach presented by Feillet et al.,[6] the root linear programs in a column-generation procedure were solved within 5% of optimality, with computation time savings of 65% to 95% for problems with randomly scattered customers.

When applied to drayage problems, a root column-generation heuristic using LSP to solve the pricing subproblem performed well when compared to optimal solutions found by the enumerative approach on small instances. First, it is important to note that for relatively small problems with twenty well-defined tasks, Namboothiri[16] shows that loosely constrained drayage problems lead to about 43,000 nondominated, feasible single-vehicle routes; note that this is a substantial increase from the results in Ileri et al.[11] for similarly sized problems. While all such routes can be enumerated in about 30 seconds, when the problem size grows to thirty well-defined tasks, the route enumeration time grows to 13 minutes. Fortunately, the root column-generation heuristic with LSP performs well on such problems. For instances with twenty to thirty tasks with the objective of minimizing the total duration of all routes, the heuristic approach achieves solutions with objective function values no greater than 2% worse than optimal solutions. For the thirty customer problems, the LSP approach resulted in average computation times of about 18 seconds, compared with 160 minutes for the enumerative approach. For larger problems with 100 tasks that could not be solved by the enumerative approach, the LSP approach required 6 to 7 minutes on average to generate solutions. Although it is not possible to assess the quality of the solutions precisely, they appear to be quite good given that vehicle utilization, measured by tasks covered per vehicle, is similar if not better than that observed for the smaller problems.

11.5.4 Variations with Flexible Tasks

The existence of flexible tasks $\mathcal{T}_f$ requires new solution methodologies to account for this additional flexibility. Francis et al.[7] developed a practical method to effectively manage the number of options considered for flexible tasks (the set ε_i) based on the geographic location of movements. As a result, options are limited in dense areas (e.g., urban locations) and expanded in isolated regions. Limiting the potential options for flexible tasks reduces the number of feasible single-vehicle routes and thus allows larger problem instances to be solved.

Researchers have developed exact approaches for solving drayage problems with both well-defined and flexible tasks. Smilowitz[19] introduces a branch-and-price approach to solve formulation (11.1) with flexible tasks, and $P_i = +\infty$ and $V_d = +\infty$ (i.e., all tasks must be served). The linear relaxation of (11.1) is solved at each node of the branch-and-price tree with column generation. Since the number of routes in $\mathcal{R}$ is often excessively large, column generation reduces the computational effort required to solve the linear relaxation by considering a reduced subset of routes. New routes are added to this subset according to a pricing problem that searches for routes with negative reduced costs. The pricing problem can be solved as a variation of the elementary shortest-path problem with resource constraints, starting and ending at the depot, with costs that are determined by the dual values from the linear relaxation of the master problem. Smilowitz and Zhang[20] identify this variation as the multiple-choice elementary constrained-shortest-path problem (MC-ECSPP). The MC-ECSPP finds the path of minimum cost that visits at most one node from each subset (representing the choice of one execution of task i from the set ε_i), constrained by the travel time of the path and time windows. Like the ESPPRC, the MC-ECSPP is NP-hard, which can be shown easily since an ESPPRC instance can be transformed in polynomial time to an MC-ECSPP instance in which each subset consists of a single node. The MC-ECSPP can be solved to optimality for small problem instances with an exact method based on k-cycle elimination from Irnich and Villeneuve[12] adapted to allow for the node subsets. Smilowitz and Zhang[20] introduce methods to obtain bounds on the optimal MC-ECSPP solution based on several aggregations of the nodes within subsets. These bounds are then used to obtain feasible solutions to the MC-ECSPP. The bounding and solution methods are incorporated into the branch-and-price method for the MRRP-FT.

Smilowitz[19] shows that provably optimal solutions to problems with flexible tasks can be generated via a branch-and-price approach when the total number of well-defined tasks and potential executions of flexible tasks is relatively small. For instances based on the geography of Chicago drayage operations with eighty-five such executions, optimal solutions can be found within 13 minutes of computation time. Much larger instances with up to 1,600 potential task executions can be addressed within the framework heuristically, where candidate routes are identified using a greedy insertion approach rather than by solving the MC-ECSPP exactly. Results of comparisons to bounds indicate that the heuristic solutions found can be of high quality.

11.6 CONCLUSIONS

This chapter has presented approaches for improving container drayage truck routing and scheduling through the use of advanced systematic scheduling approaches. We have described a flexible integer optimization model for drayage task scheduling based on ideas from set partitioning; importantly, such a model captures most of the problem complexity within the subproblem of identifying feasible routes to be conducted by single vehicles. Given this model, the chapter describes three types of approaches for its solution: (1) enumerative approaches that identify all feasible single-vehicle routes a priori; (2) heuristic approaches that use column-generation techniques to identify very good sets of candidate routes, but are not provably optimal; and (3) exact column-generation techniques that produce optimal solutions for problems that cannot be addressed by enumerative approaches. The choice of approach must be motivated by the particular circumstances faced in the specific drayage routing problem to be addressed.

ACKNOWLEDGMENTS

This work was supported in part by grants from the Federal Highway Administration and the Sloan Foundation. In addition, Dr. Smilowitz is partially supported by the National Science Foundation under grant CMII-0348622.

REFERENCES

1. S. Arunapuram, K. Mathur, and D. Solow. 2003. Vehicle routing and scheduling with full truckloads. *Transportation Science* 37:170–82.
2. O. Bräysy, and M. Gendreau. 2005. Vehicle routing problem with time windows. Part I. Route construction and local search algorithms. *Transportation Science* 39:104–18.
3. O. Bräysy, and M. Gendreau. 2005. Vehicle routing problem with time windows. Part II. Metaheuristics. *Transportation Science* 39:119–39.
4. J. F. Cordeau, G. Desaulniers, J. Desrosiers, M. M. Solomon, and F. Soumis. 2002. The vehicle routing problem with time windows. In *The vehicle routing problem*, ed. P. Toth and D. Vigo, 157–93. SIAM Monographs on Discrete Mathematics and Applications. Philadelphia: Society for Industrial and Applied Mathematics.
5. A. L. Erera, B. Karacik, and M. W. P. Savelsbergh. 2006. A dynamic driver management scheme for less-than-truckload carriers to appear in 2008. *Computers and Operations Research* New edition.
6. D. Feillet, P. Dejax, M. Gendreau, and C. Gueguen. 2004. An exact algorithm for the elementary shortest path problem with resource constraints: Application to some vehicle routing problems. *Networks* 44:216–29.
7. P. Francis, G. Zhang, and K. Smilowitz. 2006. Improved modeling and solution methods for the multi-resource routing problem with flexible tasks. *European Journal of Operational Research*, 180:1045–1059.
8. M. R. Garey, and D. S. Johnson. 1979. *Computers and intractability: A guide to the theory of NP-completeness*. New York: W. H. Freeman.
9. M. Grönalt, R. F. Hartl, and M. Reimann. 2003. New savings based algorithms for time constrained pickup and delivery of full truckloads. *European Journal of Operational Research* 151:520–35.
10. A. J. Herberger. 2001. Expanding opportunities in coastwise shipping: A multimodal, integrated coastal transportation system for the 21st century. Testimony before the House Transportation and Infrastructure Committee, May. http://www.house.gov/transportation/cgmt/05-23-01/herberger.html.
11. Y. Ileri, M. Bazaraa, T. Gifford, G. Nemhauser, J. Sokol, and E. Wikum. 2006. An optimization approach for planning daily drayage operations. *Central European Journal of Operations Research*, 14:141–56.

12. S. Irnich, and D. Villeneuve. 2006. The shortest path problem with resource constraints and k-cycle elimination for $k \geq 3$. *INFORMS Journal on Computing* 18: 391–406.
13. H. Jula, M. Dessouky, P. Ioannou, and A. Chassiakos. 2005. Container movement by trucks in metropolitan networks: Modeling and optimization. *Transportation Research Part E.*
14. A. Lowenthal. 2002. H&S 40720 marine terminal operation: Truck idling. California State Assembly. http://www.arb.ca.gov/bluebook/bb03/HS/40720.htm.
15. E. K. Morlok, and L. N. Spasovic. 1994. Redesigning rail-truck intermodal drayage operations for enhanced service and cost performance. *Journal of the Transportation Research Forum* 34:16–31.
16. R. Namboothiri. 2006. Planning container drayage operations at congested seaports. Ph.D. thesis, School of Industrial and Systems Engineering, Georgia Institute of Technology, Atlanta.
17. R. Namboothiri, and A. L. Erera. 2004. A set partitioning heuristic for local drayage routing under time-dependent port delay. In *Proceedings of IEEE International Conference on Systems, Man, and Cybernetics*, 3921–26. The Hague, Netherlands.
18. R. Namboothiri, and A. L. Erera. 2008. Planning local container drayage operations given a port access appointment system. *Transportation Research Part E* 44: 185–202.
19. K. Smilowitz. 2006. Multi-resource routing with flexible tasks: An application in drayage operations. *IIE Transactions* 38:577–90.
20. K. Smilowitz, and G. Zhang. 2006. The multiple choice elementary constrained shortest path problem. Working Paper, 06-06, Technical report, Northwestern University.
21. X. Wang, and A. C. Regan. 2002. Local truckload pickup and delivery with hard time window constraints. *Transportation Research Part B* 36:97–112.

12 Crane Double Cycling in Container Ports

Anne Goodchild and Carlos Daganzo

CONTENTS

ABSTRACT

Loading ships as they are unloaded (double cycling) can improve the efficiency of a quay crane and therefore a container port. This chapter describes the double-cycling problem, presents solution algorithms to the sequencing problem, and provides simple formulae to estimate benefits. We focus on reducing the number of operations necessary to turn around a ship. First an intuitive lower bound is developed. We then present a greedy algorithm that was developed based on the physical properties of the problem and yields a tight upper bound. The formula for an upper bound on the greedy algorithm's performance can be used to accurately predict crane performance. The problem is also formulated as a scheduling problem, which can be solved optimally using Johnson's rule. Finally, we consider the longer-term impact of double cycling on crane productivity and briefly discuss port operations. For this we use an even simpler double-cycling algorithm that is operationally convenient, easy to model, and nearly optimum. A framework is developed for analysis, and a simple formula is developed to predict the longer-term impact on turnaround time. The formula is an accurate predictor of performance. We demonstrate that double cycling can create significant efficiency gains.

12.1 INTRODUCTION

Double cycling is a technique that can be used to improve the efficiency of quay cranes by eliminating some empty-crane moves. By improving quay crane efficiency, ports can reduce ship turnaround time, improve port productivity, and improve throughput in the freight transportation system.

In contrast to other measures to increase capacity, such as terminal expansion and information technology deployments, double cycling, the method considered here, is implemented at low cost and can increase port capacity; it does not require new technology or infrastructure. Although double cycling in the long term will not solve the capacity problem, it can be more quickly implemented than other solutions and can be used to complement other strategies. Instead of using a common method, where all relevant containers are unloaded before any are loaded (single cycling), containers are loaded and unloaded simultaneously (see figure 12.1). This allows the crane to carry a container while moving from the apron to the ship (one move) as well as from the ship to the apron, doubling the number of containers transported in a cycle (or two moves). This crane efficiency improvement can be used to reduce ship turnaround time and therefore improve port throughput, and address the current capacity deficit.

The concept of double cycling is recognized by most in the industry,[1,2] and its potential to improve efficiency is intuitively understood. The technique is used to a limited extent at many ports in the United States and abroad, but a broad implementation of double cycling has not occurred. When initially considering the benefits

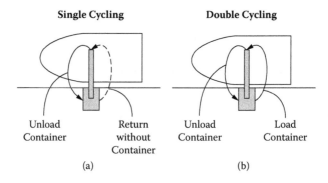

FIGURE 12.1 (a) Unloading using single cycling. (b) Unloading and loading with double cycling.

of double cycling, we assumed that existing planning tools have been used to create a loading plan, as is current practice, and that this loading plan has made no accommodations for double cycling. We therefore consider changes only to the crane's sequence of operations. In practice, this would be determined at a planning stage, and the crane operator would be given a sequence of operations to carry out, in the same way they are when performing single-cycle operations. This way we demonstrate that double cycling is feasible and beneficial without disrupting current operations.

Although problems of port design and operation are the subject of much academic research and much political attention, to date only two studies of double cycling have appeared in a scholarly journal.[3,4] Double cycling is addressed indirectly in Bendall and Stent,[6] where the authors analyze the productivity gains from hatchless ships. Actually, the hatchless vessel derives much of its benefit from the use of double cycling. Without hatches, the stacks are double cycled, and a larger portion of the vessel's containers can be unloaded and loaded simultaneously. In addition, the moves required to move the hatch covers are eliminated. In what follows we assume that only containers below deck will be double cycled when comparing hatched and hatchless ships.

Although double cycling does not appear in the academic literature, it has been a recognized concept in the industry for at least 10 years.[7] Determining the extent to which it has been used or trialed is fairly difficult, as it is not publicly available information. But we do know that double cycling, although it is not widely accepted, is used to a limited extent at some ports.

The intuition that double cycling could provide turnaround time savings is greeted with interest among the maritime community. This is tempered by significant skepticism that the technique will prove beneficial given the complicated nature of port planning and operations and the numerous parties involved in the transport of containers. This chapter provides a general analysis of double cycling and its impact on port operations.

Modern ports and vessel operators use computer programs to design loading plans, sequence loading and unloading operations, and schedule daily port operations. Double cycling could be easily incorporated into these tools. The ideas presented here are not meant to substitute for detailed terminal and vessel planning programs, which are well suited to managing a specific vessel and terminal configuration, but to provide portable insights into double cycling at a more general level. Results can be used to suggest how a port should be configured and operated when implementing double cycling.

12.2 DOUBLE CYCLING

We begin this chapter by developing a framework for analysis of the double-cycling problem. We then formulate the problem and develop algorithms to determine sequences to use when carrying out double-cycling operations. We compare the performance of these algorithms and quantify the benefits with respect to the number of cycles required to turn around a vessel.

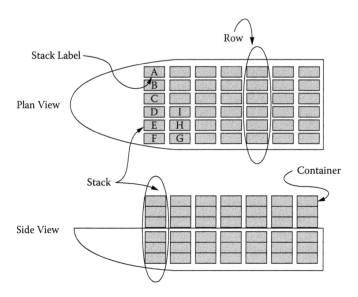

FIGURE 12.2 Plan and side views of a simplified ship (number of containers shown not representative of typical ship size).

12.3 MODELING FRAMEWORK

The layout of containers on a ship can be modeled as a three-dimensional matrix. Containers are stacked on top of one another and arranged in rows. One row stretches across the width of the ship. Large container vessels today typically hold twenty stacks of containers across the width of the ship and up to twenty stacks along the length of the ship (40-foot equivalent units). Of course, we expect these figures to increase with the penetration of 10,000 TEUs (20-foot equivalent unit) capacity Malacca-max carriers.* Figure 12.2 gives a top and side view of a typical vessel (although the number of container stacks is not representative).

It is important to point out that the number of locking and unlocking operations is not affected by double cycling. The same number of containers will need to be picked up and put down with single or double cycling. In this research we will assume that dockside containers are ready for loading when required, and containers being unloaded can be quickly removed from the immediate area.

Consider the case where a ship arrives in port with a set of containers on board to be unloaded and a loading plan for containers to be loaded. The loading plan indicates the placement of containers on the ship. Given are u_c and l_c, the number of containers to be unloaded and loaded, respectively, in each stack labeled c. Figure 12.3 is an example problem that will be used for illustrative purposes. Notice that in figure 12.3, $u_A = 3$ and $l_A = 2$. A rehandle is a container that must be moved to access containers below it, but will then be stowed again before the ship departs.

---

* Most containers in the United States are the equivalent of 2 TEUs, or 1 FEU, a 40-foot equivalent unit.

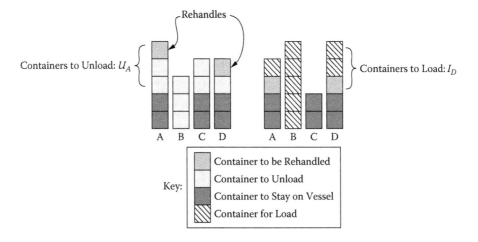

FIGURE 12.3 Detailed plan for containers to be unloaded and loaded.

Note that if any rehandles are necessary, we include these in the total number of loads and unloads. For example, if, during unloading, a container must be moved to access a container beneath it, the container moved will be counted as an unload. This container would then also be counted as a load when it is placed back in the original stack. We always assume that rehandles are replaced back into the stack from which they were removed. Note that this will overestimate the amount of work necessary to unload and load a set of containers because the container is considered moved from the vessel to the shore and back to a location on the vessel. In this work we consider a move to be between the vessel and the apron, but in reality some rehandles may only be moved between locations on the vessel, typically a shorter distance.

The set of stack labels is called S. An ordering of these stacks can be described by a permutation function, Π. A permutation is a one-to-one correspondence between the set of $n \in \{1,...,N\}$ and $c \in S$ such that $\Pi(n) = C$, or $n = \Pi^{-1}(c)$. For example, in figure 12.3, the set of stack labels is $S = \{A,B,C,D\}$. A permutation of these is $\{B,A,C,D\}$ given by the function Π_e (a permutation for our example), where $\Pi_e(1) = B$, $\Pi_e(2) = A$, $\Pi_e(3) = C$, and $\Pi_e(4) = D$.

If we consider the time it takes to unload and load a ship to be a measure of crane efficiency, then the goal of double cycling is to reduce the total turnaround time. A proxy for this is the number of cycles required to unload and load the ship. The number of cycles necessary to complete loading and unloading will be represented by the variable w. We will consider double cycling within one row of the ship. Due to the difficulty with which the crane moves laterally along the ship, it is not practical to consider double cycling across two rows. We will complete unloading and loading of one row before moving the crane lengthwise along the ship to the next row. Using this method we will not require the crane to move laterally along the ship within one cycle. We will first determine the number of cycles to complete operations on a row. We will restrict our attention to special cases of the generic double-cycling method described below:

- Choose an unloading permutation, Π'. Unload all containers in the first stack of the permutation, then all containers in the second stack of the permutation. Proceed in this fashion until all stacks have been unloaded.
- Choose a loading permutation, Π, and load the stacks in that order. Load all containers in the first stack, then in the second, and so forth. Loading can start in any stack as soon as it is empty or it contains just containers that should not be unloaded at this port. Once loading has begun in a stack, continue loading until that stack is complete.

Figure 12.4(a) is a queueing diagram for a single-cycling operation where the stacks of figure 12.3 are handled in the order $\{A,B,C,D\}$ both for loading and unloading. Time is expressed in cycles. Note that loading operations must wait until cycle $w = 10$, when unloading is finished. The process requires $w = 20$ cycles. With single cycling, the crane unloads each row of the vessel before loading any containers.

If we double cycle, we can still plot the unloading curve on the same diagram. Now, using the same sequence for unloading and loading, $\Pi' = \Pi = A,B,C,D$, we

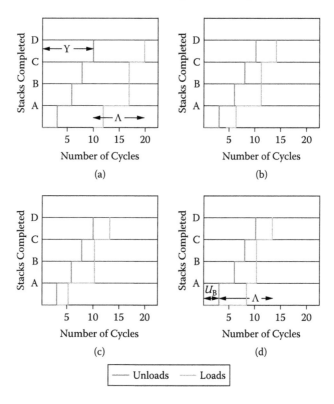

FIGURE 12.4 Turnaround time with different methods: (a) single cycling with ordering A, B, C, D unloading starts at w = 10, 20 cycles; (b) double cycling with ordering A, B, C, D unloading starts at w = 4, 14 cycles; (c) double cycling with ordering A, B, C, D unloading starts at w = 3, 14 cycles; (d) double-cycling ordering B, A, C, D unloading starts at w = 3, 13 cycles.

can shift the loading curve to the left as far as possible without overlapping the unloading curve. Figure 12.4(b) shows the maximum shift. Loading can start as early as $w = 4$, and the process would require only 14 cycles. The same number of cycles is obviously obtained if we start loading each stack as early as possible, as in figure 12.4(c). This introduces some delay as the loading operations must wait one cycle for the unloading operations to be completed in stack B, but does not change the completion time.

With single cycling one cycle is required for every container. With double cycling, however, the number of cycles will depend on the sequence. Figure 12.4(d) shows that if the loading and unloading sequence is B,A,C,D, then the completion time is $w = 13$.

This framework considers the work of one crane, working on individual rows of a vessel. This does not limit our analysis to operations where only one crane works each vessel, as it can be reproduced for each crane, assuming the working areas of the vessel can be segmented by crane.

In the next section we present a lower bound on the number of cycles using any algorithm.

12.4 LOWER BOUND

Define

$$Y = \sum_{n=1}^{N} u_{\Pi'(n)} = \sum_{c \in S} u_c, \quad \Lambda = \sum_{n=1}^{N} l_{\Pi(n)} = \sum_{c \in S} l_c. \tag{12.1}$$

Recall from figure 12.4(a) that using single cycling, the number of cycles necessary to complete a row is

$$Y + \Lambda. \tag{12.2}$$

This is intuitive: one cycle for each container. We have assumed the crane must start and finish on the dock.

For double cycling, with a specific loading permutation Π and unloading permutation Π', assuming $\Lambda \geq Y$, the number of cycles must be at least $\Lambda + u_{\Pi'(1)}$ which satisfies

$$\Lambda + u_{\Pi'(1)} \geq \Lambda + \min_c(u_c). \tag{12.3}$$

The right-hand side of (12.3) is a general lower bound that applies to all permutations when $\Lambda \geq Y$. It is also a lower bound if we force the same permutation for loads and unloads ($\Pi' = \Pi$). It is well known that there is no loss of optimality by imposing the same sequence on both machines (see Johnson[8]). It is also well known that that if the processing times (number of containers to unload and load in each stack) are interchanged, then an equivalent inverse problem results. Therefore, in the case were

$Y \geq \Lambda$, the number of cycles must be at least $Y + l_{\Pi(1)}$, which satisfies

$$Y + l_{\Pi(1)} \geq Y + \min_{c}(l_c). \tag{12.4}$$

More generally, if w is the number of cycles using any algorithm for double cycling, we can write

$$w \geq \max\{\Lambda + \min_{c}(u_c), Y + \min_{c}(l_c)\}. \tag{12.5}$$

We will now discuss an algorithm that was developed to keep the loading and unloading operations separate, but would be easy to implement, and would provide mathematical insight. It provides significant operational flexibility and an upper bound formula for the required number of cycles. The formula can be evaluated without running an algorithm and thus is useful for planning purposes.

12.5   GREEDY STRATEGY AND AN UPPER BOUND

We propose to unload and load each stack as soon as possible, assuming that the loading and unloading sequences are given by the same "greedy" permutation, $\Pi' = \Pi = G$. This greedy permutation is obtained by ordering the stacks in descending order of the variable d_c, where

$$d_c = l_c - u_c \quad \text{when} \quad \Lambda \geq Y \tag{12.6}$$

$$d_c = u_c - l_c \quad \text{when} \quad Y > \Lambda. \tag{12.7}$$

The rationale for (12.6) is that we want the unloading operations to run ahead of the loading operations as much as possible.

We will assume in this section that stacks have been labeled by position in the handling sequence with the greedy strategy. So now, u_j is the number of containers to unload in the jth stack, $d = 1...J$, or equivalently, the number of cycles required to unload the jth stack. Assume there are more loads than unloads, $\Lambda \geq Y$. Let U_j be the cumulative time (in number of cycles) at which the jth stack is finished unloading: $U_j = u_1 + u_2 + \cdots + u_j$, where the sequence is determined by the greedy strategy. Recall that l_j is the number of containers to load in the jth stack, or equivalently the number of cycles required to load the jth stack. Let L_j be the combined operational time (in number of cycles) to load j stacks: $L_j = l_1 + l_2 + \cdots + l_j$, where the sequence is determined by the greedy strategy.

Given $\Lambda - Y \geq 0$, define $d_j = l_j - u_j$ and $D_j = L_j - U_j = d_1 + d_2 + \cdots + d_j$. Notice that $d_{j-1} \geq d_j$ because our strategy is greedy. Notice also that $D_J = \Sigma_{c \in S} d_c = \Lambda - Y \geq 0$ because there are more (or equal) loads than unloads.

Lemma 12.5.1—If there are more (or equal) loads than unloads, then $D_j \geq 0 \forall j$.

Proof.   Since there are more (or equal) loads than unloads, $D_J \geq 0$. Assume now that $D_J < 0$ for some j. For D_j to be negative, one of its components, d_k, must be negative for some $k \leq j$. This, however, would mean that $d_r < 0 \forall r \geq k$, and also for $r \geq j$ (since the

d_j sequence is decreasing for the greedy strategy). Hence, $D_J = D_j + d_{j+1} + \cdots + d_J$ is a sum of negative terms and would be negative. But this is a contradiction. Thus, there cannot be a $D_J < 0$ for some j.   □

Lemma 12.5.2—If there are more (or equal) loads than unloads, the number of cycles to complete operations with the greedy strategy, $w_G \geq 0$, satisfies $w_G \leq \Lambda + \max_c u_c$.

Proof. Introduce s as the time (number of cycles) between the origin and the beginning of the loading operation. If there are no intermediate delays, then the time to begin loading the *j*th stack, B_j, is $B_j = s + L_{j-1}$, that is, the shift plus the time to load $D-1$ stacks (define $L_0 = 0$). Clearly now, if $B_j - U_j \geq 0 \forall j$, then there will be no intermediate delays to loading, and we have found a strategy with at most $s + \Lambda$ cycles. We now show that this is true. Note that $\{B_j - U_j\} = [s - u_j] + [L_{j-1} - U_{j-1}]$ and that the first term on the right side is nonnegative if we choose $s = \max_{c \in S} u_c$. We also see that the second term is nonnegative by lemma 12.5.1. Thus, $B_j - U_j \geq 0 \forall j$ as claimed, and there are no intermediate delays if $s = \max_{c \in S} u_c$. Therefore, $w_G \leq \Lambda + \max_{c \in S} u_c$.   □

Lemma 12.5.3—If $\Lambda \leq Y$, then $w_G \leq Y + \max_c (l_c)$.

Proof. G in this case is defined by (12.7). That lemma 12.5.3 is true should be obvious by symmetry. If one were to record the process of unloading and loading the row, and then play this recording in reverse, the reversed movie would display a sequence of operations with the same total time as for a problem in which the role of loads and unloads is switched. Thus, for every problem instance with loads greater than unloads, there is a dual instance where unloads are greater than loads. The greedy strategy continues to be greedy in reverse, and everything said up to this point, including the bounds, continues to hold with time running backward. The greedy strategy with time running backward implies a reverse ordering of d_c, as specified in (12.7); thus, lemma 12.5.3 holds.   □

Define $u' = \min_c (u_c)$, $l' = \min_c (l_c)$, $u^* = \max_c (u_c)$, and $l^* = \max_c (l_c)$. We can now state the following theorem:

Theorem 12.5.4—$\max (\Lambda + u', Y + l') \leq w^* \leq w_G \leq \max (\Lambda + u^*, Y + l^*)$.

Proof. See the proofs for lemmas 12.5.1 to 12.5.3.   □

For rows of large ships where $\Lambda, Y \gg (u^*, l^*)$, the greedy strategy is very close to both the upper and lower bounds since all the members of theorem 12.5.4 are relatively close to $\max (\Lambda, Y)$. More specifically, note that if u_c and l_c are bounded by a constant (stack size), then the percent optimality gap vanishes as the number of stacks (problem size) tends to infinity. Thus, the greedy strategy is asymptotically optimal. This is visually represented in figure 12.5.

If we apply the greedy algorithm to the example problem used in section 12.4, we obtain the sequence B, D, A, C. This sequence would require 13 cycles to complete unloading and loading operations. In the next section we formulate the problem of

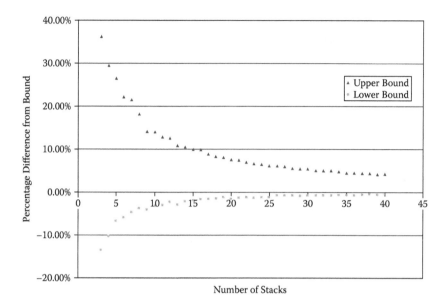

FIGURE 12.5 The greedy strategy is asymptotically optimal.

determining Π' and Π as a mixed-integer program and describe an optimal solution algorithm.

12.6 SCHEDULING APPROACH

The problem is formulated as a two-machine flow shop scheduling problem. There is one job corresponding to each stack, and each job has two operations: an unloading operation that must be completed first, and a subsequent unloading operation. In the scheduling representation, there is a machine for unloading and a separate machine for loading, but in actual implementation, the crane performs both types of operations. One cycle corresponds to one time unit in the scheduling problem. In a single cycle, the crane loads or unloads a single container. In a double cycle, the crane loads a container in one job (i.e., in one stack) and unloads a container in another job. The problem is to determine the best sequence of unloading and loading, to minimize the maximum completion time (make span). Start times are identified from the known job durations. We use the following notation:

- u_c = number of containers to unload in stack $c \in S$.
- l_c = number of containers to load in stack $c \in S$.
- FU_c = completion time of unloading $c \in S$.
- FL_c = completion time of loading $c \in S$.
- w = maximum completion time.
- X_{kj} = binary variable for ordering of unloading jobs (1 if $j \in S$ is unloaded after $k \in S$ and 0 otherwise).

- Y_{kj} = binary variable for ordering of loading jobs (1 if $j \in S$ is loaded after $k \in S$ and 0 otherwise).
- M = a large number.

The scheduling problem (SP) is to minimize the maximum completion time of all jobs subject to constraints. The result is to uniquely identify the permutations Π and Π' and a feasible set of job start and end times. It is assumed that the process starts at time zero. The formulation is

$$\text{Minimize} \quad w \qquad (12.8a)$$

$$\text{subject to} \quad w \geq FL_c \quad \forall c \in S, \qquad (12.8b)$$

$$FL_c - FU_c \geq l_c \quad \forall c \in S, \qquad (12.8c)$$

$$FU_k - FU_j + MX_{kj} \geq u_k \quad \forall j, k \in S, \qquad (12.8d)$$

$$FU_j - FU_k + M(1 - X_{kj}) \geq u_j \quad \forall j, k \in S, \qquad (12.8e)$$

$$FL_k - FL_j + MY_{kj} \geq l_k \quad \forall j, k \in S, \qquad (12.8f)$$

$$FL_j - FL_k + M(1 - Y_{kj}) \geq l_j \quad \forall j, k \in S, \qquad (12.8g)$$

$$FU_c \geq u_c \quad \forall c \in S, \qquad (12.8h)$$

$$X_{kj}, Y_{kj} = 1, 0 \quad \forall j, \ k \in S. \qquad (12.8i)$$

These constraints completely define the double-cycling problem. Constraints (12.8b) ensure that the make span is greater than or equal to the completion of loading of all stacks. Constraints (12.8c) ensure that stacks are only loaded after all necessary stacks have been unloaded. Constraints (12.8d), (12.8e), and (12.8i) ensure that every stack is unloaded after the previous one in Π' has been unloaded. This is achieved by specifying for every pair of stacks (j,k) that either stack k is unloaded before stack j (if $X_{kj} = 1$) or the reverse (if $X_{kj} = 0$) and that the time difference between the two events is large enough to unload the second of the two stacks. Constraints (12.8f), (12.8g), and (12.8i) are equivalent to (12.8d), (12.8e), and (12.8i) but for loading jobs. Constraints (12.8h) ensure that each unloading completion time allows for enough time to at least unload that stack.

It should be noted that the assumption of continuous loading of a stack and continuous unloading of a stack is without loss of optimality. In other words, preemption cannot improve the solution. Johnson[8] developed an optimal solution algorithm for the two-machine, unconstrained problem that can be used to solve the formulation above. It is well known that for two-machine problems it is sufficient to consider schedules in which the processing orders on the two machines are identical. It is also well known that if the processing times are interchanged, then an equivalent inverse problem results.

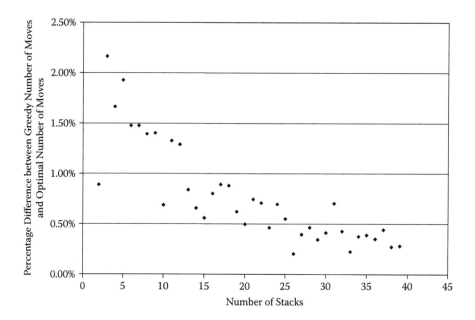

FIGURE 12.6 Performance comparison of the scheduling and greedy algorithms. Each point shows the percentage difference between the greedy solution and the optimal solution. Each point shows the average across forty generated vessels.

Using Johnson's rule on the example problem we determine the sequence that requires the minimum number of cycles: {D,B,A,C}. Notice with this sequence, which is different from the greedy sequence, we require twelve cycles to complete the unloading and loading operation. This is fewer than the thirteen cycles required with the greedy algorithm.

Figure 12.6 compares the performance of the scheduling and greedy algorithms on a larger set of data. The percentage difference between the solution using the greedy strategy and Johnson's rule is shown. Each data point shows the average difference across forty vessels. The difference is small, less than 2.5% even for three stacks. As expected, the greedy algorithm's performance improves with row size. For twenty stacks, a typical row size, the difference is between 0.5 and 1.0% for the data simulated.

Although we can solve the problem optimally with Johnson's rule, the greedy algorithm provides simple and tight bounds that can be used to estimate the turn-around time and an intuition for this result.

12.7 COMPUTER PROGRAM

A computer program was developed that generates ship data and calculates the number of moves to turn around the ship using various strategies. The purpose of the program was to provide large sets of data upon which to evaluate algorithms. The program provides turnaround times (in number of moves) using single cycling, the greedy algorithm, Johnson's rule, and the proximal stack strategy. The program can generate data for many different ship designs and market conditions (imports and exports, origins and destinations, and number of stacks). Necessary inputs into the simulator are:

- Number of stacks
- h_i, maximum stack height (in containers) for imports
- h_e, maximum stack height (in containers) for exports
- p_i and q_i, parameters for the beta distribution, applied to imports
- p_e and q_e, parameters for the beta distribution, applied to exports

The number of containers to unload and load in each stack, u_c and l_c, are assumed to be independent and generated by sampling from a beta distribution. According to the input parameters p_i, q_i, p_e, and q_e, the beta distribution is sampled twice for each stack, obtaining a value between 0 and 1, with a different distribution for imports and exports. Each sampled value is then multiplied by the maximum stack height and rounded down to the nearest integer.

The program counts the number of cycles required to complete loading and unloading operations on a vessel for the following algorithms:

- Single cycling: This algorithm is described in section 12.3.
- Greedy strategy: This algorithm is described in section 12.5.
- Proximal strategy: This algorithm is described in section 12.8.
- Johnson's rule: This algorithm is described in section 12.6.

In the next section we introduce another strategy, the proximal stack strategy. This strategy is operationally more convenient than using the sequence determined by Johnson's rule or the greedy strategy and allows for simpler mathematical analysis. It turns out the result is almost optimal.

12.8 PROXIMAL STACK STRATEGY

The proximal stack strategy is based on the strategy typically used in current operations. Let R be the number of rows on a vessel, and let C_i be the number of stacks in row i. Rows are numbered $i = 1..R$ starting from the bow. Let $\{1..C_i\}$ be the set of stack numbers, c, in one row of a vessel. The first number, $c = 1$, is the stack closest to the shore, and C_i the stack nearest the water.

Definition 12.8.1—(Proximal stack strategy) The crane processes rows one at a time in order of increasing i. For each row it:

1. Unloads all containers in the stack closest to the shore, $c = 1$, then all containers in stacks $c = 2,3$, and so on, until all stacks in the row have been unloaded.
2. Loads the stacks using the same ordering. Loading can start in a stack as soon as it is empty or contains just containers that should not be unloaded at this port. Once loading has begun in a stack it is continuously loaded until complete.

We choose to focus on this strategy because it is operationally and mathematically convenient and the strategy used in practice. For example, this is the method that was used in the real-world trial at the Port of Tacoma (see section 12.9). We again assume double cycling only takes place within one row. By operating on each row individually we do not require the crane to move laterally along the ship within one cycle.

12.9   COMPARISON OF ALGORITHMS ON LARGE DATA SETS

We use the computer program to generate problem instances and calculate the number of moves for each algorithm, to compare the benefits of the proximal stack strategy to single cycling and to those of Johnson's rule and the greedy strategy, both when double cycling above deck and below deck, or only below deck. Our comparisons always consider double cycling within one row of the ship. The results comparing the strategies are shown in figures 12.7 and 12.8. Each data point represents the average of forty generated vessels.

As expected, the benefits using the proximal stack strategy are smaller than the benefits using the greedy or optimal strategies. For a row of twenty stacks, there is a 45% reduction in number of moves over single cycling for the optimal strategy, a 44% reduction using the greedy strategy, and about a 40% savings for the proximal stack strategy. Notice that the savings range in the figures has been reduced to allow

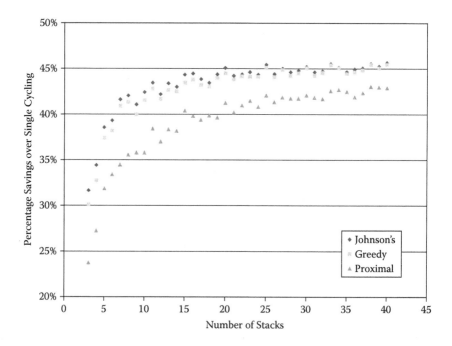

FIGURE 12.7 Performance comparison of greedy strategy, proximal strategy, and Johnson's rule to single cycling. Each data point shows the percentage savings over single cycling and is the average result for forty generated vessels.

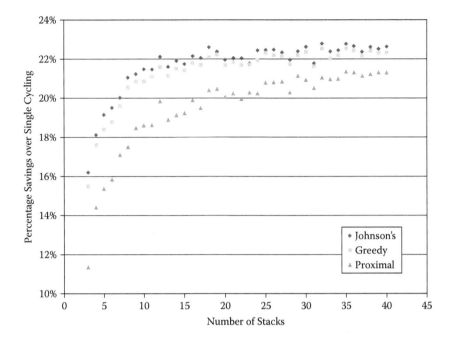

FIGURE 12.8 Comparison of the greedy strategy, proximal strategy, and Johnson's rule to single cycling when double cycling only below deck. Each data point shows the percentage savings over single cycling and is the average result for forty generated vessels.

closer comparison of the values, that benefits above 35% are commonplace, and that the benefits of using any of the three strategies are significant. These results assume there are no deck hatches (or one for each stack).

If double cycling only occurs below deck, the results from the three strategies are essentially equivalent. Figure 12.8 shows the percentage savings over single cycling of using Johnson's rule, the greedy strategy, and the proximal stack strategy to double cycle only below deck. Notice again the scale of the axis has been adjusted for closer comparison of the strategies. For a vessel with twenty stacks per row, the benefits of all three strategies are between 20 and 22%.

While the number of cycles required to turn around the vessel is a relevant metric, the real benefit comes from reducing operational time consumed by the unloading and loading process. Further, a double cycle is expected to take longer than a single cycle. Consider the time taken by the same two containers with single and double cycling. For each double cycle we save some empty-crane travel relative to the two corresponding single cycles, but we also experience a slight landside-repositioning penalty. The time penalty is incurred because after dropping a container for unloading onto a landside vehicle, the crane must wait for a container for loading to be positioned below the crane. With single cycling, this can be done simultaneously with other crane operations. The total distance traveled by the crane is reduced by one complete empty cycle between the apron and the position above either the container to load or the container to unload, whichever is closer.

After completing loading and unloading operations on a row the crane moves along the vessel to the next row. Recall that with the current method of single cycling, all relevant containers are unloaded before any containers are loaded. Thus, the crane travels the length of the vessel twice. With double cycling, each row is unloaded *and* loaded before the crane moves on to the next row; therefore, the crane travels the length of the vessel only once.

During the Efficient Marine Terminal trial at the Port of Tacoma in 2003 a trial of double cycling was undertaken. The adjusted average time for a single cycle was 1 minute 45 seconds, and for a double cycle 2 minutes 50 seconds. Estimating the time savings using the physical properties of the Tacoma trial, if the vertical and horizonal motions take place simultaneously, the difference is 25.4 seconds. If the vertical and horizontal motions do not take place simultaneously, the difference is 45 seconds. The empirical difference was 40 seconds. The upper bound is close to the empirical value. Given this, a 21% reduction in the number of cycles reduces operating time by approximately 8%. A 35% reduction in cycles would reduce operating time by 13%.

We have found the use of double cycling would provide significant reductions in operational time. From our analysis we believe the cycle time reduction will be at least 25 seconds, for every single cycle we are able to replace by a double cycle.

12.10   IMPACT ON MARINE TERMINAL OPERATIONS

This section expands the analysis from an individual vessel to an entire fleet, and develops a formula so that ports can begin planning their operations for double cycling. We compare these results to those of vessels generated by the computer program and find the formula to be a very good estimator.

Before adopting double cycling, terminal operators will need to understand its impact on requirements for landside vehicles, quay cranes, stevedores, and so on. Instead of understanding the expected turnaround for an individual vessel, they would like to base estimates on the entire fleet calling, and expecting to call, at their terminal for some relevant time horizon. In this analysis we assume knowledge only of the distribution of the number of containers to load and unload in each stack and the number of stacks on the vessel. We now develop formulae for the expected number of cycles using double cycling given just this distributional information.

Definition 12.10.1—(Demand) Introduce random variables $u_{c,i}$ to denote the number of containers to unload in stack $c \in \{1..C_i\}$ of row $i \in \{1..R\}$, and $l_{c,i}$ the number of containers to load in stack c of row i.* The random variables in sets $\{l_{c,i}\}$ and $\{u_{c,i}\}$ are assumed to be mutually independent. The variables in set $\{u_{c,i}\}$ are identically distributed with mean μ_u and variance σ_u^2, and so are the random variables in $\{l_{c,i}\}$ with mean μ_l and variance σ_l^2.

Definition 12.10.2—(Cumulative demand) Define $Y_i = \Sigma_{c=1}^{c_i} u_{c,i}$, the total number of containers to unload in row i, and $\Lambda_i = \Sigma_{c=1}^{c_i} l_{c,i}$, the total number of containers to load in row i.

---

* If a container on the vessel needs to be moved to access another container, but it is not to be unloaded at this port, this move will be considered an unload and a load if it is handled again, for simplicty.

The expected number of cycles to unload and load in row i using single cycling is equal to the expected number of containers:

$$E[\Lambda_i] + E[Y_i] \tag{12.9}$$

It should be noted that while mathematically it is satisfactory to consider the number of cycles necessary to unload and load a row, in current operations, all containers from the vessel are unloaded before any containers are loaded onto the vessel.

12.10.1 Expected Number of Cycles Using Double Cycling

In this section we use a diffusion approximation to develop an improved estimate for the proximal stack strategy. The goal is to estimate the number of cycles necessary to turn around a row with double cycling, using knowledge of the distribution of containers rather than specific information about the number of imports and exports on each ship.

Figure 12.9(a) shows a queuing diagram for the loading and unloading processes for an example problem: one row with four stacks. In this example, $u_{1,1} = u_{2,1} = 3$, $u_{3,1} = u_{4,1} = 2$, $l_{1,1} = 2$ $l_{2,1} = 5$, $l_{3,1} = 0$ and $l_{4,1} = 3$. The diagram shows two curves: one that documents the loading process and one that documents the unloading process. The figure shows the number of stacks completed for any number of cycles completed. When both loading and unloading, we operate on the stacks in the order $c = 1,2,3,4$. The loading operations begin on stack 1 as soon as the unloading operations are complete in stack 1. Note the loading operations on stack 2 are delayed by one cycle, as unloading operations on stack 2 are not yet complete. Define

$$M_i(c) = \sum_{2 \le j \le c} \{u_{j,i} - l_{j-1,i}\} \forall c \in \{2 \dots C_i\}. \tag{12.10}$$

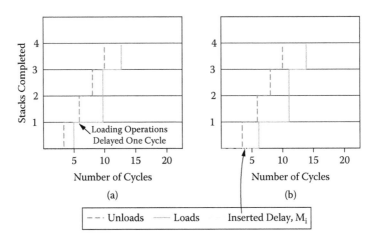

FIGURE 12.9 (a) Example queuing diagram for loading and unloading operations. Notice delay of one cycle to loading operations after completing stack 1. (b) Delay, M_i, inserted before any loading operations start to avoid later delay.

Also define $M_i(1) = 0$. Then,

$$M_i = \max_{c=\{1...C\}} \{M_i(c)\}. \tag{12.11}$$

If we delay the loading of stack 1 by M_i time units [see figure 12.9(b)], we eliminate all future delay caused by waiting, and loading operations are completed at time $\{u_{1,i} + M_i\} + \Lambda_i$. The number of cycles required to unload and load the row using the proximal stack strategy is thus

$$u_{1,i} + \Lambda_i + M_i. \tag{12.12}$$

Then on average

$$E[u_{1,i}] + E[\Lambda_i] + E[M_i]. \tag{12.13}$$

The first two terms are easy to estimate: the expected number of containers to unload in the first stack, and the expected number of containers to load in row i. In fact, a very rough estimate for the number of cycles is $E[\Lambda_i]$, but we will seek a more accurate value. Since the $u_{j,i} - l_{j-1,i}$ are independent, identically distributed random variables, $M_i(c)$ is a diffusion process with the stacks completed, c, as time. Thus, we can use diffusion formulae for $E(M_c)$. The drift is

$$d = \frac{E[M_i(c)]}{c} = \mu_u - \mu_l, \tag{12.14}$$

and the variance rate is

$$D = \frac{var(M_i(c))}{c} = \sigma_u^2 + \sigma_l^2. \tag{12.15}$$

Definition 12.10.3—(First passage time) Let $T_i(z)$ be the time (number of stacks completed) at which $M_i(c)$ first reaches z cycles, assuming the process starts from $z = 0$.

According to Feller[9] the formula for the probability density function of $T_i(z)$ evaluated at c is

$$f(c \mid z) = \frac{z}{\sqrt{2\pi Dc^3}} e^{\frac{-(z+dc)^2}{2Dc}}. \tag{12.16}$$

The cumulative distribution function is therefore

$$\Pr\{T(z) \le c\} = F(c \mid z) = \int_0^c \frac{zdy}{\sqrt{2\pi Dy^3}} e^{\frac{-(z+dy)^2}{(2Dy)}}. \tag{12.17}$$

We are interested in $F(C_i|z)$, where C_i is the number of stacks in row i. Note, however, that

$$Pr\{T(z) \le C_i\} \equiv Pr\{M_i > z\}. \qquad (12.18)$$

The expectation of a nonnegative random variable is obtained by integrating the complementary cumulative distribution function. Hence, the expected maximal excursion is

$$E[M_i] = \int_0^\infty dz \int_0^{C_i} dy \left\{ \frac{z}{\sqrt{2\pi Dy^3}} e^{\frac{-(z+dy)^2}{(2Dy)}} \right\} = \frac{2D}{d} \left[\Phi\left(\frac{d\sqrt{C_i}}{\sqrt{D}} \right) - \frac{1}{2} + \int_{\frac{-d\sqrt{C_i}}{\sqrt{D}}}^0 y\Phi(y)dy \right],$$

$$(12.19)$$

where $\Phi(x) = \int_{-\infty}^x \frac{dw}{\sqrt{2\pi}} e^{\frac{-w^2}{2}}$. An estimate for the number of cycles to unload and load a row is now given:

$$\mu_u + C_i\mu_l + \frac{2D}{d} \left[\Phi\left(\frac{d\sqrt{C_i}}{\sqrt{D}} \right) - \frac{1}{2} + \int_{\frac{-d\sqrt{C_i}}{\sqrt{D}}}^0 y\Phi(y)dy \right]. \qquad (12.20)$$

$E[u_{c,i}] = E[u_{1,i}] = \mu_u$ because $\{u_{c,i}\}$ are identically distributed and $\Phi(x) = \int_{-\infty}^x \frac{dw}{\sqrt{2\pi}} e^{\frac{-w^2}{2}}$. Notice that for $d = 0$, the value of M_i is undefined. We expect that in these cases, the value of M_i would be small. In fact, if we define $x = \frac{d\sqrt{C_i}}{\sqrt{D}}$, then the reader can verify from equation (12.11) that $\lim_{x \to 0} E[M_i] = 0$. The formula allows us to estimate the number of cycles to insert before beginning loading operations so we can avoid future delay and thus estimate the total number of cycles to complete loading and unloading operations.

Formula (12.20) tells us that only μ_u, μ_l, C_i, and $D = \sigma_u^2 + \sigma_l^2$ influence crane time; all other ship configuration data are irrelevant. Figure 12.10(a) shows the percentage reduction in the number of cycles required to complete loading and unloading operations below deck on a row for different values of C. In this example $C = 20$ and $\mu_u = \mu_l$. Figure 12.10(b) shows the same information, but for a vessel where $\mu_l = 0.5\mu_u$. As we expect, benefits decrease with increasing variance (values of D). Given that C_i is fixed, increasing values of μ_u imply a larger total number of containers, and so as a percentage, benefits are greater with increasing μ_u.

12.10.2 COMPARISON OF FORMULA RESULT AND AVERAGE OF MANY VESSELS

From the central limit theorem, we know formula (12.20) will provide a good estimate for the number of cycles for a large number of stacks, but it is unclear how well the expression will match reality for realistic numbers (up to twenty per row in current ships). We therefore compare the estimated values using the formula to

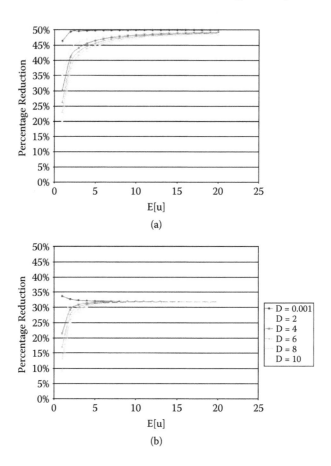

FIGURE 12.10 (a) Percentage reduction in number of cycles to complete loading and unloading operations below deck using the proximal stack strategy for varying values of D. $C_i = 20$ and $Y = \Lambda$ and (b) Percentage reduction, $Y = 0.5\,\Lambda$.

the average for a set of generated vessels. A computer program was used to generate ship data according to a distribution of imports and exports and count the number of cycles required to turn around each row. The inputs to the program are parameters p and q of the beta distribution for both the number of imports and number of exports in one stack, the number of stacks, the maximum height of an import stack, and the maximum height of an export stack. Inputs to the formula, d, D, C_i, μ_u, and μ_l, can be determined uniquely from the computer program inputs.

Figure 12.11 shows the number of moves necessary to turn around the vessel versus the number of stacks on the vessel. Each square represents the estimated number of cycles using formula (12.20). The result is equivalent to the predicted average assuming we had an infinite number of simulations. Each diamond shows the average result for thirty generated vessels. The error bars show just one standard deviation from the mean, using an estimate of the population standard deviation. We see the estimate and average are very close even for small numbers of stacks.

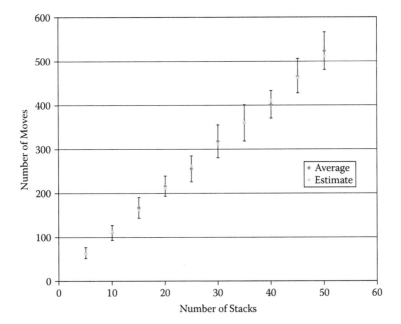

FIGURE 12.11 Squares show the number of moves as calculated by formula (12.20). Diamonds show the average of thirty simulated vessels with the same characteristics. Error bars show one standard deviation from the mean.

The estimated value from the equation is always within the error bars, and so we have demonstrated that the estimate and the average are not significantly different. The average difference between the estimate and the average was 1.13%. As we might expect, the estimate is worst for the case of five stacks, where there is a difference of 5.31%. We have thus developed a formula to estimate long-run crane and vessel productivity improvements as a result of double cycling.

12.11   IMPACT ON LANDSIDE TERMINAL OPERATIONS

In addition to improving crane and vessel productivity, the use of double cycling provides an opportunity to increase the productivity of landside equipment. In a container terminal, container handling equipment is used to transport the containers between the apron and the local storage facility. Loading and unloading the vessel simultaneously means that after a vehicle delivers a container to the apron, it can then carry a container from the apron to local storage instead of returning to the local storage without transporting a container. Container handling equipment is also used to raise or lower containers, for example, from a chassis onto the ground, or from the ground onto a stack of containers. Goodchild and Daganzo[2] provide formulae for estimating the number of landside vehicles required if double cycling; generally equipment needs are decreased. The paper also discusses techniques for developing loading plans to complement double cycling and double cycling's impact on storage equipment, yard storage locations, stevedores, and container handling equipment. Contrary to intuition, double cycling generally simplifies yard operations. It also reduces equipment needs except for the case of terminal handling equipment.

12.12   SUMMARY

This chapter provides port planners with tools and insights as to the impact of double cycling on vessel loading and unloading time. The results are general enough to allow for broad conclusions, but also allow us to quantify the benefits for specific cases. For example, our results obtained are applicable to a wide range of scenarios, including large and small vessels, those with many or few port visits, and a varying number of cranes operating per ship. An alternative approach that focuses on providing more accurate information for specific cases would require detailed information on each container. Our approach relies on only a small number of parameters, and sensitivities to these parameters are easily measured. Favorable comparisons to empirical results support our approach.

While double cycling will not eliminate current port congestion, it can be implemented quickly and, in conjunction with other measures, can ease congestion before more long-term infrastructure projects come on line.

While this chapter focuses on algorithms for double cycling, the impact of double cycling on landside operations, vessel operating time, and port economics has also been considered. See Goodchild[5] for further analysis.

REFERENCES

1. T. Ward. 2003–2005. Personal communications. JWD Group.
2. V. Yau. 2003. Personal communications. American President Lines.
3. A. V. Goodchild, and C. F. Daganzo. 2006. Double cycling strategies for container ships and their effect on ship loading and unloading operations. *Transportation Science* 40(4):473–483.
4. A. V. Goodchild, and C. F. Daganzo. 2007. Crane double cycling in container ports: Planning methods and evaluation. *Transportation Research* 41(8):875–891.
5. A. V. Goodchild. 2005. Crane double cycling in container ports: Algorithms, evaluation, and planning. Department of Civil and Environmental Engineering, Dissertation, University of California at Berkeley.
6. H. B. Bendall, and A. F. Stent. 1996. Hatch coverless container ships: Productivity gains from a new technology. *Maritime Policy and Management* 23:187–199.
7. K. H. Kim. 2004. Personal communications. Pusan National University, Korea.
8. S. M. Johnson. 1954. Optimal two- and three-stage production schedules with setup times included. *Naval Research Logistics Quarterly* 2(4):461–462.
9. W. Feller. 1950. *Probability theory and its applications*. London: John Wiley & Sons.

13 Empty Container Reuse

Hossein Jula, Hwan Chang, Anastasios Chassiakos, and Petros Ioannou

CONTENTS

ABSTRACT

Empty container repositioning is probably the single largest contributor to the congestion at and around marine ports. With a huge number of empty containers at stake at many terminals, a small percentage reduction in empty-repositioning traffic can be reflected in huge congestion reduction and improved operational cost. As a consequence of reduced congestion, container terminals can become more competitive, vehicle emissions will be reduced, and drivers will incur less congestion-related delays.

In this chapter, the *empty container reuse* concept that facilitates the interchange of empty containers outside container ports is studied. In particular, the *depot-direct* and *street-turn* methodologies are investigated, and variants of the empty container reuse problem are considered. These variants are modeled analytically, and optimization techniques are reviewed and discussed.

13.1 INTRODUCTION

As a consequence of unanticipated growth in the volume of containers at many marine container terminals, port-generated traffic has emerged as a major contributor to regional congestion. Traffic congestion and long queues at the gates of the terminals are becoming the main source of air pollution, wasted energy, driver inefficiency, and increasing maintenance cost imposed by the volume of trucks on the roadway.

For instance, over the last decade, the average annual growth of the number of containers handled in the Los Angeles and Long Beach (LA/LB) ports, the largest and busiest U.S. container port complex, has been around 9%.[1] In the year 2004, the combined ports handled about 7.3 million containers. This volume is expected to be tripled by the year 2020. In the year 2000 alone, 2.5 million out of 5.1 million containers moved in and out of the LA/LB ports were empty containers,[2] which indicates that a significant number of truck trips in the LA/LB port area involve movement of empty containers. Such trips may be reduced if more sophisticated techniques are used to deal with empty containers.

Empty container repositioning is probably the single largest contributor to the congestion at and around marine ports.[1] With a huge number of empty containers at stake, a small percentage reduction in empty-repositioning traffic can be reflected in huge congestion reduction and improved operational cost. The idea of reusing empties is an effort to reduce empty repositioning around the ports. The idea consists of using empty import containers for export loads without first returning them to the marine terminal.

In this chapter, we investigate the empty container reusing problem and its different variants. In particular, the following four variants of the problem are studied: (1) the static single-commodity empty container reuse problem, (2) the time-dependent single-commodity empty container reuse problem, (3) the stochastic single-commodity empty container reuse problem, and (4) the multicommodity empty container reuse problem. For each variant, we model the problem analytically and discuss techniques that can be used to find the optimal or near-optimal solutions.

13.2 EMPTY CONTAINER REUSE

Empty containers move both to and from the container terminals. Generally speaking, local container movements at any marine container terminal can be divided into two major categories: import and export container movements.

The import container movements, demonstrated in figure 13.1(a), can be briefly described as follows: A truck is dispatched to pick up a loaded import container from the terminal (move 1). The truck then delivers the loaded container to its designated local consignee (move 2). If an empty container is available at the time of delivery, the truck takes it back to the terminal (move 5), and then goes to its tracking company or another assignment (move 6). If an empty container is not available, the truck goes back to its trucking company or another assignment after delivering the loaded import container (move 3). When the emptied container becomes available at the local consignee, a truck is dispatched to take it back to the terminal (move 4).

Likewise, export container movements, shown in figure 13.1(b), are as follows: A truck is dispatched to pick up an empty container from the terminal (move 1).

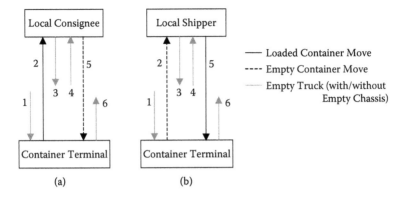

FIGURE 13.1 Container movements: (a) import movement and (b) export movement.

The empty container is trucked to a designated local shipper for loading (move 2). If a loaded container is available at the time of the empty delivery, the truck returns it to the terminal (move 5), and finally the truck goes back to its trucking company or another assignment (move 6). If the loaded container is not available, the truck goes back to its company or another assignment after delivering the empty container (move 3), and when the loaded container becomes available, a truck is dispatched to take it from the local shipper to the terminal (move 4).

The idea of reusing empties is an effort to reduce empty repositioning around the ports. It consists of using empty import containers for export loads without first returning them to marine terminals. Two major methodologies can be considered for empty reusing: depot-direct and street-turn.

Depot-direct: In addition to the marine terminals, empty containers are stored, maintained, and interchanged at off-dock container depots. The potential benefits of depot-direct are establishing a neutral supply point for reusable empties, facilitating empties' drop-off and pickup when terminal gates are congested or closed, and adding buffer capacity to the marine terminals.

The concept of off-dock empty depot may be more attractive and promising in the long term than the short term. In the long term, however, congested marine terminals and the high capital cost of expanding on-dock containers would justify the higher operating cost of empty depots.[2]

Street-turn: The empty container is directly moved from local consignees to local shippers. The potential benefits of street-turns are enormous, including (1) reducing the traffic congestion, noise, and emissions, and (2) saving driving times to and from marine terminals through avoiding the congested areas around the gates.

Considering empty container reusing, the general local container flows around the marine ports can be illustrated by figure 13.2, where the empty truck moves are eliminated for the sake of simplicity. The solid lines are loaded flows and dashed lines are empty flows.

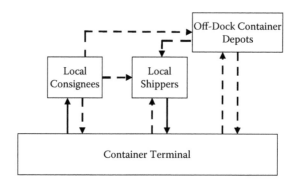

FIGURE 13.2 Local container flows. Solid lines, loaded flows; dashed lines, empty flows.

As demonstrated in figure 13.2, by forwarding empties from local consignees to local shippers directly, the empty street-turn reuse eliminates two empty moves to and from the terminals (move 5 in figure 13.1(a) and move 2 in figure 13.1[b]), and adds one off-port empty move.

It should be noted that the idea of empty reusing is very desirable, by all parties involved, yet hard to achieve.[2] Several key issues limit the ability to make the empty reuse possible, including the operational (e.g., import/export timing and location mismatch, ownership mismatch, container type mismatch) and legal (e.g., off hiring of leased containers) issues. Before the empty reuse idea is fully implemented, these barriers should be addressed and transitions should be made and encouraged to create a fast, reliable, efficient, and seamless system for empty container reuse out of the ports. It is widely expected that the optimization techniques together with the use of information technologies (IT) will play a major rule in addressing both the nonoperational and operational barriers.[2,3]

13.2.1 LITERATURE REVIEW

Surprisingly, the research efforts on empty container reusing are scant. As noted by Dejax and Crainic, even the work on developing models related to the container transportation problems is very limited.[4] Crainic et al. proposed models for the allocation of empty containers in a land distribution and transportation system.[5] The authors developed dynamic and stochastic models that incorporate the consideration of depot location problem and interdepot balancing flows. These models can be viewed as extensions to the traditional multicommodity plant location-allocation problem in which balancing flows between depots are considered.[6] The solution methods to these models vary from metaheuristic methods, such as tabu search,[7] to integer programming methods, such as dual-ascent[8] and branch-and-bound-based optimization techniques.[9]

The empty container allocation problem was also considered and studied by Cheung and Chen.[10] In their paper, the authors formulated the dynamic container allocation problem as a two-stage stochastic network model. Based on the classical restriction framework, several variants of genetic algorithm were implemented in sequential and parallel schemes that yielded comparative results.[10] The model can assist liner operators to allocate their empty containers and consequently reduce their leasing cost and the inventory level at ports.

In another related work, Choong et al. addressed the effect of the length of the planning horizon on empty containers management.[11] They used the intermodal container-on-barge operation in the Mississippi River as the case study to investigate the advantages of using a long rolling horizon. Li et al. studied the empty container allocation in a container port in order to reduce excess of empties.[12] They considered the problem as the nonstandard equipment inventory problem with positive and negative demands under a general holding-penalty cost function. Abrache et al.[13] also suggested a decomposition method that considers the substitution property of empty containers.

An elaborate survey of works related to the empty container movements and management can also be found in Boile.[14]

13.3 SINGLE-COMMODITY EMPTY CONTAINER REUSE PROBLEM

In this section, we consider single-commodity empty container movements in a deterministic and static environment. In the single-commodity problem, we assume that all containers belong to a single class of containers (see section 13.6 for more details); therefore, we do not differentiate between containers. We also assume that all information regarding the location and the time of empty requests at shippers and empty supplies at consignees are known a priori.

The single-commodity transportation problem is presented below.
Minimize

$$\sum_{i=1}^{m+p}\sum_{j=1}^{n+p} c_{ij} x_{ij} \tag{13.1}$$

subject to

$$\sum_{j=1}^{n+p} x_{ij} = s_i \qquad \forall\, i \in I \bigcup P \tag{13.2}$$

$$\sum_{i=1}^{m+p} x_{ij} = d_j \qquad \forall\, j \in J \tag{13.3}$$

$$x_{ij} \geq 0,\ \text{integer}, \qquad \forall\, i \in I \bigcup P,\, j \in J \tag{13.4}$$

where
  I = the set of consignees, $|I| = m$,
  J = the set of shippers, $|J| = n$,
  P = the set of depots and container terminals, $|P| = p$,
  C_{ij} = the cost of moving an empty container from supply node $i \in I \bigcup P$ to demand node $j \in J \bigcup P$,

x_{ij} = the number of empty containers moved from supply node $i \in I \cup P$ to demand node $j \in J \cup P$,

s_i = the number of available empties at supply node $i \in I \cup P$,

d_j = the number of empties requested by demand node $j \in J$.

Constraints (13.2) ensure that the total number of empties moved from each consignee is equal to the number of supply of empties at that location. Constraints (13.3) guarantee that the number of empties arrived at each shipper is the same as the demand of empties at that location. Finally, constraints (13.4) are the integer constraints. In the model presented in (13.1) to (13.4) depots are viewed as both dummy supply and demand nodes.

Let the total number of available containers at supply nodes $I \cup P$ be greater than or equal to the total number of requested empties by demand nodes J. In other words, we assume that all the demands can be fulfilled by internal supplies, rather than exogenous resources. Hence, we have

$$\sum_{i=1}^{m+p} s_i \geq \sum_{j=1}^{n} d_j \tag{13.5}$$

If (13.5) does not hold, to balance the flow of empty containers, the inventory and relocating of empty containers should also be included in the model presented in (13.1) to (13.4).

The problem presented in (13.1) to (13.5) can be viewed as a classical transportation problem that can be easily solved to optimality using a linear programming technique or techniques tailored for the transportation problems. Dantzig and Thapa[15] described the integer solution property of the transportation problems. Using the simplex algorithm, it follows that all the basic variables have integer values if all row and column sums of coefficients in the incident matrix are integers and the elements of the demand vector are also integers. Since all the row and column sums of coefficients in the incident matrix of the problem, presented in (13.1) to (13.5), are integers and the elements of the demand vector are also integers, all the basic variables in the problem have integer values.

13.4 TIME-DEPENDENT SINGLE-COMMODITY EMPTY CONTAINER REUSE PROBLEM

In this section, we will study the time-dependent (dynamic) empty container reuse problem in which the temporal aspects of empties availability, requests, and movements are also taken into account.

Figure 13.3 shows schematically all possible moves in the time-dependent empty container reuse scenario. The empty reuse is demonstrated in two dimensions: time and space (location). The time axis is divided into T periods, where T is the planning horizon. At each period k, $k = 1, \ldots, T$, the locations of consignees (supplies) are unfolded and shown by single circles on the space axis. For instance, at period $k = 1$, consignees 1 and 2 have supplies of empties to be picked up, while at time $k = T - 1$, only consignee i has empties available. Likewise, at each period k,

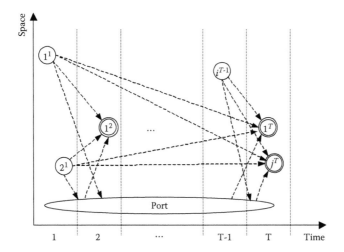

FIGURE 13.3 Dynamic empty reusing scenario.

$k = 1, \ldots, T$, the locations of shippers are unfolded and shown by double circles on the space axis. For instance, at periods $k = 2$ and $k = T$, shipper 1 has demands for empties to be delivered.

For the sake of simplicity, figure 13.3 shows only one container terminal in the region. Here, we will assume that all the depots and ports can process the empties throughout the entire time horizon, T. Dashed lines in figure 13.3 demonstrate the moves of empties from consignees to port, consignee to shippers, and port to shippers. As seen, a move may need more than one time period to be completed. That mainly depends on the distance between the geographical locations of the shippers, consignees, and the port. In congested areas, such as major cities, the moving time may also depend on the traffic condition and consequently on the instant of time when the move initiated.

We also assume that the processing of all empties occurs within the planning horizon T. In other words, we assume that no movement starts before time period $k = 1$, or ends after $k = T$.

Therefore, the time-dependent single-commodity empty container reusing problem can be modeled as follows.

Minimize

$$\sum_{k=1}^{T} \left(\sum_{k'=1}^{T} \sum_{j=1}^{n} \sum_{i=1}^{m} c_{ij}^{kk'} \cdot x_{ij}^{kk'} + \sum_{p=1}^{p} \sum_{i=1}^{m} c_{ip}^{k} \cdot x_{ip}^{k} + \sum_{p=1}^{p} \sum_{j=1}^{n} c_{pj}^{k} \cdot x_{pj}^{k} \right) \tag{13.6}$$

subject to

$$\sum_{k'=1}^{T} \sum_{j=1}^{n} x_{ij}^{kk'} + \sum_{p=1}^{P} x_{ip}^{k} = s_{i}^{k} \qquad \forall i \in I, k = 1, \ldots, T \tag{13.7}$$

$$\sum_{k=1}^{T}\sum_{i=1}^{m} x_{ij}^{kk'} + \sum_{p=1}^{P} x_{pj}^{k'} = d_{j}^{k'} \qquad\qquad \forall j \in J, k' = 1,\ldots,T \qquad (13.8)$$

$$U\left(x_{ij}^{kk'}\right) + t_{ij}^{P(x_{ij}^{kk'})} \le D\left(x_{ij}^{kk'}\right) \qquad\qquad \forall i \in I, j \in J, k, k' = 1,\ldots,T \qquad (13.9)$$

$$x_{ij}^{kk'}, x_{ip}^{k}, x_{pj}^{k'} \ge 0 \text{ and integer} \qquad\qquad \forall i \in I, j \in J, p \in P, k, k' = 1,\ldots,T \qquad (13.10)$$

where

T = the length of the planning horizon.

I = the set of consignees with excess of empties in horizon T, $|I| = m$.

J = the set of shippers with requests for empties in horizon T, $|J| = n$.

P = the set of depots and terminals, $|P| = p$.

$x_{ij}^{kk'}$ = the number of empties moved from consignee i at time k to shipper j to satisfy the demand at time k'; $x_{ij}^{kk'} = 0$ if $k' < k$.

$c_{ij}^{kk'}$ = the cost of moving an empty container from consignee i at time k to demand j in order to satisfy the request at time k'.

t_{ij}^{k} = the time needed to move an empty between consignee i and shipper j initiated at time k. This time includes the pickup time at consignee and drop-off time at shipper. We assume that this time depends solely on the time when the move started.

x_{ip}^{k} = the number of empties moved from consignee i at time k to depot or terminal p.

c_{ip}^{k} = the cost of moving an empty container from consignee i at time k to depot or terminal p.

$x_{pj}^{k'}$ = the number of empties moved from depot or terminal p to shipper j to satisfy the demand at time k'.

$c_{pj}^{k'}$ = the cost of moving an empty container from depot or terminal p to shipper j to satisfy the demand at time k'.

s_{i}^{k} = the number of supply of empties at consignee i at time k.

$d_{j}^{k'}$ = the number of empties demanded at shipper j to be fulfilled by time k'.

We also define and use the following two time index generators:

$U(x_{ij}^{kk'}) = k$ which returns the earliest pickup time of $x_{ij}^{kk'}$.

$D(x_{ij}^{kk'}) = k'$ which generates the latest delivery time of $x_{ij}^{kk'}$.

The objective function in (13.6) is to find the best matching between supplies and demands in horizon T, which minimizes the cost of dynamic empty container movements. Constraints (13.7) ensure that the total number of empties moved from a consignee at time period k is equal to the number of supply of empties at that location at the same period. Constraints (13.8) guarantee that the number of empties arrived at a shipper by time k' is the same as the demand of empties at that location. Constraints (13.9) are the time feasibility constraint. We assume that all empty moves to and from the port and depots are feasible in time. As a consequence, these moves are not included in (13.9). Finally, constraints (13.10) are the integer constraints.

To solve the time-dependent single-commodity empty container reuse problem presented in (13.6) to (13.10), Jula et al.[19] proposed a two-phase optimization technique. In phase 1, the problem is transformed into a bipartite transportation network. In phase 2, the best matching between supply and demand of empties in the transportation network is found.

Using the current and projected data from the combined Los Angles and Long Beach port area in the United States, Jula et al.[19] showed that the developed methodology can yield more than 50% reduction in the empty related truck trips to and from the container ports. The authors also demonstrated that, with a careful selection of the reuse cost function, weights can be adjusted to put more emphasis on either the street-turn or depot-direct in the time-dependent empty container reusing problem. They concluded that, when time is critical, empty reuse is shifted toward depot-direct, since the waiting time is minimal in this methodology. On the other hand, when traveling cost and traffic congestion are important factors, the street-turn methodology provides the best match between supply and demand of empties.

13.5 STOCHASTIC SINGLE-COMMODITY EMPTY CONTAINER REUSE PROBLEM

When all the information about the demand and supply of empty containers is available a priori, the empty container interchange problem can be modeled as a deterministic transportation problem. In the real world, however, one may encounter many sources of uncertainty, which could be related to spatial (location), temporal, or quantitative aspects of the future demand and supply. Broadly speaking, the stochastic empty container reusing problem incorporates one or more aspects of uncertainties.

In this section, we limit our view to the stochastic empty container interchange problem in which the demand of empties follows known probability distributions. These probability distributions can be easily extracted from available historical data.

Let the triplet (Ω, F, P) be a probability space, where F is a collection of events, $\Omega \in F$ is an event (the set of all possible scenarios), and P is the probability measure. We view ω as an outcome (i.e., a scenario) of event Ω which is a random experiment.

Let us consider the single-commodity empty reusing problem presented in (13.1) to (13.4) in its matrix vector form as follows:

Minimize

$$\mathbf{c}^T \mathbf{x} \qquad (13.11)$$

subject to

$$\mathbf{A}_U \mathbf{x} = \mathbf{s} \qquad (13.12)$$

$$\mathbf{A}_L \mathbf{x} = \mathbf{d}(\omega) \qquad (13.13)$$

$$\mathbf{x} \geq \mathbf{0} \qquad (13.14)$$

where c is the cost vector, $\mathbf{s}$ is the supply vector, $\mathbf{d}(\omega)$ is the stochastic demand vector of all scenarios ω, and $\mathbf{A}_U$ and $\mathbf{A}_L$ are to represent equations (13.2) and (13.3) in matrix form, respectively.

As seen, constraints (13.13) depend on the decision vector $\mathbf{x}$ and the specific realization ω. To compensate for any constraints violation, we will provide a recourse vector $\mathbf{y}$ that, after observing the realization of ω, will affect the choice of $\mathbf{x}$. In other words, since the decision vector $\mathbf{x}$ must be made before the realization of ω is known, a second-stage linear program is introduced, whose values are uncertain but will influence the choice of $\mathbf{x}$.

Since every scenario may involve a different set of constraints, a more reasonable objective is to choose the decision variables so that the expected cost of the following recourse formulation is minimized:

$$\text{minimize } \mathbf{c}^T\mathbf{x} + E_\omega[\mathbf{q}^T\mathbf{y}]$$

$$\text{subject to } \mathbf{A}_U\mathbf{x} \qquad\qquad = \mathbf{s}$$

$$\mathbf{A}_L\mathbf{x} \quad + \mathbf{W}\mathbf{y} = \mathbf{d}(\omega) \qquad \omega \in \Omega \tag{13.15}$$

$$\mathbf{x}, \mathbf{y} \geq 0$$

where E_ω stands for the mathematical expectation that is the weighted average over all ω and $\mathbf{W} = [\mathbf{I}_m - \mathbf{I}_m]$ is the recourse matrix, $\mathbf{I}_m$ is an identity matrix, and m is the number of shippers.

In the above formulation, exogenous variable $\mathbf{y}$ is used as the second stage variable and the recourse cost vector $\mathbf{q}^T$ is introduced to penalize constraints violations.

13.5.1 MONTE CARLO SAMPLING AND VSS

Using the deterministic equivalent, the two-stage stochastic program can be solved by linear programming techniques. However, even for a moderate number of possible scenarios, the deterministic equivalent could result in a huge linear programming problem. For example, in the case of the stochastic transportation problem in (13.11) to (13.14), we have assumed demands in the form of finite sets of scenarios. Since the problem includes a set of depots, with a large amount of supplies, every realization scenario ($\omega \in \Omega$) will be feasible and hence the number of feasible scenarios will be

$$|\Omega| = \prod_{j=1}^{n} |d_j(\omega)| \tag{13.16}$$

where n is the number of shippers, $d_j(\omega) = \{d_{j1}, \ldots, d_{js_j}\}$ is the finite set of demands (scenarios) at shipper J, and s_j is the cardinality of the set $d_j(\omega)$.

In other words, although Ω is a finite set, the number of realizable scenarios is too many to enumerate. One way to overcome this problem is to use decomposition methods, such as Bender's decomposition method.[16] Unfortunately, although a decomposition method could reduce the number of variables substantially, the method still generates an extremely large number of constraints. For this reason, a sampling technique called the Monte Carlo simulation method can be used to estimate the expected value of the stochastic program.

Consider the stochastic transportation problem in (13.11) to (13.14). Let v^* be the optimal expected value of the problem, which can be expressed as the following two-stage stochastic program:

$$v^* = \min_{x \in S}\{f(\mathbf{x}) \equiv E_\omega[g(\mathbf{x}, \mathbf{d}(\omega))]\} \tag{13.17}$$

where

$$S \equiv \{\mathbf{x}|\mathbf{A}_U\mathbf{x} = \mathbf{s},\ \mathbf{x} \geq 0\} \text{ and } g(\mathbf{x}, \mathbf{d}(\omega)) \equiv \mathbf{c}^T\mathbf{x} + \min_{y \geq 0}\{\mathbf{q}^T\mathbf{y}|\mathbf{W}\mathbf{y} = \mathbf{d}(\omega) - \mathbf{A}_L\mathbf{x}\}.$$

Let random samples $\mathbf{d}^1, \ldots, \mathbf{d}^N$ be N realizations of the random vector $\mathbf{d}(\omega)$, and

$$\hat{f}_N(\mathbf{x}) \equiv N^{-1}\sum_{k=1}^{N} g(\mathbf{x}, \mathbf{d}^k) \tag{13.18}$$

be the sample average approximation (SAA) of $f(\mathbf{x})$. By replacing $f(\mathbf{x})$ with $\hat{f}_N(\mathbf{x})$ in (13.17), we find the optimal expected value of the *approximated* stochastic problem by

$$\hat{v}_N = \min_{x \in S}\left\{\hat{f}_N(\mathbf{x}) \equiv N^{-1}\sum_{k=1}^{N} g(\mathbf{x}, \mathbf{d}^k)\right\} \tag{13.19}$$

Since the random realizations $\mathbf{d}^1, \ldots, \mathbf{d}^N$ have the same probability distribution as $\mathbf{d}(\omega)$, it follows that $\hat{f}_N(\mathbf{x})$ is an unbiased estimator of $f(\mathbf{x})$ for any $\mathbf{x}$.[17] By generating M independent sample sets $\mathbf{D}^j = \{\mathbf{d}^{1,j}, \ldots, \mathbf{d}^{N,j}\}, j = 1, \ldots, M$, each of size N, and solving the corresponding SAA problems in (13.19), the sample average of the optimal expected values of the approximated stochastic programs can be computed by

$$L_{N.M} = M^{-1}\sum_{j=1}^{M} \hat{v}_N^j \tag{13.20}$$

where $\hat{v}_N^j$ is the optimal expected value of the SAA problem in (13.19) for each sample set $\mathbf{D}^j$. It can be shown that $L_{N.M} = M^{-1}\Sigma_{j=1}^{M}\hat{v}_N^j$ is an unbiased estimator of $E[\hat{v}_N]$.[17]

In stochastic optimization problems, the value of the stochastic solution (VSS) is defined as the difference between the optimal expected value of the stochastic problem and the solution of the deterministic equivalent computed by replacing stochastic variables by their mathematical expectations. The former is called the stochastic solution (SS), which is the solution of (13.20), and the latter is called the expected value (EV) solution. VSS indicates the benefit of knowing the distributions of the stochastic variables.[18]

In particular, EV can be computed by taking the following procedure. During the first stage, the mean-value solution is computed by replacing stochastic demands in (13.15) by their expectations. Subsequently, with the first stage values fixed, each submodel solution is independently computed and averaged over all sampled scenarios.

Let $\bar{\mathbf{x}}$ be the solution to the mean-value problem, which is constructed by replacing random variables by their expectations. Hence, EV is obtained by

$$v_E = \min_{x \in S}\{f(\mathbf{x}) = g(\mathbf{x}, E_\omega[\mathbf{d}(\omega)])\} = g(\bar{\mathbf{x}}, \mu) \tag{13.21}$$

where $\mu = E_\omega[\mathbf{d}(\omega)]$.

We generate N independent random samples $\mathbf{d}^1, \ldots, \mathbf{d}^N$ of $\mathbf{d}(\omega)$. For each $\mathbf{d}^k$, $k = 1, \ldots, N$, we compute

$$v^k = g(\bar{\mathbf{x}}, \mathbf{d}^k) \tag{13.22}$$

where $g(\bar{\mathbf{x}}, \mathbf{d}^k) \equiv \mathbf{c}^T\bar{\mathbf{x}} + \min_{y \geq 0}\{\mathbf{q}^T\mathbf{y} | \mathbf{W}\mathbf{y} = \mathbf{d}^k - \mathbf{A}_L\bar{\mathbf{x}}\}$. The expectation of the EV (EEV) can be estimated by obtaining the sample average of $v^k = g(\bar{\mathbf{x}}, \mathbf{d}^k)$ over all the sampled scenarios, that is,

$$E[\hat{v}_E] = N^{-1}\sum_{k=1}^{N} g(\bar{\mathbf{x}}, \mathbf{d}^k) \tag{13.23}$$

Usually, if the difference

$$EEV - EV = E[\hat{v}_E] - v_E \tag{13.24}$$

is small, EEV is said to be a reasonably good solution to the stochastic program.[18] Furthermore, the value of the approximated stochastic solution (VSS) is computed by

$$VSS = EEV - ESS = E[\hat{v}_E] - E[\hat{v}_N] \tag{13.25}$$

VSS in (13.25) indicates the price of using a naive EV model instead of SS.

13.6 MULTICOMMODITY EMPTY CONTAINER REUSE PROBLEM

In this section, we investigate the possibility of fulfilling the request of one type of containers with another type in an effort to further reduce the cost of empty container reusing. This problem is referred to as the multicommodity empty container substitution problem. For the sake of simplicity, in this section, we consider the deterministic, single-period multicommodity empty container substitution problem. In other words, we assume that all the information regarding the number of supply and demand of empties over the entire single planning horizon is fully available in advance.

13.6.1 TYPES OF CONTAINERS AND SUBSTITUTION RULES

Containers can be classified into different types (classes) according to their intended use, external dimension, ownership, and so on. For instance, if the types of containers can be determined by only three attributes—purpose, dimensions, and ownership—a type t container can be expressed as

$$\text{type } t = \{\text{purpose, dimension, ownership}\}$$

The purpose indicates the intended use of containers such as general-purpose (dry cargo) or specific-purpose (refrigerated, hazmat, etc.) containers. Most containers are sized according to the International Standards Organization (ISO). Based on ISO, containers are described in terms of 20-foot equivalent units (TEUs) in order to facilitate comparison of one container system with another. A TEU is an 8-foot-wide, 8.5-foot-high, and 20-foot-long container. An FEU is an 8-foot high, 40-foot long container and is equivalent to two TEUs. Containers with a height of 9.5 feet are usually referred to as high-cube containers. The most widely used containers are general-purpose FEU containers. Without loss of generality, in the rest of this section, we will only consider dry cargo containers with the following standard dimensions:

$$D_1: 40' \times 8' \times 8.5', \quad D_2: 20' \times 8' \times 8.5', \quad D_3: 40' \times 8' \times 9.5' \qquad (13.26)$$

13.6.2 SUBSTITUTION RULES

The substitution conforms to the certain rules specified for substituting each ordered pair of container types. These rules may be symmetric or asymmetric. For instance, suppose that there are three types of containers:

$$t_1 = \{\text{General, } D_1, \text{Hanjin}\}, \ t_2 = \{\text{General, } D_2, \text{Hyundai}\}, \ t_3 = \{\text{General, } D_3, \text{Maersk}\}$$

$$(13.27)$$

The possible asymmetric container substitutions between t_1, t_2, and t_3 could be as follows:

Asymmetric in type for (t_1, t_3): One request for t_3 could be fulfilled by one supply of t_1, but the reverse substitution is not permitted. This case may happen when certain customers do not accept the high-cube containers t_3 due to their facility limitations.

Asymmetric in number for (t_2, t_1): One request for t_1 can be fulfilled by two supplies of t_2. However, sometimes two requests for t_2 can only be satisfied with two supplies of t_1. This case happens when export cargos in a shipper location have two different destinations or when it is desirable to have two small containers because of the weight limitation.

Generally speaking, the substitution rules come from differences in the handling capacity of the loading/unloading facilities, the destination of cargos, the weight of cargos, or the nature of cargos. Other factors may include operational regulations or limitations set forth by freight liners or trucking companies.

13.6.3 PROBLEM MODELING AND SOLUTION METHODS

The structural model for the multicommodity empty container reuse problem is represented in figure 13.4. Similar to the interconnecting flow variables, dummy sets of demand nodes are introduced to represent substitution flow variables. These spawned sets of demand nodes complicate the problem structure and deteriorate the running time of a solution procedure. However, they provide a scalable structure to deal with substitution between multiple commodity types.

The multicommodity empty container reuse problem can be modeled as follows:

$$\text{Minimize} \quad \sum_{t \in T}\sum_{i=1}^{m+p}\sum_{j=1}^{n+p} c_{ij}^{t} x_{ij}^{t} + \sum_{t \in T}\sum_{i=1}^{m+p}\sum_{k=1}^{\bar{n}} c_{ik}^{t} x_{ik}^{t} \tag{13.28}$$

$$\text{subject to} \quad \sum_{j=1}^{n+p} x_{ij}^{t} + \sum_{k=1}^{\bar{n}} x_{ik}^{t} = s_{i}^{t}, \qquad \forall\, i \in I \bigcup P,\ t \in T \tag{13.29}$$

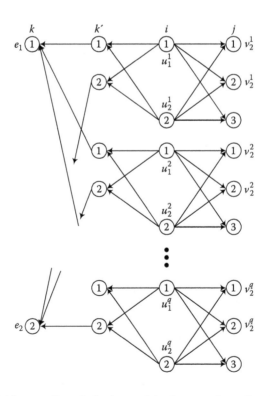

FIGURE 13.4 Multicommodity substitution model, where $m = 1$, $n = 2$, and $p = 1$

$$\sum_{i=1}^{m+p} x_{ij}^t = d_j^t, \qquad \forall\, j \in J,\, t \in T \qquad\qquad (13.30)$$

$$\sum_{t\in T}\sum_{i=1}^{m+p} r_{ik}^t x_{ik}^t = e_k, \qquad \forall\, k \in \bar{J} \qquad\qquad (13.31)$$

$$x_{ij}^t,\ x_{ik}^t \geq 0,\ \text{integer}, \qquad \forall\, i,\, j,\, k,\, t \qquad\qquad (13.32)$$

where

T = the set of container types, $|T| = q$,

$\bar{J}$ = the subset of shippers that allow substitution, $\bar{J} \subseteq J$ and $|\bar{J}| = \bar{n}$,

c_{ik}^t = the cost of transporting a type t container from supply node $i \in I \cup P$ to demand node $k \in \bar{J}$,

x_{ik}^t = the number of type t containers transported from supply node $i \in I \cup P$ to demand node $k \in \bar{J}$,

e_k = the sum of extra requests in demand node $k \in \bar{J}$, which is unspecified by a certain type.

Constraints (13.29) specify that the supply s_i^t can be shipped to satisfy both the exact type of requests from real nodes and the extra requests from dummy nodes. Constraints (13.30) indicate that the exact requests should be met. Finally, constraints (13.31) indicate that the extra requests should be met without violating the substitution rule constraints. In this multicommodity substitution problem, the overall demand on a demand node $k \in \bar{J}$ can be expressed as

$$d_k = \left(d_k^1;\ d_k^2;\ldots;\ d_k^q;\ e_k \right).$$

Therefore, the substitution coefficient r_{ik}^t represents the number of containers of type t to satisfy one extra request in demand node $k \in \bar{J}$.

Chang et al.[20] investigated the multicommodity empty container reuse problem that integrates substitutions of container types. They developed a decomposition methodology that divides the problem into dependent and independent parts and applies a branch-and-bound (B&B) method to the dependent part. The methodology yields a suboptimal solution relatively fast.

Chang et al.[20] also tested their developed methodology against the current and projected data obtained from the combined Los Angles and Long Beach port area in the United States.[19] The authors concluded that when the substitutions between different types of containers are allowed, additional cost reductions (compared to Jula et al.[19]) ranging from 4 to 47% can be obtained. Furthermore, by allowing substitution, the number of empty related truck trips to and from the ports can be reduced substantially.

13.7 CONCLUSION

In this chapter, the empty container reuse concept, in which the emptied import containers are recycled for export loads without first being returned to marine terminals, was studied. The two major methodologies for reusing, depot-direct and street-turn, were discussed in detail, and pros and cons of each methodology were outlined. Furthermore, four different variants of the empty container reuse problem were investigated, and each variant was modeled analytically and optimization techniques were reviewed and discussed.

It was found that by implementing the empty container reuse idea a substantial reduction in the empty related truck trips to and from the container ports as well as in the empty container moving related costs can be obtained. Moreover, by allowing substitution between different types of empty containers, we can further reduce the number of the truck trips and decrease the empty related costs. This indicates that by implementing the empty container reuse idea the traffic congestion around the container terminals can be improved, and as a consequence, noise and emissions can be reduced significantly.

ACKNOWLEDGMENTS

This work is supported by the National Center for Metropolitan Research (METRANS) located at the University of Southern California (USC) and the California State University at Long Beach (CSULB).

REFERENCES

1. Mallon, L. G., and Magaddino, J. P. 2001. *An integrated approach to managing local container traffic growth in the Long Beach–Los Angeles Port Complex, Phase II.* METRANS technical report 00-17. California, United States.
2. The Tioga Group. 2002. Empty ocean logistics study. Technical report, submitted to the Gateway Cities Council of Governments. California, United States.
3. Hanh, L. D. 2003. *The logistics of empty cargo containers in the Southern California region.* Technical report, Metrans Report.
4. Dejax, P. J., and Crainic, T. G. 1987. A review of empty flows and fleet management models in freight transportation. *Transportation Science* 21:227–47.
5. Crainic, T. G., Gendreau, M., and Dejax, P. 1993. Dynamic and stochastic models for the allocation of empty containers. *Operations Research* 41:102–26.
6. Crainic, T. G., Dejax, P., and Delorme, L. 1989. Models for multimode multicommodity location problems with interdepot balancing requirements. *Annals of Operations Research* 18:279–302.
7. Crainic, T. G., Gendreau, M., Soriano, P., and Toulouse, M. 1993. A tabu search procedure for multicommodity location/allocation with balancing requirements. *Annals of Operations Research* 41:359–83.
8. Crainic, T. G., and Delorme, L. 1993. Dual-ascent procedures for multicommodity location-allocation problems with balancing requirements. *Transportation Science* 27:90–101.
9. Crainic, T. G., Delorme, L., and Dejax, P. 1993. A branch-and-bound for multicommodity location with balancing requirements. *European Journal of Operations Research* 65:368–82.
10. Cheung, R. K., and Chen, C. Y. 1998. A two-stage stochastic network model and solution methods for the dynamic empty container allocation problem. *Transportation Science* 32:142–62.
11. Choong, S. T., Cole, M. H., and Kutanoglu, E. 2002. Empty container management for intermodal transportation networks. *Transportation Research* 38E:423–38.

12. Li, J. A., Liu, K., Leung, S. C. H., and Lai, K. K. 2004. Empty container management in a port with long-run average criterion. *Mathematical and Computer Modeling* 40:85–100.
13. Abrache, J., Crainic, T. G., and Gendreau, M. 1999. *A new decomposition algorithm for the deterministic dynamic allocation of empty containers.* CRT-99-49, University of Montreal.
14. Boile, M. P. 2006. *Empty intermodal container management.* Technical report FHWA-NJ-2006-005, Federal Highway Administration.
15. Dantzig, G. B., and Thapa, M. N. 1997. *Linear programming 2: Theory and extensions.* New York: Springer.
16. Bertsimas, D., and Tsitsiklis, J. N. 1997. *Introduction to linear optimization.* Belmont, MA: Athena Scientific.
17. Linderoth, J., Shapiro, A., and Wright, S. 2002. *The empirical behavior of sampling methods for stochastic programming.* Optimization technical report 02-01, University of Wisconsin–Madison.
18. Birge, J. R., and Louveaux, F. 1997. *Introduction to stochastic programming.* Springer Series in Operations Research and Financial Engineering. New York: Springer.
19. Jula, H., Chassiakos, A., and Ioannou, P. 2006. Port dynamic empty container reuse. *Transportation Research* 42E:43–60.
20. Chang, H., Jula, H., Chassiakos, A., and Ioannou, P. 2008. A heuristic solution for the empty container substitution problem. *Transportation Research E* 44E:203–16.

14 Port Labor
The Effects of Competition, Devolution, Labor Reform, Trade, and Technology

Kristen Monaco, Christine Ann Mulcahy, and Lindy Helfman

CONTENTS

ABSTRACT

Containerization facilitated increased globalization, competition between ports, and port devolution. These concurrent changes had deleterious impacts on port labor worldwide. In this chapter we provide a cross-country analysis of how labor was affected by changes in the structure of water transportation, using a case study of West Coast port workers to contrast the outcomes under different economic, legal, and structural regimes. We conclude with a discussion of current issues facing European and Asian ports and hypothesize their potential impacts on labor markets.

14.1 INTRODUCTION

Trade liberalization and economic growth have led to ever-increasing reliance on ocean transportation to facilitate commerce. This, in turn, has led to changes in the structure of ocean carriers, terminal operators, and ports themselves. Whereas port operations and longshore labor conditions were fairly similar across developed countries in the late 1800s and early 1900s, key differences in port governance, national economic performance, and labor bargaining power have led to radically different outcomes for port labor across developed and developing countries. This chapter provides a comparison of port labor conditions across countries, emphasizing how the evolution of the water transport industry and economic performance across nations have transformed labor markets, to the benefit of workers in some countries (for example, the West Coast of the United States) and the detriment of others. As has been pointed out in prior studies, it is critical to examine interconnectedness of technology, trade, geography, government, and labor-management relations when comparing the cross-country labor outcomes.

This chapter starts with a brief overview of international port rankings and changes in the water transport internationally to provide context for the cross-country labor comparison. We then analyze the evolution of labor demand and earnings of workers at major ports, with a focus on market and government forces that have shaped variations in labor outcomes. The current trends and outlooks for labor in Asian and European ports are described. Though dockworkers are the focus of much labor research in water transport, seafarers are also integral to water transportation. We conclude with a brief description of the labor market conditions of these workers.

14.2 PORT RANKINGS AND PORT GOVERNANCE

Currently, Asian ports dominate water transport, measured by container volume. Of the top fifteen container ports in 2004, nine were located in Asia, three in the United States, and three in Europe (see table 14.1).

These trends in port rankings reflect many things, including the growth of the Chinese manufacturing centers and the importance of West Coast ports in U.S. trans-Pacific trade. The competition among ports varies, primarily due to geography. As the ports of Hong Kong, Singapore, and Busan handle a great deal of transshipping, they compete with one another and potential competitors (Japan, Malaysia) that seek to increase their share of transshipping volume. Given the considerable competition among these ports, there is pressure for ports and terminal operators to invest in productivity-enhancing capital. Likewise, the ports of Europe (Rotterdam, Hamburg, the UK ports) are in sufficiently close proximity to one another that ships could be diverted from one to another.

Transshipping is nearly nonexistent in the United States, and the competition between ports is less marked due to infrastructure and geography. For example, though Long Beach and Los Angeles are competitors, the two ports are located adjacent to one another and have cooperated on some programs, including their new environmental initiatives. These two ports have a major advantage in attracting trans-Pacific trade, as they have infrastructure that can accommodate the largest ships in operation. New York and New Jersey are at a disadvantage in attracting this trade due

TABLE 14.1
2004 Container Port Rankings (TEU Measured in Thousands)

Rank	Port	Country	TEUs
1	Hong Kong	China	21,984
2	Singapore	Singapore	21,329
3	Shanghai	China	14,557
4	Shenzhen	China	13,615
5	Busan	South Korea	11,430
6	Kaohsiung	Taiwan	9,714
7	Rotterdam	Netherlands	8,281
8	Los Angeles	United States	7,321
9	Hamburg	Germany	7,003
10	Dubai	United Arab Emirates	6,429
11	Antwerp	Belgium	6,064
12	Long Beach	United States	5,780
13	Port Kalang	Malaysia	5,244
14	Quingdao	China	5,140
15	New York/New Jersey	United States	4,478

to the constraint on ship size moving through the Panama Canal and the accompanying increase in transport time to send Asian goods to the East Coast of the United States using all-water transport.

One component of labor demand, and therefore labor bargaining power, is the traffic entering the port; however, the positive effect of increased trade on labor may be mitigated if traffic could be easily diverted to another port (which is a function of infrastructure and geographical proximity). The relative advantage of West Coast longshoremen in the United States versus those in Asia or Central Europe is caused in part by the lack of alternatives for trans-Pacific trade entering the United States.

Port governance is often categorized by four models. The service port is a fully public port, where the port provides all capital and services. The tool port is characterized by a public port authority that is responsible for providing capital to the private firms that service the stevedoring activities at the port. Landlord ports are ports where the government owns the land and might provide some limited capital, but private stevedoring companies are responsible for the bulk of goods movement capital. Most container ports in the United States meet this latter model, where the ports provide some dredging and other infrastructure; however, the terminal operators invest in cranes, yard hostlers, and other capital required for their operations. Finally, in a fully privatized port model, private firms own the port complex and provide all services. An example would be the ports of Great Britain.[2] Though there has been a general trend in ports moving from primarily service ports to landlord or private ports, many ports still do not fall neatly into one of the four categories above.

There were 112 port privatization projects undertaken between 1990 and 1998 involving U.S.$9 billion in total investment.[1] Though there has been a general move to devolution, this has not been a smooth transition for many countries. Devolution need not mean strict privatization. Indeed, full privatization where the private company owns the land on which the port operates is rare and may not be an optimal choice for a country. For example, in the case of the United Kingdom, it is estimated that the land was undervalued by approximately 50–75% due to lack of prior sales of this nature.[2,3]

Rather than full privatization, many countries opt for a hybrid approach, including commercialization, build operate transfer (BOT), or joint venture. Under commercialization, the government is still responsible for port operations, but is made more accountable for port performance and may issue concessions for stevedoring services. An example of this would be changes in Morocco in 1984. Joint venture activities include the ports of Shanghai and Saigon.[3] In Shanghai, Hutchison Ports (a Hong Kong–based terminal operator) took over terminal operations in 1993. The first movers in privatization were Kingston Port, Jamaica, in1967, Port Klang, Malaysia, in 1986, and Manila Harbor, Philippines, in 1988. Ports in Colombia and Mexico are run as concessions.[1]

As ports move to privatization, there is increased pressure to cut costs, which may not be felt by service ports. This generally leads to decreased labor demand and loss of earnings or employment for workers. The World Bank encourages ports, especially ports in developing nations, to enact port reforms and move toward privatization, as is documented in their *Port Reform Took Kit*.[4] The World Bank recommends that governments consider how displaced workers will be handled. In many cases, this requires governments themselves to pay for lost income. For example, the government of Chile spent U.S.$30 million in severance for displaced workers in 1981; however, it is estimated that the country saved $96 million in 1996 due to port reform.[3,4] It is estimated that reform cost $50 million in Mozambique, $50 million in Colombia, and $182 million in Venezuela, where there were approximately 13,000 dockworkers laid off.[4]

Generally, the move to a for-profit model of port operations requires strategic planning and is made more successful if preceded by the implementation of port reform authorities, as seen in Thailand and Australia. It is notable that the World Bank recommends that port reform be accompanied by provisions for increased competition between terminal operators. Colombia, for example, specifically issued concessions that would result in competition between ports and stevedoring companies.

14.3 SHIPPING LINES AND TERMINAL OPERATORS

The move to privatization puts more power in the hands of stevedoring companies—a sector that has undergone considerable consolidation as a result of many factors, including devolution. In this section, we briefly describe the role of ocean shipping lines and stevedores/terminal operators.

We rank the top ten ocean carriers below by their share of imports to the United States in 2005 (see table 14.2).

Ocean carriers have been characterized in the past by conferences that set rates and capacity along trade lanes. As countries have moved to deregulate ocean carriers, these conferences have largely been replaced by alliances that serve to share information and perhaps capacity. The composition of these alliances has changed over the past 12 years, reflecting the great deal of consolidation that has occurred in the industry. (Tables 14.3 and 14.4 present the changes in alliances between 1994 and 2006.)

TABLE 14.2
Ranking of Ocean Carriers by Share of U.S. Import Traffic, 2005

Shipping Line	Country Headquartered	Share of U.S. Imports
Maersk	Denmark	15.1%
Evergreen	Taiwan	8.6%
Hanjin	Korea	6.6%
APL	Singapore	6.4%
Mediterranean Shipping	Switzerland	6.3%
Hapag-Lloyd	Germany	5.2%
China Shipping	China	4.6%
NYK Line	Japan	4.6%
Cosco	China	4.4%
OOCL	Hong Kong	4.1%

An example of the degree of consolidation activity is the mergers of Maersk and Sealand, and P&O and Nedlloyd. Subsequently, Maersk Sealand purchased P&O Nedllyod, resulting in considerable market power for Maersk.

The emergence of "pure" stevedoring companies that specialized in operations across ports occurred subsequent to the emergence of containerization in the 1960s. Many of these companies have increased their port operations business since then, but stay primarily specialized in stevedoring. These include (headquarter country in parentheses) Hutchison Ports Holdings (Hong Kong), SSA (United States), Dubai Ports World (Dubai, which recently purchased P&O Ports, United Kingdom), and Port of Singapore Authority (Singapore). It is notable that these stevedoring companies have operations in a number of countries.

The vertical integration of shipping lines into stevedoring companies evolved primarily after the development of post-Panamax ships (ships that are too wide to fit through the Panama Canal). Some examples of vertically integrated shipping lines and stevedores include Maersk and APM, Evergreen, K-Line, and Hanjin. These "integrated global carriers" benefit primarily through improved coordination between

TABLE 14.3
Shipping Alliances, 1994

Alliance	Members
Global Alliance	APL, Mitsui, OOCL, Nedlloyd
Grand Alliance	NYK Line, Hapag-Lloyd, Neptune Orient Line, P&O Line
Hanjin/Tricon	Hanjin, DSR-Senator, Cho Yang
Maersk/Sealand	Maersk, Sealand
Unnamed	"K" Line, COSCO, Yang Ming

TABLE 14.4
Shipping Alliances, 2006

Alliance	Members
Grand Alliance	OOCL, NYK, Hapag-Lloyd
New World Alliance	APL, Mitsui OSK Lines, Hyundai
CKYH	COSCO, "K" Line, Yang Ming, Hanjin

ship and landside operations and guaranteed infrastructure and capacity in terminals.[5] While it has been argued that privatization has led to increased concentration among stevedoring companies, it is clear that containerization also contributed to the concentration of ocean shipping lines and, in turn, to the privatization movement. We turn next to a discussion of the industry before and after containerization.

14.4 IMPACT OF CONTAINERIZATION

Prior to containerization, there was no standardized method for handling freight stowed on ships. The handling of break-bulk freight was labor intensive. "Gangs" of workers loaded and unloaded freight from the bodies of ships, which was ultimately assisted by cranes and "slings" that were loaded on the dock and then unloaded on the ship. Freight entering or exiting the port was typically handled both on the ship and at warehouses and other facilities adjacent to the docks. The jobs of dockworkers were similar across countries—arduous, potentially unsafe work that paid well, but was irregular, resulting in considerable uncertainty in earnings.

Ships typically stayed in port until fully loaded to take advantage of economies of scale, leading to irregular ship schedules. This variability of work led to potential corruption on the docks, with workers bribing employers for work when the ship was in port. For example, in Shanghai, workers often paid stevedoring companies to receive work assignments.[6] Often workers were required to "shape up" multiple times per day to get assigned work.

Firms benefited from having a pool of "casual" available labor to service the ships; however, this resulted in income uncertainty for dockworkers. Workers pushed for methods of restricting labor supply (through registering workers) and a predictable way of allocating work when it was available (through worker control of hiring halls). In many countries this led to unionization of workers at the ports, with unions achieving control of the hiring hall in several countries, while in others the hiring hall was jointly run by labor and management.

These unions fought for jurisdiction over hiring and registration of labor as well as work rules that ostensibly were designed to protect labor. Examples include spelling, continuity, fixed manning, and no transfer. Under spelling, workers would alternate performing tasks (for example, 2 hours working, followed by 2 hours off). Continuity required that gangs not be altered until work was completed. Fixed manning (which still exists today in many ports) fixed a minimum gang size. No-transfer provisions meant that workers could not be switched between tasks. In Venezuela workers would become registered upon working 60 consecutive days, and 200 workers were allocated to a ship, regardless of the actual need for labor. In Puerto Limon, Costa Rica, two gangs of thirty-two men each were assigned to each ship.[7] Clearly there was a tension between these work rules and employers' desire for increased productivity.

Additionally, the rotation of labor led to less attachment between workers and a particular stevedoring firm, which was to the advantage of workers, as it made it more difficult for firms to discipline them.

The concept of containerization was developed by Malcolm McLean in the 1950s. He saw the potential for standardized shipping with improved efficiency due to reduced labor demand and standardized freight handling. Though containerization began in the 1960s, it did not emerge as the primary form of ocean transportation until the 1980s. Containerization radically redesigned the business model of ocean shipping, as the industry moved from labor intensive to capital intensive. Containers were ultimately standardized, with 20-foot (TEU) and forty-foot (FEU) equivalent unit containers used most frequently. Containers were loaded to and unloaded from ships using gantry cranes (rather than the original model of sling loads). Yard hostlers were used to move containers around the terminal facilities, and containers could be taken to their destination via rail or truck without the freight being touched. The combination of new capital technology to move freight and the reduction of freight handling at the ports led to a marked decrease in the demand for labor. Additionally, the existent work rules were particularly ill-suited to containerized freight. Gang sizes would need to be reduced, and while firms preferred casual labor when handling break-bulk freight, they demanded regular labor for containerized cargo. The reason for the latter was twofold: first, containerization and the concurrent increase in waterborne trade led to more regular service than in the past; second, firms preferred a workforce that had been trained to use the firm's capital. Generally, the use of more steady labor has been paid for by employers through guaranteed income programs. Following containerization, some form of a guaranteed income program was established in the United States, the Netherlands, the United Kingdom, Belgium, Australia, and New Zealand. In Belgium and the Netherlands, this was jointly funded by employers and the state (though in the Netherlands, the state suspended payments in 1995). In the balance of the countries, the programs were funded by the employers (and, in some cases, directly by shippers).[8]

The ability of labor to effectively insulate themselves from this change was a function of several variables. West Coast longshoremen in the United States are typically thought to have had the confluence of conditions that led to an advantage for these workers. We will present a brief case study of the West Coast longshore workers and then contrast this to the outcomes experienced by workers in other countries and on the East Coast of the United States.

14.5 LABOR ON THE WEST COAST: SUCCESS IN THE FACE OF CONTAINERIZATION

In 1902 the West Coast unions joined with those on the East Coast (the International Longshore Association [ILA]). Except for a strike in 1916, local unions struck alone, reducing the power of labor as freight could be diverted to nearby ports. There were few work rules, employers controlled hiring, and bribing for dispatch was common.

The local unions of the West Coast joined forces in 1933 and obtained a charter from the ILA as a single unit. Negotiations ensued between the Pacific Coast ILA and the employers (Waterfront Employers' Union [WEU]), with the union demanding a coast-wide agreement and union-run hiring hall in each port.[9] Unable to reach an agreement with the WEU, the workers went on strike, which ultimately was settled by the National Longshoremen's Board. The new contract included a joint employer-union hiring hall with a dispatcher selected by the union and provisions for overtime pay after 6 hours of work. The union also implemented "low man out" dispatching, the practice of giving the worker with the fewest accumulated hours of work the first opportunity at a job, intended to provide a stable level of income for all longshoremen. The effect was that by 1937 "casuals" (workers who were not full-time) were a supplementary workforce. During this period, the International Longshore and Warehouse Union (ILWU) split from the ILA and Harry Bridges became its first president and presided over the ILWU from its inception in 1937 until his retirement in 1977.[10]

On September 2, 1948, a strike between the ILWU and employers broke out and ended after employers agreed to accept the union-controlled dispatch hall and low-man-out dispatching in return for the union's acceptance of a no-strike, no-lock-out, no-work-stoppage clause, except when health and safety were at risk.[11] In effect, the employers gave up control over hiring and dispatching, and in return there were fewer work stoppages.

The union recognized early that containerization would lead to problems for labor. In 1956 the Longshore Caucus created a committee to examine potential problems, which resulted in the Coast Committee Report (October 1957), which advocated union flexibility in negotiations with the employers (the Pacific Maritime Association [PMA]), since "acting otherwise could provoke an already frustrated employer into a fight the union might find difficult to win."[12]

Negotiations between the two parties concerning technology were finalized on October 18, 1960. The agreement can be summarized on two fronts: mechanization (employers could introduce new technology and the union would share in the returns from this technology) and modernization (specification of work rules)—M&M. A fund was created and used to guarantee longshoremen income equivalent to 35 hours of work per week and provide pensions for voluntary early retirement.[12] This period was marked by cooperation between the ILWU and the PMA (the employers association) with no coast-wide strikes from 1948 to 1971.

In 1966, the M&M agreement and contract were renewed for 5 years (the new pension agreement would be unchanged for 10 years), but a few amendments were made. The wage guarantee fund (amounting to two-fifths of the total employer contribution fund, or $13 million) would be distributed to eligible registered men. The PMA won the right to "steady men," skilled men that they could employ on a monthly basis.

It is important to note that in the 1950s the PMA's approach to mechanization was to share the productivity gains with the union, with the understanding that gains would be split 50–50. By 1966 this had changed to a model where the PMA "bought out" work rules to ensure increased productivity, which undoubtedly saved it considerably as the amount of containerized freight increased substantially.[11] In fact, the payout total for early retirement and loss of hours was $29 million, while the estimated productivity gains were estimated at $120 million.[13]

Between 1960 and 1980, the number of registered longshore workers dropped from 14,500 to 8,400.[10] During the same period, productivity, measured in tons per hour, increased from 0.837 to 5.498, and the cost per ton decreased from $4.94 to $3.60 despite a fourfold increase in the longshore hourly wage rate.[14] Despite these gains, there are still considerable inefficiencies in the West Coast system. First, workers were still dispatched daily from the hiring hall, which the PMA estimated reduced productivity 7% per year.[15] Second, technology that would reduce the need for some manual data entry has been resisted by the ILWU since it would reduce the need for clerk labor. The employers and the union have agreed to phase in these technological changes in a way that would not displace clerks.

The ability of the ILWU to sustain high incomes for its workers and stable employment since the initial displacement following containerization stems from several sources. First, starting in the 1930s, the local unions along the West Coast decided to bargain as one unit. This led to an advantage for labor since employers (who also cooperated with one another) could not simply divert freight from one port to another during a dispute. Second, rather than simply resisting the impending changes due to containerization, Harry Bridges, the long-time union leader, realizing the inevitability of containerization as well as the potential disastrous effects on labor, decided on a strategy to dealing with containerization in a way that could benefit union members. This strategy included establishing an employer-funded trust for workers and reducing the register of longshoremen by encouraging early retirement. Third, the increase in trade between Asia and the United States led to an increased demand for water transport workers, which favored the West Coast, due to geographical proximity and, ultimately, the fact that port-Panamax ships reduced the ability to divert the freight to the East Coast without adding considerably to transit time. Fourth, the landlord model of ports in the United States meant that private firms were competing with each other prior to the rise in containerization; therefore, there was not a double shock to labor from containerization and privatization.

Generally, the impact of containerization on labor is a function of union strength, port structure, trade flows into a port, competition between ports, and a nation's legal structure.

14.6 INTERACTION OF UNION STRENGTH, PORT STRUCTURE, AND PORT COMPETITION

Union power is dependent on the nation's legal structure and the ability of employers to undermine worker cohesion. Though both the ILA union on the East Cost of the United States and the ILWU on the West Coast face a similar national labor relations climate, the ILA was undermined by the corruption of its leadership, which ultimately made the union less effective. In New York, the union lost the right to dispatch workers when the Waterfront Commission assumed control of the hiring hall. A weakened ILA was unable to secure a coast-wide labor agreement, which meant that workers at different ports were in competition with one another—freight from New York could be diverted to Hampton Roads in the case of a labor dispute.

In New Zealand, the longshoremen's union that represented all workers was abolished in 1951 and replaced by twenty-six unions bargaining at the port level, reducing the power of labor.[8] This power was further reduced when its labor commission (Waterfront Industry Commission [WIC]) was eliminated in 1989, and New Zealand eliminated some strike actions. A similar situation arose in France, whose Les Bureaux Centraux de la Main-D'Oeuvre Portuaire (BCMO) labor commission was abolished in 1989. Further hampering the bargaining power in these two countries is the fact that in France and New Zealand crane operators were employed by the port authorities and dockworkers by stevedores.[16]

In the Port of Rotterdam, there was one association of employers (the SVZ, founded in 1908) that bargained with several different longshore unions.[6] The centralization of the management unit and decentralization of the labor units gave more bargaining power to the employers. The employer-run hiring hall led to considerable variations in work hours among longshoremen, unlike West Coast workers, who had more homogeneous hours and earnings. Finally, in 1960 the European Container Terminus argued that it should be able to hire its own specialized workers to handle its freight and should not have to accept workers from the general pool. The union acquiesced, and the terminal hired its own workers. It is estimated that containerization caused the loss of 7,000 jobs at the Port of Rotterdam (falling from 17,000 in 1966 to 10,000 in 1985).[6]

Much like the Gulf ports of the United States, where the ILA has been weakened due to the existence of nonunion terminals, labor at the Port of Genoa was undermined by the ability of shipping lines to divert freight to the port at La Spezia, which was privatized and did not have unionized labor.[6]

Under the National Dock Labor Scheme (NDLS) in the United Kingdom, dockworkers in Liverpool, London, and Southampton were employed by the ports authorities. The decrease in labor demand due to containerization was accommodated through both a severance program and a program that distributed labor across ports. As this led to increased costs of operating these ports, ports that were not included in the NDLS became more desirable for the shipping lines, and thus traffic was diverted to Dover and Felixstowe. Felixstowe, omitted from the NDLS due to its small size, actually grew significantly as a result.[8,16]

Changes in the containerized freight business undoubtedly led to devolution in many ports, not necessarily as a direct effect of containerization, but as a result of the technology made possible by the use of containers. Shipping lines were able to realize economies of scale from using ever-larger ships. Introducing large ships quickly adds capacity to the network, which should drive shipping rates down. Thus, shipping lines are forced to keep costs down. In turn, terminal operators are required to increase capital investment to accommodate these large ships and speed loading and unloading time. This requires terminal operators to keep labor costs down, which is made more difficult under a service port model, thus pressuring service ports to move to a landlord model or outright privatization.[16,17]

Privatization in Chile was preceded by a law eliminating dockworkers' "right to work," decreasing labor power.[7] Privatization at Port Klang in Malaysia in 1980 was also preceded by changes in national law that made it more difficult for unions to collectively bargain. Workers could join "in house" unions at their firms, but not national unions. Following the privatization of the port, dockworkers were offered jobs at the private firm and guaranteed 5 years of employment.[18] The fairly favorable terms for labor, somewhat surprising given the lack of union power, were largely obtained as a result of the revenue generated by an increase of Malaysian trade concurrent with privatization.[3] Even in the face of ample trade volumes, devolution may not lead to favorable outcomes for workers. In Taiwan, workers at the Kaohsiung International Shipowners Association had earnings cut by 40%, a number of workers were laid off, and income guarantees were cut.[16] Wages were also cut when Hutchison Ports took over operations in Felixstowe (United Kingdom) and in Panama.

14.7 CURRENT ISSUES IN PORT LABOR: EUROPE AND ASIA

14.7.1 EUROPEAN UNION

In early 2006, the European Parliament rejected the European Commission's second proposal for liberalizing Europe's ports by a count of 532 to 120. One key facet of the proposal was the "ports package," a directive on market access to port services, aimed at imposing conventions for the management of the ports, differentiating the regulatory and operational dealings of the port authority, and allowing for competition in the supply of port services. The principle of self-handling, which would allow vessel operators to hire their own labor to load and unload cargo, was opposed by the trade unions. Moreover, some port operators also disapproved of the directive, arguing that the regulatory burden was too great.[19]

Proponents of the proposal argue that liberalization will result in a higher level of operating efficiency at the ports. The industry coalition and the commission consider current port operations to be monopolistic in nature; therefore, liberalizing port operations would result in increased competition that could potentially save importers and exporters millions of euros.[20] Despite this defeat, there is a movement in the EU to develop more business-friendly policies, particularly in the area of flexible labor markets, in hopes of strengthening the job sector and promoting economic growth.[21]

The unions opposed the commission's proposal and conducted widespread disruptions at some of Europe's largest ports. Well-orchestrated strikes were conducted across Europe, with as many as 40,000 participants from 12 countries turning out, including 2,000 to 3,000 workers from Germany's Port of Hamburg.[21] The main motivation for the strong opposition to market liberalization reforms by dockworkers originates from the likely scenario that shipping companies would hire cheaper, nonunionized labor to load and unload cargo rather than having to hire certified union workers, as the current law stipulates.[22]

If the legislation were to pass, the increase in the supply of available workers would likely result in either large job losses of unionized labor or significant reductions in the wages companies would be willing to pay dockworkers. The dockworkers also argue that liberalization will ultimately lower safety standards, which the commission refutes with a provision stating that safety and training standards would not change under such a proposal. Moreover, equipment owners might be hesitant to allow noncertified port workers to use the highly technical and expensive equipment required to load and unload cargo.[22] Although the unions were able to rally against this directive, their strength against future market-liberalizing legislation may be questionable, as unions have witnessed their influence on EU policies diminish over recent years.[19]

14.7.2 SOUTHEAST ASIA

In Asia, the rapid economic and export growth of China has spurred other Asian ports to lower costs in order to compete. Hong Kong, Taiwanese, South Korean, and Japanese ports are taking steps to differentiate their services to gain market share. The minister of the Chinese Ministry of Communications, Li Shenglin, announced a plan to build five big port clusters within the next 5 years. The expansion will increase China's throughput capacity to 7.5 billion tons, with 115 million TEUs in container volumes.[23]

With China growing as rapidly as it is, Hong Kong, which once held the title as the "port of entry for capital" into China, must find a way to compete. The average wage for unskilled labor in Hong Kong is $1,000 a month, whereas in the Guangdong province of China, workers earn $135 a month. Furthermore, in 2003, Hong Kong's port was charging the highest handling fees in the world, compared with the Chinese ports of Shenzhen, located only 20 miles away, charging 25% less, or Shanghai, whose fees are 75% lower but is over 700 miles away.[24]

In nearby Taiwan, the Kaohsiung Harbour Bureau is working on a plan to expand the busiest Taiwanese port. The deputy director of the bureau, Teng Yu, reports that growth in container volumes is about 10% a year, and the medium-term plan to build four additional berths by 2009 will help meet the future demand. The Taiwanese port can handle about 9.9 million TEUs annually, of which half is estimated to be in transit. This plan is expected to cost approximately $262 million. In 2002, Shanghai jumped pass Kaohsiung's port to become the fourth busiest in the world. Because of the competition from China, the Taiwanese government passed a Free Trade Harbour Zone Statute (FTHZS), which "features favorable tariff rates, free flow of products, processing for export and a one-stop administrative process."[25]

In 2005, the Japanese government announced that it would reduce the cost of using its ports by 30%–40% in an effort to compete with other Asian ports, namely, China, South Korea, and Taiwan. In 2005, Japanese ports held the title of the most expensive in Asia, resulting in a decrease in container volume and competitiveness. The Transport Ministry plans to lower unloading costs by 30%, allow ports to operate 24 hours a day, and streamline the port-entry process. Japan's major port, Kobe, was ranked the twenty-seventh busiest in 2002.[26]

Across the Sea of Japan, South Korea is committed to becoming an economic hub in Northeast Asia. In 2004, South Korea opened a free trade zone at the southern Port of Kwangyang that exempts shippers from custom duties and clearance procedures. The port expanded to include 12 berths in 2006 and has plans to expand to 19 berths by 2008 and 33 in 2001. South Korea's Port of Busan, the third busiest container port in 2003 (behind

Hong Kong and Singapore), handled over 9 million TEUs, of which nearly 4 million TEUs were in transit, further making its way toward the role of an Asian hub. The Busan port is also expanding by adding six berths and 5,500 feet of dockside and intermodal facilities at a cost of over $476 million.[27]

In 2004, the South Korean dockworkers were serious about competing with other Asian ports and signed an agreement with the government to ensure labor peace for the rest of the year in exchange for a 4.5% wage increase. The Ministry of Maritime Affairs and Fisheries hopes the agreement will lead to investment from foreign companies. This comes at a time when China's Shanghai and Shenzhen ports surpassed Busan, which now ranks fifth, on the list of world's busiest ports.

The inter-Asian port competition leads to increased questions about what will happen to dockworkers' earnings. The pressure to decrease costs will bring downward pressure on wages, while the increased labor demand from increased traffic (especially in Chinese ports) has the potential to increase earnings.

14.8 SUMMARY AND CONCLUSIONS

The wages and working conditions of dockworkers have changed as a result of the changing climate of water transport resulting from containerization. Containerization reduced the demand for labor substantially and transformed the jobs from labor-intensive irregular work to capital-intensive steady employment. The increased globalization facilitated by containerization ushered in an era characterized by port reform and increased competition across and within ports.

The ability of unions to withstand the resulting shock to the labor market was a function of union bargaining power, port traffic, the structure of port operations, and national law. Even in successful cases, such as that of the ILWU workers on the U.S. West Coast, the past 20 years has seen the roles of labor shrinking considerably as workers are replaced by technology.

While privatization placed additional downward pressure on wages and employment of dockworkers, it has been deemed a success in countries such as Malaysia, where the workers, whose union had relatively little power due to decentralized bargaining, saw the effects of privatization softened by a large increase in trade volumes. Many terminal operations have been assumed by large private companies who have continued to expand through new contracts and merger. In Taiwan, though trade volumes are high, the competition with Asian transshipment ports and the entry of private terminal operators has led to significant decreases in wages and employment.

While not the focus of this chapter, it must be noted that dockwork is only a portion of the labor market of workers involved in international freight movement. There has been relatively little research on the working conditions of individuals operating the ships. Often these workers are from developing countries and receive relatively low pay. They are most typically heard of when there are incidents involving ships. In September 2006 a ship was held in the Port of Long Beach due to the crew complaining to the International Transport Workers Federation that they had not received payment in over a year. In 2006, the International Labor Organization pushed for countries to ratify changes in the working conditions of the estimated 1.2 million individuals who work as seafarers, including a minimum age of 16 years for employment. Those countries ratifying the 2006 Maritime Labor Convention will have the authority to examine ships for compliance.[28]

REFERENCES

1. Sommer, D. 1999. *Private participation in port facilities: Recent trends.* Note 193, Public Policy for the Private Sector, Washington, DC.
2. Brooks, M. R. 2004. The governance structure of ports. *Review of Network Economics* 3:168–83.
3. Haralambides, H. E., Ma, S., and Veenstra, A. W. 1997. *World-wide experiences of port reform.* In H. Meersman and E. v. d. Voorde (eds.) Removing the Port and Transport Business. Leuven, Belgium: Acco Publishing.
4. World Bank. 2001. *Module 7, Port reform tool kit: Labor reform and related social issues.* Washington, DC: World Bank.
5. Midoro, R., Musso, E., and Parola, F. 2005. Maritime liner shipping and the stevedoring industry: Market structure and competition strategies. *Maritime Policy and Management* 32:89–106.
6. Kagan, R. A. 1990. How much does law matter? Labor law, competition, and waterfront labor relations in Rotterdam and US ports. *Law and Society Review* 24:35–70.
7. Harding, A. S. 1990. *Restrictive labor practices in seaports.* Policy, research, and external affairs working paper, WPS 514, World Bank.
8. Turnbull, P., and Sapsford, D. 2001. Hitting the bricks: An international comparative study of conflict on the waterfront. *Industrial Relations* 40:231–57.
9. Pilcher, W. W. 1972. *The Portland longshoremen.* New York: Holt, Rinehart and Winston.
10. Finlay, W. 1988. *Work on the waterfront.* Philadelphia: Temple University Press.
11. Killingsworth, C. C. 1962. The modernization of West Coast longshore rules. *Industrial and Labor Relations Review* 15:295–306.
12. Fairley, L. 1979. *Facing mechanization: The West Coast longshore plan.* Los Angeles: Institute of Industrial Relations Publications.
13. Martin, D. L. 1990. Some economics of job rights in the longshore industry. *Journal of Economics and Business* 93–100.
14. Pacific Maritime Association. 1990. Annual report.
15. Leung, S. Dock workers nix bid to update dispatch system. *Wall Street Journal,* December 15, 1999, p. CA2.
16. Turnbull, P. 2000. Contesting globalization on the waterfront. *Politics and Society* 28:367–91.
17. Fusillo, M. 2003. Excess capacity and entry deterrence: The case of ocean liner shipping markets. *Maritime Economics and Logistics* 5:110–15.
18. Hill, D. 2004. *Globalization and labour relations: The case of Asian ports.* Working paper, University of Wollongong.
19. European Union. Ports move no threat to liberalization. *OxResearch,* January 15, 2006.
20. Position on port services directive. *Financial Times,* February 17, 2003, p. 12.
21. Minder, R., and Parker, G. Ports braced for strikes as dock workers step up action. *Financial Times,* January 12, 2006, p. 6.
22. Miller, J. European ports overhaul likely won't be delivered. *Wall Street Journal,* January 18, 2006, p. A9.

23. China to build 5 big port clusters in next five yrs. *SinoCast China Business Daily News*, April 21, 2006, p. 1.
24. Meredith, R. 2003. Crash! *Forbes* 171:56.
25. Yapp, J. 2003. Major expansion for port of Kaohsiung. *Asia Today International* 21:41.
26. Sanchanta, M. Japan ports to cut fees to boost competitiveness. *Financial Times*, January 6, 2005.
27. Bangsberg, P. T. 2003. Korea port lines up financing for expansion. *Journal of Commerce*, May 30, 2003.
28. Labor standards proposed in ILO seafarer charter. 2006. *Journal of Commerce*, on-line edition, February 23, 2006.

15 Intelligent Freight Technologies

A Longshore Perspective

Domenick Miretti

CONTENTS

ABSTRACT

The rapidly expanding number of cargo containers moving through California ports has encouraged marine terminal operators to expand their use of new waterfront technology and challenged longshore labor to deal with new and innovative ways of moving cargo.

In this chapter we will study longshore labor, members of the International Longshore and Warehouse Union (ILWU), and their involvement with goods movement technologies in West Coast seaports. More specifically, we will delineate and discuss the union's involvement in three technological phases, the events of which have been instrumental in molding, coloring, and tempering their view of changing goods movement practices. Such an understanding is essential to the future introduction of intelligent transportation technologies designed to make seaports more productive while at the same time maintaining the integrity of the longshore workforce.

The three technological phases will include:

1. The union's formative years and the break-bulk era, circa 1930–1959
2. Mechanization and modernization, circa 1960–2001
3. The years of expanding technology and the globalization of trade, circa 2002 to the present

The findings of this study are based on a review of primary and secondary sources, related mainly to the longshore industry and a review of the oral history of the ILWU. Personal interviews were also conducted with key union and industry leaders. From a rather unique perspective, study findings and conclusions are drawn from personal observations and experiences gained as a member of both ILWU Locals 13 and 63 for more than five decades and as the union's senior liaison to the ports of Los Angeles and Long Beach for more than two decades.

It should be noted that the opinions and conclusions expressed or implied in this chapter are those of the author. They are not necessarily those of the International Longshore and Warehouse Union, its officers, or individual members.

15.1 INTRODUCTION

The purpose of this study is to provide a broad understanding of longshore labor's perception of and involvement with waterfront technology as tempered by the events of three technological eras spanning a period of more than seven decades.

The terms *waterfront technology* and *intelligent freight technologies* will be used interchangeably in this study. In general, both terms refer to fast, efficient, and more secure ways of moving cargo.

Traditionally, waterfront technology refers to a variety of dock-side machines, trucks, trains, and cranes that initially modernized longshore work and are still operational and effective today.

On the other hand, intelligent freight technologies are newer, smarter, and more sophisticated goods movement techniques. They include a wide range of electronic-based devices such as cameras, scanners, transponders, optical readers, and computers; many are satellite assisted.

They are designed and used to improve freight system efficiency and productivity, increase global connectivity, and enhance freight system security against common threats and terrorism.[1]

The study's technological phases include:

1. The union's formative years and the break-bulk era, circa 1930–1959. This period spanned an era of moving cargo by primitive means under deplorable working conditions, severe labor management strife,[2] the Big Strike of 1934, the formation of the ILWU, its right to bargain collectively, and the establishment of a hiring hall.

2. Mechanization and modernization, circa 1960–2001. This phase was initiated by the signing of the Mechanization and Modernization (M&M) contract by the ILWU and the Pacific Maritime Association (PMA), the waterfront employers' bargaining agent.[3] The contract ushered in the age of containerization and intermodalism, the concept of moving goods interchangeably in cargo containers. This accord gave longshore some assurances that men and machines could coexist. The period's relative calm was punctuated by longshore's longest strike in 1970. This event strengthened the union's solidarity and reaffirmed its commitment to job creation, work preservation, expanding its jurisdiction, and control of its workplace. While enjoying the economic benefits of this era, the union also became actively involved in questioning the adverse affects of the new technology and seeking solutions to related problems.
3. The years of expanding technology and the globalization of trade, circa 2002 to the present. In 2002 the ILWU and PMA secured a 6-year labor contract.[4] After much deliberation the two sides agreed to a comprehensive process that would create a twenty-first century waterfront with modern tools and efficient practices.[5]

With the right to bring technology to the waterfront the employer began the process in 2003 of researching and developing systems that would best fit his needs.[6] With increased cargo volume in 2004 the employer pushed for fully integrated technology systems on marine terminals.[7] The year of technology, 2005, saw the accelerated implementation of technology in all major West Coast ports.[8] The union saw that the proliferation of technology created a number of longshore problems. New technology will surely initiate a reduction in job categories, blur longshore's division of labor, raise the question of job jurisdiction within the longshore locals and among the increasing number of marine terminal superintendents, and reopen the question of increasing the steady workforce and how all of this will be influenced by the globalization of trade.[9]

In this discourse, longshore labor will refer to waterfront workers that belong to the ILWU, Locals 13, 63, and 94, based in the Southern California region.

15.2   ILWU WORKFORCE

ILWU dockworkers facilitate the flow of cargo throughout U.S. West Coast seaports, stretching from San Diego, California, to Alaska, British Columbia, Canada, and the Hawaiian Islands.[10] In the East and Gulf Coast ports of the United States most of the longshore work is done by union members who belong to the International Longshore Association (ILA).[11]

As of December 2005 the ILWU registered workforce stood at 14,000. These numbers were complemented by thousands of additional "casual" workers who typically work part time.[12]

The ports of Los Angeles and Long Beach have the largest number of registered longshore and casual workers.[13] This segment of the ILWU was responsible for moving more than 14 million containers (20-foot equivalent units) through these twin ports, the busiest seaports in our nation, during the year 2005.[14]

15.3 ILWU DIVISION OF LABOR

There is a distinct division of labor within longshore ranks. Workers can be divided into three broad work categories; longshore, clerks, and walking bosses and foremen.

The detailed language describing a worker's specific functions can be found in the ILWU-PMA coast-wide agreements for longshore and clerks. The agreement also includes walking bosses and foremen and spells out their specific duties.[15]

A somewhat more generic, vivid, and personal description of longshore's division of labor follows.

Those longshore members who traditionally use muscle to move cargo belong to ILWU Local 13. They lash containers to ships' decks, ferry containers around container yards, and operate massive machines that facilitate the flow of cargo. Perched in a lofty position the gantry crane operators choreograph proper vessel discharge and loading operations.

ILWU marine clerks, Local 63, identify, verify, account for, and count every item of cargo that passes through a marine terminal. They receive cargo for export and release imported goods. Hatch clerks check every container during vessel discharge and loading operations. Clerk planners plan the stowage of cargo on vessels and on dock.

Marine clerk computer operators interface with truckers at terminal gates, interchanging thousands of truckers per day. Marine clerks, or "checkers" as they are called on the docks, perform their duties armed with computers, counters, scanners, cameras, and pencils, and use reams of paper to document and record their daily activities hour by hour and often minute by minute.

Walking bosses and foremen belong to ILWU Local 94. They supervise the work activities of all Local 13 longshore workers. Foremen are highly visible as ship, dock, and yard bosses. They interface with lashers, crane operators, and heavy equipment operators, and hold and dockworkers, and assign and coordinate the work activities of all longshore gangs. Foremen are in a unique position. They represent management in a supervisory capacity yet are union and belong to the ILWU.

15.4 EVOLUTIONARY STAGE OF LONGSHORE TECHNOLOGY

A review of the literature suggests that advancing marine terminal technology, specifically affecting longshore workers, can be traced through three evolutionary stages:

1. The union's formative years and the break-bulk era, circa 1930–1959[16]
2. Mechanization and modernization, circa 1960–2001[17]
3. The years of expanding technology and the globalization of trade, circa 2002 to the present[18]

From personal workplace observations, experience, and a review of the literature the author concludes that the events of each era have profoundly influenced longshore's view of newer and more efficient ways of moving cargo.

15.5 UNION'S FORMATIVE YEARS AND THE BREAK-BULK ERA, CIRCA 1930–1959

This period spanned an era of moving cargo break bulk or piece by piece under deplorable working conditions, the Big Strike of 1934, the formation of the ILWU, its right to bargain collectively, and the establishment of a hiring hall.

Containers, cranes, massive machines, and electronic devices were nowhere to be found during the break-bulk era. Instead cargo was manipulated, piece by piece, by muscle, a longshoremen's hook, rope slings, four wheelers, the lever, the pulley, the inclined plane, or any means possible.[19]

In this era, dockworkers struggled to secure work and physically survive under some of the most deplorable working conditions to be found in any seaport. Work was casual, containing the evils of employer favoritism and discrimination. Events included the shape-up, kickbacks, favored hiring practices, the speed-up (men working faster and faster, pushing their physical limits in an effort to curry the bosses favor), unmercifully long work shifts, extremely dangerous working conditions, and labor and management always at odds.[20] These working conditions are what old-timers talk about and young longshore workers find hard to comprehend.

The daily congregation of individual dockworkers at specified places and times to be selected by a foreman for work has been referred to by Keller as the notorious shape-up or shape (known in Great Britain as "calling on"). He goes on to say: "It is a system which has propagated favoritism, bribery and demoralization."[21]

Along much of the West Coast longshoremen were hired on the docks. In San Francisco the Embarcadero was known as the "slave market." Men hung around the docks all day waiting for a work opportunity.[22]

A small minority of men who successfully worked the system were privileged to work themselves to death, while a majority were reduced to the level of casual labor.[23] There were no safety rules or safe working standards.[24] Ships' riggings were rarely maintained in proper working order and, straining under the weight of unrestricted load limits, often gave way, dumping their loads on workers below. Longshore workers were commonly injured and often killed.[25]

The speed-up was a common work practice on the West Coast waterfront.[26] Designed to avoid overtime payments, it also encouraged a frenzied work pace, pitting longshore workers and gangs against each other. Those with the highest performance records were accorded work.[27] The 8-hour workday was unheard of. Men worked until a ship was finished. Shifts of 24 to 36 hours without sleep were not uncommon.[28] In an effort to put a stop to worker abuses, ILWU founder Harry Bridges issued a call for a coast-wide convention.[29] Demands by convention delegates were clear. They demanded a coast-wide contract, a 6-hour day, wage increases, and, most importantly, union hiring halls.[30] A strike was authorized if delegate demands were unmet. Employers quickly refused worker demands. Tensions rose along the West Coast as a strike deadline approached.

On May 9, 1934, the start of the Big Strike, approximately 12,000 longshoremen and seafarers walked off their jobs along the entire U.S. West Coast. A violent struggle and bloody battle ensued. "In San Francisco July 1934, the laboring population laid down its tools in a general strike."[30] San Francisco's governor Frank Merream ordered 4,000 National Guard troops to the docks in an effort to control the city.[31] As the strike continued, hundreds of strikers were injured and many lost their lives. In San Francisco alone, 2 strikers were killed and 109 injured in what has become known as Bloody Thursday.

The 82-day-old strike resulted in a National Arbitration Award handed down by the National Longshoremen's Board appointed by President Roosevelt, which on October 12, 1934, ruled in favor of striker demands.[32] According to Fairley, the strike was an overwhelming union victory, with all major demands being met.[33] A coast-wide contract defined longshore jurisdiction and work. Hiring halls, jointly operated by labor and management but largely union controlled, were established to ensure equal work opportunity.[34] Workforce registration providing for a stable labor pool, the 6-hour day (designed to spread out the work and reduce the workday), higher wages, and safer working conditions were all a part of the National Arbitration Award.[35] Initially a part of the ILA, the ILWU was formed in 1937 under the leadership of Harry Bridges.[36]

The difficult years on the West Coast waterfront during the early 1930s created a unique longshore character.[37] Fairley describes in part the character of the ILWU. He sees them as being "progressive." They tend to be outspoken on social, economic, and political issues.[38] At the same time, Fairley describes the union as extremely militant and tough, united by years of struggle against equally tough employers.[39] A hallmark of the ILWU since its beginning, Fairley concludes, is its unusual degree of internal democracy, which helps to explain its differing and apparent divergent characteristics.[40]

Based on personal experience as an ILWU member and Fairley's findings, the author suggests that the ILWU character has colored the union's perception of labor management relations, its view of the role it plays on the waterfront, and how it may be affected by changing goods movement technology.

The hiring hall was the first and single most important factor that has influenced longshore's perception of waterfront technology. Keller explains that "the union clung tenaciously to one principle, namely that without union control of the hiring halls the right to organize was meaningless."[41] The union's house of labor, its hiring hall, is considered by the ILWU to be its heart, its unifying force, and its strength.

The hall must be protected and preserved at all costs. Any new work practice that would threaten the integrity or existence of the hiring hall would be immediately suspect and surely rejected. When the California Marine International Transportation System Advisory Council (CALMITSAC) in its 2006 interim report, "Growth of California Ports: Opportunities and Challenges," suggested dispatching workers directly to terminals and bypassing their daily visits to dispatch halls, it raised considerable ire and rankled union officers and rank-and-file members.[42]

Deeply engrained in the ILWU character is a reverence for its past and a deep belief that aside from its leaders, the rank and file is the union's ultimate strength. Harry Bridges had an uncompromising devotion to his union.[43] His faith in the ultimate wisdom and triumph of his organization's rank and file was shown in the 1934 strike and demonstrated the might of the rank and file. Bridges stated, "Economists, lawyers, financial advisors, and even the officers they elected to lead them, while valuable and truly important skilled tools, ran second to the strength of the workers. Rank-and-file strength is shown for what it is—indispensable. This principle remains eternally sound.'[44] Bridges along with his early followers are respected, honored, and revered for the creation of a union that has provided longshore workers with unprecedented economic and social gains. Union traditions must be practiced, protected, and preserved. To do anything less would be disrespectful and against good union principles.

Given the past, it is not surprising that the union approaches new technological ideas with concern, apprehension, caution, and a certain amount of pessimism. This era came to a close at the end of the 1950s. The union was now faced with a new and formidable challenge—mechanization, the use of the machine.

15.6 MECHANIZATION AND MODERNIZATION, CIRCA 1960–2001

This phase was introduced by a new and revolutionary way of moving goods (containerization), labor/management agreement on the use of the machine, the signing of the M&M contract, and some union assurance that men and machines could coexist.

The new technology and the 1970 strike reaffirmed union solidarity and provided longshore an incentive to more carefully manage and control its workplace. The period also served as a catalyst for greater longshore involvement in seeking solutions to growing goods movement problems while at the same time sharing in the era's economic benefits.

During the latter part of the 1950s the maritime industry faced the challenge of introducing new waterfront technology. Obviously, continuing to move cargo by old break-bulk methods was no longer a viable option. Finlay cites a consultant from the American Hawaiian Steamship Company as saying: "The so-called break-bulk method of handling cargo used by the shipping industry for centuries is obsolete in an economy with high labor costs."[45]

The idea of putting cargo in huge steel boxes (containers) and handling them interchangeably by ship, truck, rail, and plane ended the break-bulk era for the movement of most waterborne goods. This new method of cargo conveyance ushered in the age of containerization and intermodalism, by all accounts a revolutionary and creative way of doing business. Longshore viewed this new form of waterfront technology with great concern, growing fear, and initial resistance to its introduction and use.[46]

Almost 50 years ago, mechanization, use of the machine, was the most talked about topic on the waterfront. At the time dockworkers speculated on the impact machines would have on the longshore industry and raised many questions concerning their use. Would massive machines replace workers, create a job speed-up, create unsafe working conditions, reduce wages and welfare benefits, and challenge the equal work opportunity of the hiring hall?[47] Others speculated on the union's loss of hard-earned collective bargaining gains it had achieved since the Big Strike.[48] These were questions of paramount importance then and still have relevance today.

In 1960 it was the modernization and mechanization agreement (M&M contract) that would serve as a guide for the introduction and implementation of new and revolutionary types of waterfront technology. In retrospect, the contract would be viewed as a unique collective bargaining pact between management and labor. It was said to be "epoch-making."[49] The agreement provided for more than putting cargo in big steel boxes. More important, it created an economic and social compact, a labor-management bill of rights. The union conceded to the use of the machine. Working rules the employer found restrictive were relaxed, and manning was significantly adjusted.[50] In return there would be job security, no lay-offs, no employer speed-up; job safety would command the highest priority.[51] Wages, welfare, and working conditions would be improved. Pensions and a pay guarantee plan were put in place.[52]

Of greater significance than the economic and social provisions of the contract, and paramount to its success, was the understanding that when new waterfront technology was introduced it would be done cooperatively by both labor and management, for its time a simple yet far-reaching idea that encouraged contract ratification.

Successful implementation of the contract was the legacy of the rank and file. They were a dedicated, yet adventurous group of workers willing to test and implement newfound ideas that many felt threatened their very existence. With the implementation of the M&M contract, workers speculated that the new technology might empower employers, giving them greater control of the workplace and reducing longshore's control of production.

In retrospect, the author suggests, based on research findings and personal on-the-job work experience, that the contract gave longshore an opportunity to improve its work skills, gain greater control of its workplace, and become more involved in on-the-job decision making in the face of changing technology.

Finlay provides an interesting comparative analysis of how postwar technological change has affected workers on the job. He points out two different perspectives on the impact of technological change on work and workers with special reference to longshore.[53]

One view focuses on subordination of workers and the other on empowerment of workers. Examples of the former include the works of Braverman,[54] Clawson,[55] and Edwards.[56] Examples of the latter position are presented by Blauner,[57] Sabel,[58] and Thurow.[59]

In the subordination view workers experience loss of control at work along with a depreciation of job skills. According to the empowerment view, new technologies increase worker influence over the process of production by creating the opportunity for new skills and new responsibilities.

Finlay agrees with the empowerment of worker view, as does the author. He argues that mechanization and modernization in the West Coast longshore industry have not deskilled workers or weakened their autonomy or job control. In some respect, workers have gained increased skills and strengthened control of the work process.

Starting in the early 1970s mechanization on the waterfront encouraged the PMA (employers) to make increased demands on worker availability and the use of "steady men." These are ILWU workers who report to a work site on a regular basis rather than being dispatched from the longshore hiring hall.

From the union's point of view the need for job security and wage maintenance was greater than ever and there was a concern over job jurisdiction. From an ILWU perspective these issues were the modernization part of mechanization and needed to be adequately addressed. Neither the ILWU nor PMA could agree on the issues. The result was the 1971–1972 strike. Longshore workers stayed off the job for 134 days. The longest longshore strike in ILWU history reaffirmed rank-and-file solidarity and committed the union to a proactive role in confronting and dealing with the effects of any new waterfront technology.

Labor, along with many others, has come to realize that moving goods interchangeably through marine terminals coupled with labor management cooperation has not only increased port productivity but also, in large part, benefited labor, port cities, their regions, states, and the nation as a whole. In the job category nearly 30,000 jobs in the city of Long Beach are supported indirectly and directly by the port, more than 315,000 jobs in the port's five-county metropolitan area (region), and approximately 371,000 jobs in the state of California.[60]

The impact of port operations on the nation is considerable. About 1.4 million jobs are supported directly and indirectly by port operations.[61] The multiplier effect of port production also creates more jobs and generates needed municipal tax revenue. Reviewing Port of Los Angeles statistics, O'Brien found that the port alone generated $1.4 million in state and local tax revenue for Southern California.[62] Erie suggests that international trade through our ports generates widespread benefits emphasizing increased economic productivity and lower-priced imported goods for consumers.[63]

These economic gains have not been without serious side effects. Air quality has been substantially compromised in the Los Angeles basin, traffic congestion is a serious problem in the region, and port communities are experiencing a decline in their neighborhood quality of life. The ports of Los Angeles and Long Beach contribute to the single largest source of air pollution in Southern California.[64] The entire port complex creates about 25% of the diesel pollution in the region.[65] Diesel exhaust is a known carcinogenic agent. The MATES II study found that 71% of all cancer risks from air pollution are derived from diesel exhaust.

A map from the report delineates a "diesel death zone" existing in and around the ports.[66] Research has shown that children living in the area can suffer from asthma, have reduced lung capacity, along with higher rates of school absences.[67] The additional number of trucks needed to move increasing cargo volume is adding to already serious traffic congestion in the Southern California region.[68]

On a daily basis approximately 35,000 trucks move in and out of the San Pedro Bay complex. That number could increase to 55,000 by the year 2020.[69] Aside from contributing to traffic gridlock and higher incidents of freeway accidents, trucks add a significant amount of diesel exhaust in the Los Angeles Basin.[70]

Port communities have expressed concern about noise and visual blight in and around the ports. They see empty container storage yards, tall container cranes, vehicle emissions, and small neighborhoods being traversed by trucks as serious environmental impacts that reduce their quality of life.[71]

The author contends that the union sees these negative impacts as an incentive to become actively involved in the technological decision-making process and seek solutions to relevant goods movement problems. The union, along with the author, has taken a proactive role in seeking ways to reduce truck traffic on local transportation arteries and improve air quality in and around West Coast seaports.

The ILWU has a three-point plan. It provides for greater use of on-dock rail, the creation of near- and off-dock container transfer facilities, and the development of inland ports. The union suggests its plan could speed the flow of cargo through marine terminals, substantially reduce port-related truck traffic, and cut down on harmful diesel truck exhaust.

Most recently the ILWU announced a "Saving Lives" campaign to reduce air pollution in seaports from Seattle to San Diego. Its clean air initiative calls for a 20% reduction in emissions by 2010 for all ships calling at West Coast ports. It also hopes to reduce pollution from trucks and cargo handling equipment on the docks.[72]

Another attempt by the union to become actively involved in goods movement technology is presently being proposed by an ILWU member. The guideway system is a possible solution to the pollution and infrastructure problems for the San Pedro Bay ports and could provide additional off-dock jobs for union workers.[73] The electronically operated rail system is designed to move containers from the ports to satellite terminals for later truck pickup.

Given the union's efforts to help solve goods movement problems, one can assume that it is not totally opposed to intelligent transport technologies. For example, this book's research proposals designed to make our ports safer, speed the flow of cargo, and make our ships and ports greener would be viewed by longshore as having merit and would generate further longshore interest. It appears that the union's involvement in the assessment of goods movement technology serves as a central theme in its perception of new and emerging ways of moving cargo. The next longshore phase presents the union with a set of new technological challenges.

15.7   INTELLIGENT TRANSPORT TECHNOLOGIES AND THE GLOBALIZATION OF TRADE

For the union and its involvement with new and smarter technologies this is a period of renewed concern, questioning, involvement, and problem solving.[74] The events of this period that have colored and will continue to color the union's perception of the use of new waterfront technology include but are not limited to present and proposed new technology, its impact on security-related issues, greater political involvement by stakeholders in labor/management issues, and the globalization of trade.[75]

Almost 50 years ago longshore workers questioned the concept of mechanization and its possible impact on their workplace. Today these workers are coming to grips with intelligent freight technologies and attempting to assess how these more sophisticated ways of moving cargo will impact their workplace.

Methods to improve freight system efficiency and productivity, increase global connectivity, and enhance freight systems' security against common threats and terrorism are broadly defined as intelligent freight technologies.[76] These technologies are currently deployed in several areas: (1) asset tracking, (2) on-board status, (3) gateway facilitation, (4) freight status information, and (5) network status information.[77]

They include but are not limited to:

1. Optical character recognition (OCR). Cameras and scanners that read container and chassis information, increasing the number of gate moves at marine container terminals while reducing labor requirements.
2. Radio frequency identification (RFID). This system tracks trucks, containers, and cargo on or off dock. When used at marine terminals it can pass information from one piece of equipment to another, thus eliminating the need for an ILWU marine clerk to locate and record a container's location.
3. Equipment positioning system (EPS). A satellite-based system tied to marine terminal receivers and used to locate and monitor the flow of cargo during ship and dock operations. EPS can and have reduced longshore clerk jobs.
4. Real-Time Location Systems (RTLS). Used to mitigate terminal gate congestion and help draymen and terminals operate more efficiently. The system can combine accurate near-real-time information on queues and traffic delays with terminal and delivery scheduling.
5. Application service provider software (ASPS). Provides a software system designed in part to support interoperable machine-to-machine interaction over a network. The software functions as a gateway between proprietary trading systems.[78]

Variations in the application of terminal technology affect longshore personnel.

Working at a mega container terminal in the Port of Los Angeles using Navis SPARCS, a highly sophisticated planning and central system for container planning and central handling, is a worker challenge.[79] In comparison, dockworkers find a nearby terminal more user friendly. It relies on OCR and GPS along with in-house proprietary measures and a less complex cargo handling system.[80]

Longshore workers understand that working marine terminals smarter, coupled with new technology is a key element needed for future port growth and expansion. As such, the union is not opposed to technology. In fact, it has agreed contractually that the employers have the right to implement technology subject to a number of controlling and procedural principles.[81]

First, employers must discuss with the union, at the local level, their intent to introduce new technology. They then must submit to the union at the coast level a "technology letter" that describes the new technology and proposed impacts on marine clerks. The union responds by letter to the employer stating its position. In turn, the employer responds by letter to concerns raised by the union. If all parties agree, the employer shall have the right to implement new technology 35 days after the coast technology letter is submitted to the union. If issues are raised by either party, they may be presented to the area arbitrator. If agreement is not reached at the local level, the matter shall immediately be referred to the coast arbitrator for final resolution.

The author contends that technology in itself is not the issue. Concerns center around the impact new technology will have on jobs. On the positive side, longshore job numbers have risen substantially. Statistics reveal that the workforce has increased by 34% since June 2002, the date for the last labor contract.[82] In 1994 the registered workforce was 8,000. Today PMA members employ 14,000 registered workers and thousands more part-time or casual workers.[83]

As encouraging as growing the workforce is, it clouds the issue of reduced job options, blurs longshore division of labor, raises the question of job jurisdiction within the union and among company superintendents, and increases the possibility of a steady workforce. Marine clerks serve as an excellent example of loss of job opportunity. Ober's findings substantiate this claim. He states that "although it may be true that the volume of labor to be performed is a changing quantity, it is equally true that in individual instances employment opportunities have declined as a result of the introduction to technological changes."[84]

The number of marine clerks who normally worked out-gates at container terminals has declined by 80%.[85] One terminal reported that its "in-gate" crew dropped from 27% to 7% because of the use of OCR and related technology.[86] Except for marine clerk computer operators interacting with truckers, gate jobs are for all practical purposes nonexistent.

Marine clerk yard jobs have also been significantly reduced. Clerks are no longer assigned to each yard crane to assist in yard delivery and receiving of containers. A limited number of "rovers," a new job description, are assigned to a group of machines interacting with their operators only when needed.

A major objective of trade unions, Ober states, has been the control over definite categories of work.[87] Traditionally longshore workers physically move cargo, marine clerks expedite its flow, while ILWU foremen supervise Local 13 workers.

The author contends that technology is blurring longshore division of labor. What work responsibility or job jurisdiction each longshore local now has will ultimately be altered. The problem will arise as to which group of workers should be assigned to the changed jobs:[88] longshore, clerks, foremen, union, or nonunion.

Increasingly, crane and heavy equipment operators are using on-board computers and OCR technology to assist them in completing their work assignments. Many of their tasks are completed without interacting with a marine clerk. Assisting and directing the flow of cargo is contractually defined as marine clerk work.

Newly created on dock technology has initiated a rapid increase in the number of marine terminal superintendents. The author fears that these lower-level management personnel will play a greater role in supervising longshore workers, presently the union's exclusive responsibility, and act as a barrier filtering out newly generated electronics data that could serve as a basis for areas of new longshore jurisdiction. Rank-and-file members speculate that superintendents could serve as replacement workers hoping to keep marine terminals functioning on a limited basis in the event of a major labor management dispute.

An added union concern is that as technology becomes increasingly more complex, employers will seek out the most highly qualified and motivated individuals and recruit them as steady employees. The union fears that increasing the steady workforce could philosophically divide workers and jeopardize the existence of the hiring hall.

The lack of longshore involvement on technological issues will surely have a negative effect on its perception of new goods movement practices. The ILWU Clerks'

Technology Committee, made up of union officers and appointed union members, feels that employers are not involving the union in the installation and testing of new technology, and that they are using vendors along with their own management to do work considered to be clerks' work.

The committee sees off-dock terminal control centers (TCCs) as the clerks' biggest challenge. At these inland centers much of the same scope of work will be performed by nonunion people that now takes place at marine terminal control centers presently employing ILWU marine clerks.[89]

The events of 9/11 have introduced a new form of waterfront technology: workers' identification cards. All union members seek a safe and secure work environment free of terrorist activity. However, the procedural aspects designed to accomplish that goal, using the Transportation Worker Identification Credential (TWIC) system, is being questioned by the ILWU. Their concerns center around worker background checks, who and how data will be collected, and, once compiled, where will it be housed and under whose control. Some workers fear that an individual would not be eligible for employment in a security-sensitive position if the Department of Homeland Security finds the individual has been convicted of a disqualifying crime during the past 10 years.[90]

The author, along with many other dockworkers, see new technology as a double edged sword. On the one hand, OCR's and related devices have increased container throughput, while at the same time their use has eliminated ILWU marine clerks who traditionally checked and inspected all empty export containers and checked the integrity of seals on loaded outbounded containers. This changed post 9/11 work practice has seriously compromised the level of security at marine terminal gates and has made our ports vulnerable to terrorist attack.

The ports of San Pedro Bay are an integral part of the global economy. Erie points out that Los Angeles serves as the chief hub for the U.S. waterborne commerce and has become the nation's leading Pacific Rim gateway.[91] Given this global connection, there is an increased demand for high productivity through marine container terminals that serve as hubs of international trade.

One highly automated concept is the Gottwald Automated Transport System.[92] Automated guided vehicles (AGVs) are suitable for unmanned container transport from quay to the stack and railway terminal areas. The Port of Rotterdam is presently using the system.

A major West Coast carrier is planning to construct and operate such a terminal on the U.S. East Coast. This new facility will surely impact longshore labor along the eastern seaboard. It could adversely affect longshore manning should a similar facility be built at a major West Coast port.

In the global scheme of things recent actions under the auspices of the North American Free Trade Agreement (NAFTA) and the European Union have raised ILWU concerns and will surely color the union's perception of and involvement with the activities of international goods movement.

The NAFTA Super Highway serves as a good example. The project raises many unanswered questions.[93] Will the plan designed to allow containers from the Far East to enter the United States through the Mexican port of Lazaro Cardenas cost West Coast ports market share, resulting in a loss of West Coast longshore jobs? Will

Mexican trucks traveling the NAFTA Super Highway from Mexico to Canada replace American union truckers, present an open-border security issue, and why haven't affected unions been consulted or involved in such a massive project?

European port reform, a controversial plan to liberalize port services throughout the twenty-five nations of the European Union, is opposed by West Coast dock-workers. One official called it a smoke screen for union busting.[94]

The most controversial part of the plan is the so-called self-handling clause. This practice would allow seafaring crewmen to carry out work traditionally performed by longshore personnel. Dockworkers fear the program could cause widespread job loss, reduce worker pay, and lower job security, especially in the cargo handling sector.

Both the ILWU commitment to international worker solidarity and a strong belief in the slogan "An injury to one is an injury to all" will continue to color the union's perception of any and all goods movement technology that affects working people.

If called upon by workers seeking help in addressing technological issues that might threaten them economically, politically, or socially, the union will respond and become actively involved. The rather recent ILWU delegation sent to Europe to demonstrate against government and employer attempts to de-unionize European ports and the 2005 Mining and Maritime Conference held in Long Beach, California, bringing together Pacific Rim maritime workers, serve as examples of ILWU global involvement in worker solidarity and support.

It seems evident that longshore firmly believes that all workers must be considered an integral part of the technological equation. The author feels that workers who implement new technology are the real source of its success. Without proper and skilled application the technology in itself is an inert force, simply a potential energy source.

Benedetti embraces this idea in an article that explains how the Port of Cartagena, Venezuela, has dramatically increased port productivity by using new technologies. However, he goes on to say that "the high productivity levels would not have been reached without the human factor—the ingredient that assures true comparative advantages are sustainable over time."[95]

Fernandez states that the essential factor in port productivity increases worldwide has been technology. He adds, "The human factor unites all other elements because without it a complex industry like the maritime transport industry could not be understood."[96] The Fernandez assumption leads us to a question of paramount importance ripe for research and further discussion. What role do members of the transportation sector see workers playing in the planning, implementation, and use of new water-front technology given the fact that labor is an essential part of every aspect of goods movement? The answer to that question will have a profound influence on longshore's perception of intelligent transport technologies as colored by the events of its past.

15.8 SUMMARY

It has been suggested that the International Longshore and Warehouse Union has progressed through three technological phases spanning more than seven decades of union involvement in moving goods through West Coast seaports.

The first phase was the union's formative years and the break-bulk era. During this period cargo was moved by primitive methods under deplorable working conditions. The Big Strike of 1934, the formation of the ILWU, its right to bargain collectively, and the establishment of a hiring hall were important era events.

Mechanization and modernization marked the second phase. Significant here was the signing of the M&M contract that ushered in the age of containerization and intermodalism. The union accepted the use of the new technology with some assurances that men and machines could coexist. The strike of 1970 strengthened the union's solidarity and furthered its commitment to job creation and preservation, while at the same time it sought to expand its jurisdiction and control its workplace. Near the end of this phase the union became actively involved in questioning the adverse effects of new waterfront technology.

The years of expanding technology and the globalization of trade marked the third and final technological phase. This period saw increased cargo volumes passing through West Coast ports. Employers pushed for sophisticated and fully integrated technological systems on marine terminals and the accelerated implementation of their use in container handling facilities. The union questioned the proliferation of expanding technology and saw its use creating a number of longshore issues. Paramount among these concerns is a reduction in job categories, a blurring of longshore's division of labor, the steady workforce question, and how all of this will be influenced by the globalization of trade and its impact on workers in general. The events unfolding in this phase of longshore's involvement with smarter marine terminal technology will continue to color its perception of all goods movement practices.

15.9   CONCLUSION

As the transportation industry comes to grips with the expanding and complex technologies of the twenty-first century, it will face challenges far greater than its predecessors. Hopefully all parties involved will apply a humanistic approach in solving technological issues.

Industry must be cognizant of and sensitive to economic, social, and environmental problems caused by new technology, and be ready to mitigate such impacts. The author believes that the measure of industry's success in dealing with twenty-first-century technology will be gauged not so much by the kind of technology but by the kind of joint decisions made as an industry for the application of that technology.

REFERENCES

1. FHWA Freight Management, The Freight Technology Story, Report documentation page, p. 1, http://www.frieght_tech_story/from/form.htm.
2. For a vivid accounting of labor relations up to 1934, see Quin, M., *The Big Strike* (New York: International Publishers, 1979).
3. Fairley, L., *Facing Mechanization: The West Coast Longshore Plan* (Los Angeles: Institute of Industrial Relations, University of California, 1979), 1.
4. Pacific Maritime Association, *2005 Annual Report* (San Francisco: Pacific Maritime Association, 2005), 28.

5. Ibid.
6. Ibid.
7. Ibid.
8. Ibid.
9. Assumptions made by Domenick Miretti were agreed to by Mark Mendoza, Joe Gasperov, and Daniel Miranda, presidents of ILWU Locals 13, 63, and 94.
10. Pacific Maritime Association, *2005 Annual Report*, 21.
11. Grobar, L., *METRANS 2005 Annual Report* (Long Beach: University of Southern California and California State University, Transportation Center).
12. Pacific Maritime Association, *2005 Annual Report*, 31.
13. Ibid.
14. Ibid.
15. Ibid., 32.
16. Fairley, *Facing Mechanization*, 9.
17. Ibid., introduction.
18. Pacific Maritime Association, *2005 Annual Report*, 28.
19. *Men and Machines: A Story about Longshoring on the West Coast Waterfront*, a photo story of the mechanization and modernization agreement between the International Longshoremen's and Warehousemen's Union and the Pacific Maritime Association, printed in United States by Phillips and Van Orden Company.
20. Fairley, *Facing Mechanization*, 9.
21. Works Progress Administration, *Decasualization of Longshore Work in San Francisco: Methods and Results of the Control of Dispatching and Hours Worked, 1935–37*, National Research Project, Report L-2 (Philadelphia: Works Progress Administration, 1939), chap. 1, p. 2.
22. Quin, *The Big Strike 31*.
23. Ibid., 32.
24. Ibid.
25. Bridges, H., *A Centennial Retrospective: An Oral History of the Origins of the ILWU and the 1934 Strike*, edited and with introduction by Harvey Schwartz. San Pedro, CA: Harry Bridges Institute (2001), 6.
26. Quin, *The Big Strike*, 32.
27. Ibid.
28. Ibid.
29. Local 13 Education Committee, International Longshore and Warehouse Union, *ILWU Caretakers of a Great Inheritance* (ILWU, 2005), 6, 7.
30. Quin, *The Big Strike*, 3.
31. *ILWU Caretakers of a Great Inheritance*, 7.
32. Ibid., 8.
33. Fairley, *Facing Mechanization*, 9.
34. Ibid.
35. Ibid.
36. Pacific Maritime Association, *2005 Annual Report*, 34.
37. Fairley, *Facing Mechanization*, 2, 3.
38. Ibid.
39. Ibid.
40. Ibid.
41. Works in Progress Administration, 12.
42. Barton, M., "State Tackles Growth at California Ports." Center for International Trade and Transportation, California State University–Long Beach, *Building Bridges*, 5(3), 2006, 1.
43. Harry Bridges, *A Centennial Retrospective*, 7.

44. Quin, *The Big Strike,* xi.
45. William Finlay, *Work on the Waterfront: Worker Power and Technological Change in a West Coast Port* (Philadelphia: Temple University Press, 1988), 9.
46. Fairley. A brief overview of the M&M contract is presented based on Fairley's findings in *Facing Mechanization,* 112–138.
47. Ibid., 1.
48. Ibid.
49. Ibid.
50. Ibid.
51. Ibid.
52. Ibid., 2.
53. Finlay, *Work on the Waterfront,* 181.
54. Harry Braverman, *Labor and Monopoly Capitol* (New York: Monthly Review Press, 1974).
55. Dan Clawson, *Bureaucracy and the Labor Process* (New York: Monthly Review Press, 1980).
56. Richard Edwards, *Contested Terrain* (New York: Basic Books, 1979).
57. Robert Blauner, *Alienation and Freedom* (Chicago: University of Chicago Press, 1964).
58. Charles Sabel, *Work and Politics* (Cambridge: Cambridge University Press, 1982).
59. Lester C. Thurow, *Generating Inequality* (New York: Basic Books, 1975).
60. Port of Long Beach, *Economic Imports, Contributing to the Local State and National Economies* (Port of Long Beach, 2006).
61. Ibid.
62. Tom O' Brien, *Quality of Life and Port Operations: Challenges, Successes and the Future,* Sixth Annual CITT State of the Trade and Transportation Industry Town Hall Meeting, A White Paper, August 30, 2004.
63. Steven P. Erie, *Globalizing L.A. Trade Infrastructure and Regional Development* (Stanford, CA: Stanford University Press, 2004).
64. O'Brien, *Quality of Life and Port Operations,* 11.
65. Ibid., 11.
66. South Coast Air Quality Management District, *Multiple Air Toxic Exposure Study on the South Coast Air Basin: MATES II Final Report and Appendices* (Diamond Bar, California SCAQMD, 2000).
67. O'Brien, *Quality of Life and Port Operations,* 11.
68. Ibid.
69. Ibid.
70. Ibid.
71. Ibid.
72. Barbara Schoch, Labor Lends Its Clout to Port Pollution Battle, *Los Angeles Times,* January 28, 2006.
73. Alexia Terzopoulos "Longshoreman Designs Transportation System to Shoulder Weight of Port Problems." Long Beach Business Journal July 31, 2007, 1. 22(15) Jim Whelan, *The Guideway System.*
74. Mark Mendoza, president, ILWU Local 13.
75. Joe Gasperov, president, ILWU Local 63.
76. FHWA Freight Management, Freight Technology Story.
77. Ibid.
78. Ibid.
79. Joe Gasperov, president, ILWU Local 63.
80. Frank Pisano, vice president, Trans Pacific Container Service Corporation, Wilmington, CA.

81. *Pacific Coast Contract Document ILWU*, PMA 2002-2008.
82. Pacific Maritime Association, *2005 Annual Report*, 31.
83. Ibid.
84. Harry Ober, *Trade-Union Policy and Technological Change* (Report L-8, Works Projects Administration, Philadelphia, Pennsylvania, April 1940), 5.
85. Joe Gasperov.
86. Frank Pisano.
87. Ober, *Trade-Union Policy and Technological Change*, 10.
88. Ibid.
89. *ILWU Clerks Technology Committee Report* (Committee of the Whole, ILWU International Caucus, April 4–6, 2005).
90. *Port Security Task Force Meeting* (Port of Los Angeles Section 8, June 10, 2003).
91. Erie, *Globalizing L.A. Trade.*
92. Guided Vehicles for automating container terminals.
93. Jerome R. Corsi, Bush Administration Quietly Plans NAFTA Super Highway, June 12, 2006, http://www.humanevents.com/article.php?id=15497.
94. Alan M. Field, "Steering a New Course, *Journal of Commerce*" 7(5)(2006): 42.
95. Organization of American States, *Productivity Today* (Inter-American Committee on Ports Publication) 3 (2005): 36.
96. Ibid., 40.

16 Environmental Management of the Logistic Chain
Concepts and Perspectives

Antonis Michail and Christopher F. Wooldridge

CONTENTS

ABSTRACT

The logistic chain may be defined as the network of successive links involved in the transport and placement of goods. Intermodality, the process of transporting freight by means of a system of interconnected networks involving various combinations of modes of transportation, lies at the heart of the concept of the logistic chain. The chain consists of successive links between patterns of movement (transport modes) and nodal points (logistic nodes) in an integrative, intermodal concept from point of origin to point of consumption.

The environmental management of the logistic chain addresses the functional organization and use of appropriate response options necessary to minimize the environmental impacts arising from the operation of intermodal transport systems, including all its activities, operations, and services. The concept recognizes the physical and economic links associated with an integrated transport system, or chain, specifically designed to transport or deliver goods in a cost-effective and sustainable manner. The challenge is to minimize the environmental impact of the chain by adding the environmental component into the decision making and management of its operation.

The chapter examines the interaction between transport systems and the environment and highlights the significance of modal choices. Taking a European perspective, the extent to which policies of environmental protection are actually implemented is discussed with regard to the European transport system, and outstanding issues are discussed. In addition, the chapter identifies the responses of the major players and operators in intermodal transport chains as they face the challenge of sustainable chain operation. This includes related policies, recent initiatives, and examples of good practice. Finally, the focus is centered on the role of the major logistic nodes operators with special reference to seaports.

16.1 INTRODUCTION

Freight transport makes a vital contribution to the world's economy and to the well-being and sustenance of society in general and is at the heart of globalization. However, the trend of its dramatic growth over recent years, especially in the road sector, is widely perceived as having produced unacceptable effects on the environment through such impacts as congestion, noise, air pollution, and demand on energy. It is timely and topical to examine the current practices in Europe with regard to the environmental management of transport because of the density of traffic and the significance of environmental issues in terms of politics, planning, and the whole debate on sustainable development. In this context, managerial actions are focused on the environmental dimension of the holistic concept of the logistic chain and its included operations. The chapter may be seen as a generic study of the perspectives of the different stakeholders in the European transport sector in responding to the challenge of striking a balance between the socioeconomic benefits of transport growth and the related environmental constraints. The case of European seaports is investigated as an exemplar of response options.

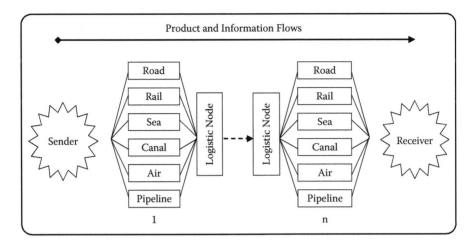

FIGURE 16.1 The logistic chain.

16.1.1 DEFINITIONS AND BACKGROUND

The widely recognized concepts of supply chain and logistic chain management were mostly developed and driven by financial imperatives. With business management entering the era of internetwork competition, individual businesses were no longer competing as solely autonomous entities, but rather as supply chains.[1] Transportation services play a central role in supply-chain operations, moving materials from supply sites to manufacturing facilities, repositioning inventory among different distribution centers, and delivering finished products to customers. Research in the field of supply-chain management argues that transportation often represents one of the chain's weaker elements.[2]

Focusing on transport as a critical element of the supply chain, the logistic chain may be defined as the network of the successive physical and conceptual links involved in the transport and placement of goods. Intermodality, the process of transporting freight by means of a system of interconnected networks involving various combinations of modes of transportation, lies at the heart of the concept of the logistic chain. The chain (figure 16.1) consists of successive links between movement patterns (transport modes) and nodal points (logistic nodes) in an integrative, intermodal concept from point of origin to point of consumption.

The environmental management of the logistic chain addresses the required functional organization and the available management options in order to minimize the environmental impacts arising from the operation of intermodal transport systems. The concept recognizes the physical and economic links associated with an integrated transport system, or chain, specifically designed to transport or deliver goods in a cost-effective and sustainable manner.

16.1.2 MAIN ENVIRONMENTAL IMPACTS OF TRANSPORT

Transport systems play a major role in the economic life of industrialized countries and in the daily life of their citizens. An efficient transport system is a crucial precondition for economic development and an asset in international competition. In the European Union, the transport service industry accounts for about 7% to 8% of the GDP.[3] Apart from its economic and social significance, transport is a major contributor to various environmental problems. "The dramatic increase in transport demand, and in particular for road transport and aviation, has made the sector a major contributor to several health and environmental problems in Europe (71)."[4]

Transport operations have a significant impact on the natural environment and are main contributors to local and global environmental problems. Emissions of air pollutants such as CO, NO_x, SO_2, Hydrocarbons (HCs), Volatile Organic Compounds (VOCs), lead, and particulates contribute to local air pollution, endangering human health. Carbon dioxide (CO_2) emissions from transport are a major contributor (29% of man-made CO_2 is emitted by transport[5]) to the greenhouse effect and global warming. The construction of transport infrastructure can result in the modification of water systems and the disruption of hydrological processes. Furthermore, transport infrastructure covers an increasing amount of land to the virtual exclusion of other uses, cuts through ecosystems, and spoils the view of natural scenery and historic monuments. Runoff from roads caused by vehicles leads to surface and groundwater pollution, while at sea, routine and accidental releases of oil by shipping contribute further to the pollution of the seas. Additionally, accidents associated with transport produce a heavy social cost, and nuisances from traffic noise, congestion, and the consumption of nonrenewable natural resources also represent major environmental issues.

16.1.3 ENVIRONMENTAL REFERENCES OF THE DIFFERENT TRANSPORT MODES

There are six main modes of transportation: road, rail, marine shipping, inland water shipping, air, and pipelines. Although all the transport modes impact on the environment to a certain extent, modal choices have a significant influence on the environmental performance of transport systems. A table summarizing the significant environmental effects of the different transport modes can be found at the end of the chapter in the appendix. In terms of energy efficiency and emissions of greenhouse gases and other air pollutants, road and air transport may be considered less environmentally friendly modes than shipping and rail transport. Table 16.1[6] presents some key figures with regard to the environmental performance of the different transport modes in selected areas.

Figure 16.2 presents a comparative example of road transport and short sea shipping in terms of energy efficiency and CO_2 emissions.[7] The comparison is based on fuel consumption and the emitted CO_2 of a full load by truck in a journey from Dortmund to Lisbon. The figures for short sea shipping take into consideration the road precarriage to the Port of Rotterdam and on-carriage from the Port of Lisbon.

The fuel consumption and the emissions of carbon dioxide appear to be more than three and a half times higher in the case of road transport.

16.1.4 ADDING THE ENVIRONMENTAL COMPONENT IN POLICY AND DECISION MAKING

Transport responds to society's demands, and regardless of the efficacy of the planning and operation, a certain level of environmental impact arising from the

TABLE 16.1
Environmental References by Mode of Transportation

Energy efficiency: In terms of ton-kilometers, shipping consumes 0.12–0.25 MJ, rail transport 0.60 MJ, and road transport 0.70–1.20 MJ. Concerning the efficiency of energy consumption, 1 kg of oil for 1 km can transport 50 tons by truck, 97 tons by rail, and 127 tons by water.

Air pollution: Carbon dioxide emissions in the European Union area are 30 g per ton-kilometer in short sea shipping, 41 g in rail transport, and 207 g in road transport. Of the total amount of nitrogen oxide emissions, 51% originates from road transport vehicles and 12% from other traffic. The majority of sulfur dioxide emissions from transportation originate from shipping.

External costs: These are the costs that traffic causes society, such as expenses connected with air emissions, climate change, infrastructure, noise, accidents, and congestion. The total amount of external costs incurred in EU countries, Norway, and Switzerland is 134.3 million euros per year. Of these expenses, 92% are caused by road transport, 2% by rail transport, and 0.5% by shipping.

Source: Reference 6.

operation of the system is inevitable. The concept of sustainability embraces the economic, sociopolitical, and environmental considerations. Sustainable transport can be seen as "satisfying current transport and mobility needs without compromising the ability of the future generations to meet these needs (151)."[8] From the perspective of environmental policy, dedicated environmental management of the logistic chain offers an opportunity to move toward a more sustainable, more efficient, and less polluting transport system that makes less demand on resources.

In launching Part D (*Transport and the Environment*) of *Transportation Research* in 1996, the editor-in-chief states in the preface: "The environment is one of those

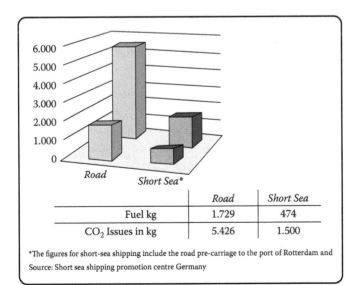

	Road	Short Sea
Fuel kg	1.729	474
CO_2 Issues in kg	5.426	1.500

*The figures for short-sea shipping include the road pre-carriage to the port of Rotterdam and
Source: Short sea shipping promotion centre Germany

FIGURE 16.2 Comparison of a full load by truck from Dortmund to Lisbon.

TABLE 16.2
Modal Split of Freight Transport in Europe

	Road (%)	Rail (%)	Inland Waterways (%)	Pipelines (%)	Sea (%)
1970	34.7	20.0	7.3	4.5	33.5
1980	36.3	14.6	5.3	4.3	39.4
1990	41.9	10.9	4.6	3.0	39.6
1991	42.3	9.8	4.5	3.3	40.0
1995	43.0	8.5	4.4	3.1	41.0
2000	43.2	8.2	4.2	2.8	41.6
2001	44.0	7.9	4.1	2.8	41.1
2002	44.7	7.7	4.1	2.8	40.8

topics that refuse to go away. Meanwhile there is mounting concern that with the current rapid expansion of traffic, sustainable development will not be possible without major changes in transport policy and technology." In this context the challenge for transport policy is to strike a balance between the economic and social benefits of transport and its negative impacts on society and the environment.[4] From an operational perspective the challenge could be rephrased so as to add the environmental component into the decision making and management of the logistic chain.

16.2   EUROPEAN POLICY PERSPECTIVE

16.2.1   EUROPEAN TRANSPORT SYSTEM: MODAL SHARE AND GROWTH TRENDS

The general picture with regard to the European transport system and the actual performance of the different transport modes are shown in the following tables. Table 16.2 gives an overview of the modal split in European freight transport from 1970 to 2002.[5] Figure 16.3 compares the tonne-kilometer growth of each transport mode in Europe for the period from 1995 to 2004.[9]

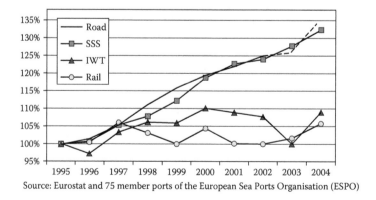

Source: Eurostat and 75 member ports of the European Sea Ports Organisation (ESPO)

FIGURE 16.3 Tonne-kilometer growth by mode of transportation in Europe.

It can be observed that road transport and short sea shipping are by far the most commonly used modes for freight transportation in Europe. Additionally, road transport and shipping are the modes that present the highest annual growth rates. Rail transport and inland water transport currently fail to reach their full potential.

16.2.2 Main EU Policies and Programs

Taking into account the realities and trends in the European freight transport system and especially the imbalanced growth in favor of road transport in land transportation, in 2001 the European Commission launched its White Paper on transport: "European Transport Policy for 2010: Time to Decide."[10] The central idea of the paper with regard to freight transportation is that while the amount of transported goods increases, distributions between modes of transport should be balanced so that the relative share of road transportation does not grow further. In line with the central idea, the main policy objectives of the White Paper are aimed at:

1. Shifting the balance between the modes of transportation by rebalancing the modal shift. This includes a series of proposed measures such as revitalizing the railways, promoting short sea shipping and intermodal transportation, and improving services in road transport and aviation.
2. Eliminating bottlenecks in the transport system. The proposed measures include infrastructural and procedural interventions in the European transport network, but also promoting research in technical and conceptual areas in order to tackle inefficiencies in unimodal and intermodal transport operations.

The modal shift from road transport to cleaner transport modes such as short sea shipping, inland shipping, and rail transport is one of the main aims of the European transport policy. Intermodality is of fundamental importance for developing competitive alternatives to road transport,[11] and therefore rebalancing the modal split in the European transport system could be achievable through the development of door-to-door, integrated, and intermodal transport chains. Intermodal transport chains are characterized by a much higher degree of complexity than single-transport-mode solutions in terms of procedures, administration, and technology in use, as shown in figure 16.4.[12] Those complexities have an impact in terms of economics and efficiency and also play a part in the restriction of the growth of intermodal transportation. In a few important European corridors, intermodal transport has the potential to reach a market share of 30%, while currently it represents between 2 and 4% of freight transport.[13] New logistics concepts, innovative technologies, and holistic chain management tools are needed before intermodal transport can really fulfill its potential by offering competitive alternatives to road transport in terms of cost and quality on an integrated, door-to-door basis.

In line with the policy objectives of its White Paper on transport, the European Commission launched a series of policy initiatives and programs aimed at rebalancing of the modal split and the growth of intermodal transportation. Selected initiatives and programs are presented in table 16.3.

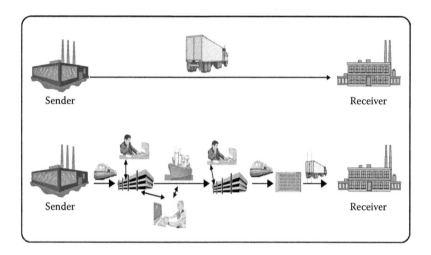

FIGURE 16.4 Administrative and operational complexities in intermodal transport chains.

TABLE 16.3
Selected EU Policy Initiatives and Programs

EU Policy Initiatives	Description
White Paper	Modal shift from road and air transport to cleaner transport modes "The growth in road and air traffic must therefore be brought under control, and rail and other environmentally friendly modes given the means to become competitive alternatives."
Marco Polo program	To help shift the expected increase of international road freight to short sea shipping, rail, and inland waterway.
Short sea shipping promotion	Establishment of a network of national short sea shipping promotion centers in all EU countries.
Motorways of the sea	To offer real competitive alternatives to land transport. The "motorways of the sea" concept aims at introducing new intermodal maritime-based logistics chains in Europe.
Standardization and harmonization of intermodal loading units	To reduce inefficiencies in intermodal transport resulting from various sizes of containers circulating in Europe. Furthermore, the measure will help to better integrate short sea shipping into the intermodal transport chain.
Freight Integration Action Plan	To improve the organization of intermodal freight transport. With this initiative, the commission intends to help improve freight-forwarding practices to boost intermodal transport.

16.3 PERSPECTIVE OF THE MAJOR PLAYERS

16.3.1 MAJOR PLAYERS IN THE LOGISTIC CHAIN

The following parties play a significant role in the freight transport market: shippers, carriers, and transport operators of all modes, third-party intermediaries, and logistic node operators. The shippers are the owners of the cargo to be transported. Shippers do not operate transport but generate and guide the transport demand from point A to point B. The carriers and transport operators of the different transport modes are the owners of the means of transportation (trucks, locomotives, vessels, barges, aircrafts, and pipelines). They are the parties that physically perform the transportation of cargo. The third-party intermediaries (freight forwarders, logistic service providers, intermodal transport operators) are working on behalf of the shippers and are responsible for the management of the logistics associated with the transport of goods. Finally, the logistic node operators are the parties responsible for the handling, warehousing, or forwarding of goods inside the physical limits of a given logistic node (for example, seaport, inland port, airport, dry port, warehouse, logistic center).

The business relationship between the major players and their position in the chain is shown in figure 16.5. It is a genuine complex of interrelated interests and responsibilities. It highlights the difficulties of gaining consensus on environmental

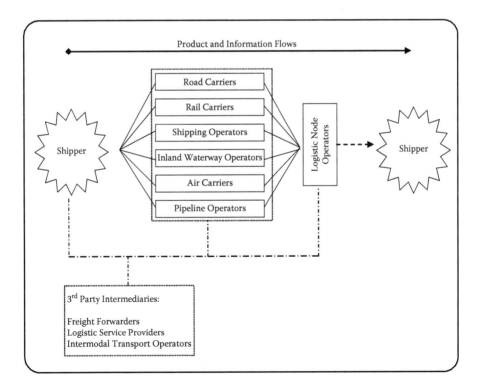

FIGURE 16.5 The major players in the logistic chain.

issues, achieving comprehensive implementation of policies, and demonstrating compliance with legislation.

Nowadays, a clear distinction in the roles and responsibilities of the major players in the transport market is becoming an increasingly challenging exercise due to the different possible arrangements for transporting goods, but also due to the dynamic nature of the transport market and the horizontal and vertical integrations that occur throughout the logistic chain. Horizontal merging occurs between port operators and between shippers in order to create economies of scale. It is a common practice for carriers to invest in terminals and for unimodal carriers to invest in other transport modes and logistic service providers offering intermodal and integrated door-to-door services.

16.3.2 Interest and Practice with Regard to the Environmental Management of the Logistic Chain

The interest and practice of the major players with regard to the environmental management of the logistic chain has been researched as part of the EC Project ECOPORTS, with particular reference to sustainable, logistic chain operation.[14] The study of the logistic chain addressed the main policies, objectives, actions, and initiatives of shippers, carriers, and third-party intermediaries (logistic service providers and intermodal transport carriers) concerning the environmental management of their transport operations. Information was obtained based on sources that included online websites and publications such as newsletters, environmental reports, and other related studies of fourteen selected companies acknowledged as carrying out best practices with regard to the environmental management of transport. Although it is acknowledged that the sample cannot be deemed representative of the transport industry as a whole, the results serve as a summary of current best practices and indicate progress on selected issues.

The assessment was based on a checklist of information summarized in table 16.4.

TABLE 16.4
Information Checklist

Checklist	Explanation
Policy statements	How companies perceive their role and responsibilities in relation with the environmental management of transport
Objectives and reported actions and solutions	Main policy objectives aiming to reduce the environmental impact arising from transport operations and related actions taken and solutions implemented
Benefits	Resulting benefits for the environment and for the companies
Initiatives	Projects and initiatives that address health, safety, environmental, and security aspects of transport in a chain concept

TABLE 16.5
Policy Statements

Shippers

"We believe that our social and environmental responsibility covers the whole supply chain." —IKEA

"We leverage our communication opportunities to convince suppliers and consumers alike of the importance of environmental protection, and to strengthen the role of the environment as a factor in the supply-and-demand equation." —Otto Versand

Carriers and Logistic Service Providers

"Since we are one of the leading logistics service providers in Europe, the company has to carry part of the burden to reduce environmental impacts of transport." —Schenker AG

The observations that can be drawn from the information available on policies, objectives, benefits, and initiatives may be summarized as follows.

16.3.2.1 Policy Statements

With regard to the policy statements, selected examples of which are presented in table 16.5, the following observations can be made:

Although the shippers are not physically involved in transport operations, they actually generate and guide the transport demand. In line with the concept of corporate social and environmental responsibility, good shippers' practice starts by accepting environmental responsibility for their entire supply chain and of their generated transport as part of it. A further step in good shippers' practice in this context is using their influence as customers toward the outsourced transport and logistic services in order to ensure the operation of an efficient logistic chain in both economic and environmental terms.

Concerning the carriers and logistic service providers, the picture is more straightforward, as their core business is directly related to transportation. Good practice could not be considered anything less than committing to carry part of the burden to reduce the environmental impacts of transport operations.

16.3.2.2 Policy Objectives, Reported Actions, and Solutions

Table 16.6 categorizes and summarizes the policy objectives and reported actions and solutions implemented by the researched companies with regard to the environmental management of the logistic chain.

Main shippers' policy objectives referred to the modal shift from road transport and aviation to more environmentally friendly transport modes, to the reduction of the CO_2 emissions of their transportation, to monitoring and reporting environmental performance of their logistic chain, and to cooperating with other partners in the chain undertaking common initiatives to tackle common challenges. In terms of actions and implemented solutions, some major shippers are using their influence as customers while outsourcing transport and logistic services to ensure

TABLE 16.6

Policy Objectives and Implemented Solutions

Policy Objectives	Reported Actions and Solutions	
Reduce greenhouse gas emissions	Emission control technology	
Increase efficiency	Efficient engine technology	
Reduce energy use	State-of-the-art vehicles, vessels, aircrafts	Technical
State-of-the-art technology		
Alternative fuels	Alternative fuels	
	IT systems	
Improve the environmental performance of transport operations	Performance criteria on carriers	
Consolidating goods	Cooperation with other partners in the chain	Procedural
Coordinating transports		
Avoiding empty positioning		
Modal shift: from road and air to rail, sea, and inland waterways	Initiatives toward the modal shift	
Monitor transport impacts	ISO 14001 certification	Managerial
Report performance	Training in eco-driving	

an efficient and sustainable distribution of their products. Good-practice examples include applying specific criteria for the evaluation of the environmental performance of the transport operators and demanding commitment toward continuous environmental improvement. "We require that our transport service suppliers inform us about, and continuously improve their environmental performance."[15] The policy objectives of carriers and logistic service providers are oriented toward reducing air emissions and increasing the energy efficiency of transport operations. Therefore, they are investing in the implementation of state-of-the-art innovative technical solutions, including emission control and engine technology, vehicle technology, and alternative fuel technology.

16.3.2.3 Benefits

The reported benefits for both the environment and industry from the applied policies, measures, and solutions are presented in table 16.7.

16.3.2.4 Existing Initiatives

In addition to the individual transport policies and objectives of each company, scrutiny of the companies' websites provides details of collaborative initiatives bringing together shippers, carriers, and logistic services providers designed to deliver a more sustainable transport system. IKEA, Hewlett Packard, Maersk, and P&O Nedlloyd participate in the Business for Social Responsibility (BSR) Clean Cargo working group that aims to promote sustainable product transportation at sea and in the port. IKEA, again, together with Stora Enso, is part of the Business Leaders Initiative on

TABLE 16.7
Reported Benefits for the Industry and the Environment

Reported Benefits	
For the Environment	**For the Industry**
Reduction in CO_2 emissions	Differentiation from competitors
Reduction of air emissions	Proactively alleviate negative publicity
Decrease in fuel consumption per ton of goods	Improvements in efficiency and processes
Energy savings	
Lorries off the road	Increased trust
Safer distribution	Cost savings (efficiency, insurance rates)

Climate Change (BLICC), which aims to reduce CO_2 emissions arising from manufacturing and transport operations. Shell Chemical integrates the Responsible Care management code on product distribution. The code is designed to reduce the risk that the transportation and storage of chemicals poses to the public, carriers, customers, contractors, company employees, and the environment.

16.4   PORT SECTOR'S PERSPECTIVE

16.4.1   SIGNIFICANCE OF SEAPORTS AS MAJOR LOGISTIC NODES

Logistic nodes are the nodal points in the logistic chain where the functions of cargo handling, warehousing, and modal transferring take place. Logistic nodes can be dry ports, seaports, inland ports, airports, warehouses, stores, and production and manufacturing sites. Seaports and their port areas are highly significant nodes within the whole chain complex through virtue of concentration, diversity of operation, and critical connectivity of the chain operation. Seaports are characterized by a higher degree of complexity and variety of operations in comparison with other logistic nodes. Many port areas demonstrate intense intermodal concentration as several transport modes coalesce to form the functioning node (figure 16.6).

In its most complex form, the transport network of seaports may include sea routes, inland waterways, roads, railways, and pipelines. Airports may be situated in close proximity or even within some seaport areas, and they are commonly linked to major ports. Additionally, in most cases port areas are situated in close juxtaposition with urban areas, and may even be bounded by, or include, areas of special environmental significance due to the presence of protected habitats and ecosystems. The diversity of types of cargo, the range of activities, products, and services conducted within the port area, the multiple use of the land and sea areas, and the physical impact of the associated infrastructure all identify seaports as major logistic nodes.

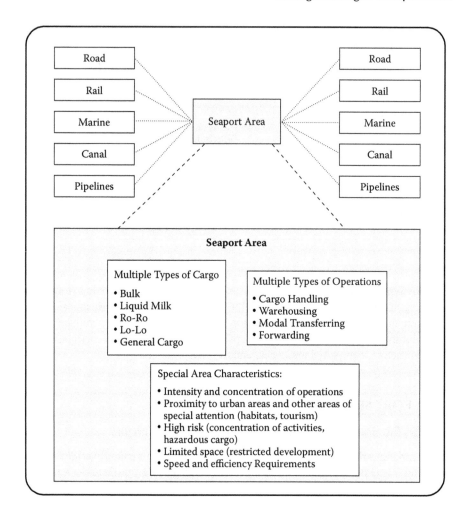

FIGURE 16.6 The seaport area as a major logistic node.

16.4.2 ROLE OF PORTS IN THE ENVIRONMENTAL MANAGEMENT OF THE LOGISTIC CHAIN

Many port authorities are increasingly active in applying environmental manage-ment to their port area, not just to the immediate vicinity of the waterfront or areas devoted solely to port-related activities. They have been driven by their liabilities and responsibilities as landlords insofar as that in the interpretation of some envi-ronmental legislation port authorities may reasonably be expected to bring some influence to bear on the environmental performance of their tenants and operators. In identifying their significant environmental aspects, elements of the authority's activities, products, or services that can interact with the environment,[16] ports should

TABLE 16.8
Progress in Implementation of Key Environmental Management Activities by ESPO Members Based on EPF/ESPO Surveys

Environmental Management Component	1996	1999	2004	2006	%ᵃ + or –
Environmental plan?	45	52	62	82	+37
Plan aims for compliance plus?	32	41	48	72	+40
Does plan aim to raise environmental awareness?	44	62	69	68	+24
Is environmental monitoring carried out?	53	60	65	72	+19
Does plan involve community and stakeholders?	53	60	39	78	+25
Is ESPO code available?	41	48	53	53	+12
Designated personnel?	55	65	67	88	+33

ᵃ Given the vagaries of questionnaire survey returns in terms of the extent to which respondents are truly representative of the sector, it may be suggested that the trends are more relevant than the absolute percentage values. In this case, the sector can demonstrate continual improvement, which in itself is a positive attribute of an environmental management system. ISO 14001 (1996) defines such progress as the process of enhancing the environmental management system to achieve improvement in overall environmental performance in line with the organization's (ports sector) environmental policy. It notes that the process need not take place in all areas of activity simultaneously.

take into account aspects for which they are legally liable, those of their tenants and operators over which they could bring some influence, and issues of national or local significance pertinent to the port area. Functional organization of an environmental program for the port area ipso facto implies influence or involvement with environmental facets of the logistic chain. Table 16.8[17] demonstrates the progress made by members of the European Sea Ports Organization (ESPO) during the period 1996–2006 in implementing key components of an environmental management program. The involvement of the local community and other stakeholders is particularly apposite in the context of the logistic chain.

The dilemma for the port authority is that, as in the case of identifying significant environmental aspects, it may not necessarily be directly, legally responsible for the activities, products, and services of the components of the logistic chain, but its overarching administrative role, ownership of the estate (land and sea), and permanency of operational presence mean that the port is the obvious point of contact and readily identifiable player for any issues related to the environment in the whole port area. This is a situation similar to various fiscal, health, safety, and waste issues. The emerging role of port authorities with regard to the environmental management of the logistic chain is therefore that of facilitator. The concept of ports as facilitators refers to the contribution that ports can make in helping the whole port community (including partners in the logistic chain) to deliver compliance with legislation, prevention of pollution, reduction and mitigation of environmental impacts, sustainable development, and evidence of satisfactory performance. Positive steps are being made to achieve these objectives by the development and implementation of appropriate procedures for the exchange of information and cooperation among the different players in the logistic chain, collaboration on research and development of practicable tools and methodologies, and identification of best practices solutions to common challenges. This approach by the port authority includes providing the necessary communication platforms and coordinating the exchange of safety, health, and environmental information among the different port commercial visitors, and between the port and other authorities. It also entails working with other parties in tackling the informational, technical, and procedural bottlenecks restricting the efficient operation of intermodal transport chains.

Some selected good-practice examples concerning the active role of ports in tackling the information, technical, and procedural bottlenecks of intermodal transport include:

- **Port Infolink**—Port of Rotterdam: Port Infolink is a port-wide IT platform used in the port of Rotterdam aiming toward one single port community system. The port community system enables all the links within the port of Rotterdam's logistics chain to efficiently exchange information with one another.[18]
- **Port Railway Information and Operation System (PRIOS)**—Port of Hamburg: A state-of-the-art IT system used in the port of Hamburg aiming to optimize rail transport operations.[19]
- **AMSbarge**—Port of Amsterdam: AMSbarge is an inland navigation ship that has its own heavy container crane and can load and unload containers independently of terminals or cranes on the quay—provided there are mooring facilities. The concept has been developed by the Port of Amsterdam in cooperation with a number of large shippers.[20]

16.5 DISCUSSION

The holistic management of the logistic chain in terms of efficiency and cost effectiveness is a reality in modern business management. The challenge now is to add the environmental component to the decision-making process and the day-to-day management of the logistic chain. The transport sector is a major contributor to various environmental problems, and therefore there is an absolute need to manage the environmental impacts arising from transport operations.

The balance between the different transport modes has a significant impact on the environmental performance of the transport system. The current balance in Europe and the trends in transport growth constitute a serious threat to the sustainable operation of the European transport system. The policies of the European Commission focus on rebalancing the modal split in Europe and improving the environmental performance of transport by promoting the implementation of technical, procedural, and managerial solutions throughout the logistic chain.

With regard to the impact of the EU transport policies, the following remarks can be made:

- Several research and development projects, supported by the European Commission, have been undertaken during the last 5 years. Those projects have produced valuable tools, methodologies, and innovative technical, procedural, and managerial concepts contributing to the improvement of the environmental performance of the European transport system.
- The trends concerning the envisaged modal shift from road transport to cleaner transport modes do not appear to be optimistic. Road transport is still gaining modal share and is presenting the highest growth rate among the different modes of transportation.
- The communication between the European Commission and the major transport players involved in intermodal transport needs to be strengthened. In many cases, gaps can be observed between the industry's practice and the objectives of the European transport policy. An example is the envisaged harmonization of container sizes in Europe. The European Commission has been considering since 2001 to introduce the European intermodal loading unit, a container unit 44 feet in length, to replace current containers and to cope with inefficiencies occurring due to the presence of different container sizes throughout Europe. The practice, though, is currently oriented toward the use of 45-foot containers, which are increasingly being introduced to the market as a result of the competition between Ro-Ro and Lo-Lo carriers.

Concerning the transport industry's perspective toward the environmental management of the logistic chain, it should be noted that industry's best practices are in line with the main policies of the European Commission. Such practices include efforts toward the modal shift from road transport and aviation to cleaner transport modes and the implementation of state-of-the-art technologies and IT systems. Port environmental management has progressed steadily during the last decade, and further progress is achievable through the proactive, collaborative programs of the sector itself aimed specifically at compliance with legislation and the attainment of high standards of environmental protection through voluntary self-regulation. The role of port authorities as facilitators within the logistic chain is poised to make a substantive contribution toward the goal of effective environmental management throughout the transport system.

As the concept of environmental management of the logistic chain evolves, there is likely to be growing awareness among all the stakeholders involved that increased collaboration based on the free exchange of information and experience will produce mutual benefit in terms of cost and risk reduction, and evidence that the goals of environmental protection and profitable commercial activities are not necessarily mutually exclusive. Effective environmental management of the logistic chain has the potential to deliver credible sustainable development in practice.

APPENDIX

TABLE 1
Significant Environmental Aspects of the Different Transport Modes

Transport Modes	Air	Water Resources	Land Resources	Solid Waste	Noise	Risk of Accidents	Other Impacts
Marine transport	Emission of sulfur dioxide (SO_2)	Modification of water systems during port construction and dredging. Water pollution due to routine and accidental releases of oil	Land taken for infrastructure. Dereliction of obsolete facilities	Vessels withdrawn from service		Bulk transport of fuels and hazardous substances	
Inland water transport		Modification of water systems during canal cutting and dredging operations	Land taken for infrastructure. Dereliction of obsolete facilities	Vessels and barges withdrawn from service		Bulk transport of fuels and hazardous substances	
Rail transport			Land taken for rail infrastructure. Dereliction of obsolete facilities	Abandoned lines equipment and rolling stock	Railway noise and vibration	Transport of hazardous substances	Spatial separation effects (wildlife habitats, farmlands, urban areas) Amenity and severance

Mode	Air pollution	Water pollution	Land/resource use	Solid waste	Noise	Safety/hazards	Spatial effects
Road transport	Local air pollution (CO, NOx, HCs, VOCs, Pb, particulates) Global air pollution (CO₂, CFCs)	Pollution of surface and groundwater due to runoff from roads. Modification of water systems, disruption of hydrological processes	Runoff from roads Land taken for infrastructure Erosion of exposed soil surfaces due to the extraction of road building materials	Discarded vehicles Abandoned spoil tips and rubble from road works	Noise and vibration	Transport of hazardous substances Deaths, injuries, and property damage Risks of structural failure in old facilities	Congestion Spatial separation effects (wildlife habitats, farmlands, urban areas) Amenity and severance
Air transport	Air pollution (CO, NOx, HCs, VOCs) Global air pollution (CO₂) Depletion of ozone due to aircraft emissions in the stratosphere	Modification of water tables, river courses, and field drainage due to airport construction	Land taken for infrastructure Dereliction of obsolete facilities	Aircrafts withdrawn from service	Noise around airports		
Pipelines		Pollution of surface and groundwater due to oil leakages				Transport of fuels	Barrier to wildlife migration in the case of aboveground pipelines

Sources: References 3, 21–29.

REFERENCES

1. Lambert, D. M. 2001. The supply chain management and logistics controversy. In *Handbook of logistics and supply-chain management*, ed. A. M. Brewer, K. J. Button, and D. A. Hensher, 99–126. Amsterdam: Pergamon.
2. Stank, T. P., and T. J. Goldsby. 2000. A framework for transportation decision making in an integrated supply chain. *Supply Chain Management: An International Journal* 5:71–77.
3. Stanners, D., and P. Bourdeau, eds. 1995. *Europe's environment, The Dobris assessment*, 676. Copenhagen: European Environmental Agency (EEA).
4. European Environmental Agency. 2003. *Europe's environment: The third assessment*, 344. Copenhagen: European Environmental Agency.
5. European Commission Directorate-General for Energy and Transport. 2003. *Energy and transport in figures: Statistical pocketbook 2003*, 198. Brussels: European Commission Directorate-General for Energy and Transport.
6. Shortsea Promotion Centre Finland. 2003. Shortsea shipping and the environment. Theme bulletin. http://www.shortsea.fi/cutenews/data/upimages/Environment_2003.pdf (accessed January 12, 2005).
7. Shortsea Promotion Centre Germany. 2005. Environmental data. http://www.shortseashipping.de/eng/umwelt/umwelt.html (accessed January 12, 2005).
8. Black, W. R. 1996. Sustainable transportation: A US perspective. *Journal of Transport Geography* 4:151–59.
9. Vanderhaegen, M. 2006. Motorways of the sea in the European Short Sea policy. In *RoRo 2006*. Ghent.
10. European Commission. 2001. *European transport policy for 2010: Time to decide*, White paper, 1–119. Luxemburg: European Communities.
11. European Commission. 2004. *Transport research in the European research area: A guide to European, international and national programs and other research activities*, ed. Directorate-General for Energy and European Communities, 78.
12. Logit Systems. 2005. Conference presentation. http://www.logit-systems.com/.
13. European Commission. *Freight intermodality: Results from the transport research program*, ed. Directorate-General for Energy and Transport, 20. European Communities.
14. ECOPORTS Foundation. 2005. *Short feasibility study on how to integrate these new modalities in the ECOPORTS project*, 64. ECOPORTS project D23a, ECOPORTS Foundation, Amsterdam.
15. Stora Enso. 2005. www.storaenso.com (accessed February 21, 2005).
16. International Organization for Standardization. 1996. EN ISO 14001.
17. Wooldridge, C. 2006. Promoting sustainable development: Progress through partnership. Paper presented at European Sea Port Organization's Conference, Stockholm.
18. Port infolink. 2005. Towards one single port community system. http://www.portinfolink.com/english/content/informatie/over_portinfolink.asp accessed February 2, 2005).
19. ECOPORTS Foundation. 2005. *Concise plan for the implementation of environmental solution in port area*, 25. ECOPORTS project D182005, ECOPORTS Foundation, Amsterdam.
20. Port of Amsterdam. 2006. RePort overview 2006. http://www.portofamsterdam.com/smartsite14601.dws (accessed May 1, 2006).
21. Rothengatter, W. 2003. Environmental concepts: Physical and economic. In *Handbook of transport and the environment*, ed. D. A. Hensher and K. J. Button, 9–35. Amsterdam: Elsevier.

22. Hunter, C., J. Farrington, and W. Walton 1998. Transport and the environment. In *Modern transport geography*, ed. B. Hoyle and R. Knowles, 374. Chichester, UK: John Wiley & Sons.
23. Linster, M. 1990. Background facts and figures. In *Transport policy and the environment*, ed. European Conference of Ministers of Transport, 9–45. Paris: ECMT-OECD Publications Service.
24. Houghton, J. 1996. Sustainable transport: How the Royal Commission sees the future. In *Transport and the Environment*, ed. B. Cartledge, 23–44. The Linacre Lectures 1994–95. Oxford: Oxford University Press.
25. Handy, S. 2003. Amenity and severance. In *Handbook of transport and the environment*, ed. D. A. Hensher and K. J. Button, 117–40. Amsterdam: Elsevier.
26. Owens, S. 1996. "I wouldn't start from here": Land use, transport, and sustainability. In *Transport and the environment*, ed. B. Cartledge, 45–61. The Linacre Lectures 1994–95. Oxford: Oxford University Press.
27. Reid, B. 1996. Railways and sustainable development. In *Transport and the environment*, ed. B. Cartledge, 81–99. The Linacre Lectures 1994–95. Oxford: Oxford University Press.
28. Somerville, H. 1996. Airlines, aviation, and the environment. In *Transport and the environment*, ed. B. Cartledge, 110–31. The Linacre Lectures 1994–95. Oxford: Oxford University Press.
29. Graham, B. 1998. International air transport. In *Modern transport geography*, ed. B. Hoyle and R. Knowles, 311–36. Chichester, UK: John Wiley & Sons.

17 Green Ports and Green Ships

Christopher F. Wooldridge, Thomas H. Wakeman, and Sotiris Theofanis

CONTENTS

ABSTRACT

The chapter examines the apparently discrete areas of port and environmental management makes the case for practicable integration of their environmental implementation strategies through collaborative effort to mutual advantage of their commercial interests and the environment as a whole. The global, strategic significance of the wide range of activities and operations facilitated through the shipping industry and port sector is established in terms of world trade, importance to the nation-state, and the increasingly significant, timely, and topical environmental imperative. With examples drawn from the United States and Europe, the major issues are identified, and the various management response options are examined in terms of policy, environmental management systems, implementation, and case studies. The legislative and regulatory regimes are summarized, stakeholder pressure is identified by interest area, and the benefits of a proactive response to environmental liabilities and responsibilities are tabulated. The challenge of filling the gap between policy and actual delivery of continuous improvement of environmental quality is addressed through examination of the role and implementation of environmental management systems. Baseline and benchmark performance is established with reference to surveys of port sector achievements over the last 10 years of collaborative research and development of practicable tools and methodologies. Finally, an overview of options, instruments, and approaches for encouraging and facilitating further advances in the effective environmental management of port and shipping activities is set out for consideration. In closing, a paradigm shift is proposed and a call for a sustainable environmental philosophy is set forth as a framework for future maritime actions.

17.1   INTRODUCTION

Since the 1950s, the world's population has doubled, and the global marketplace has emerged to serve this growing population's desire for manufactured products and commercial goods. International freight transportation has expanded to meet the world's trading partners' demands for connectivity. Much of this trade moves in containers with the volume more than tripling from 1995 to 2006.[1] Simultaneously during this expansion period, there was an awakening by the world's citizens to the environmental impacts of these industrial and transportation activities. It is estimated that the volume of imported cargo into the United States will triple between 2000 and 2020, with further environmental scrutiny by special-interest groups and community advocates because of their placement near urban centers and in sensitive environmental areas.[2] In addition, the number of national and international environmental laws and regulations has grown and expanded as rapidly as world trade, creating a legal framework for these groups to act against maritime activities. The collision of these two trends is nowhere more apparent than at the water's edge, where today's maritime industry's dynamic operation and development activities and environmental pressure oppose one another. The tension between these two societal forces can cause economic stagnation, transportation inefficiencies, and human health impacts if not dealt with in a positive manner.

The global marine transportation system is critical to the continued economic viability of the United States, Europe, and most other nations. No state in the world can sustain its industries and commerce without recourse to sea transport.[3] Ports have long been the gateways for trade to pass between coastal nations. They also continue to serve the interests of landlocked states in their role as strategic nodes in the international logistic chain. The oceans, coastal seas, and inland waterways of the world provide the most economical, cost-effective mode of transportation for goods to and from the global marketplace. Maritime transportation and the trading it supports cause landside activity that stimulates local economic activity, nurtures the emergence of civic centers, and ultimately feeds the prosperity of surrounding urban areas. The marine environment also provides a wide range of other significant benefits to humans, including food resources, recreation opportunities, and, indeed, life support. With the expansion of maritime activities on the world's oceans and waterways for bulk and containerized cargo transportation, there has been an increased potential for maritime operations to interfere with the flow of these other benefits to coastal populations. Similarly, as the volumes of cargo have grown, environmental impacts associated with this terrestrial movement have created negative land quality and air quality conditions. These conditions are becoming more pronounced over time. Accordingly, public concerns have mounted with each passing year. Recently there has been increasing interest in ensuring that these activities are "green" or environmentally sensitive. The situation calls out for proactive environmental management actions by the maritime community for both green ports and green ships. What are green ports and green ships? Are they part of a sustainable development initiative?

17.1.1 GROWING PRESSURES

For hundreds of years, ports were nodes that provided the critical interfaces between the sea and land. Ships were important, but it was the seaport that was at the core of the urban waterfront. It was the place where people gathered to talk and await the paddle wheelers along the Mississippi, the dock where men waited for building materials and women for cloth in London, and where bankers and brokers made their money in New York. Today's port is different. It is often isolated from the public and unnoticed. Many people have the attitude that the seaport is an anachronism. Nothing could be further from the truth. In fact, the seaport was never more important to the seafaring nations of the world than it is today. It is the gateway for the international trading of 95% of all manufactured products, and some believe container traffic to be the lynchpin of today's global marketplace.[1] Port operations are much more complex today than in earlier days, when all that was needed was "a basket, a couple of ropes, and three or four strong men." Now, ports are highly organized systems not only for cargo handling at the wharf and onto the terminal, but also as critical nodes in the whole logistic chain. In addition, ports have emerged as major industrial centers in their own right.[4] Shippers are concerned with dwell time, reliability, cargo transparency, and transportation costs. The port operations manager must be concerned with these issues as well as port safety, health, security, paving, traffic, customs, channel navigability, and crisis management, just to name a few of the many categories. Port managers must also be environmentally aware. They must be concerned not only because of regulatory oversight and community relations, but also because the potential costs that can follow an environmental mishap may be disastrous to financially strapped terminals.

Global trade and economic expansion are creating pressures that require ports to expand their infrastructure to handle the increasing volumes of cargo. This expansion, unless it is managed correctly, is potentially damaging to the adjacent land and waterside environments and to public relations with the surrounding communities. For example, factors on the landside may include transportation networks, local industries, urban development and zoning, governmental consents, permit requirements, and intermodal connections. Waterside factors are different but no less difficult to address. Factors may include ship design, number of ship calls, channel and berth dimensions, handling techniques, and international regulations. The driving force for today's port authority is not only how it deals with all these industrial factors, but also how it deals with growing environmental and community pressures.

17.1.2 EMERGING POLICY PARADIGM

Beginning in the 1960s and 1970s, incorporating environmental considerations and impacts analyses into transportation sector activities became a necessity by law. In order to complete the mandated analyses, the ecosystem was broken into components and individual modal impacts were evaluated. The environmental was divided into media categories (air, water, soil, and sediment) in order to analyze impacts. This formulation also led to the development of laws, regulations, and management strategies to tackle each media separately. Unfortunately this approach provided a sense of separation between media that does not exist in the environment. The media boundary is artificial and the system is actually closed, requiring an integrated approach if impacts and solutions are to be holistically dealt with by society.

Likewise, the transportation system was also seen until recently as a mix of several different and dissimilar modes. Ships, trains, and trucks were considered as separate modal and environmental stovepipes. Recognizing that transportation is now used and perceived as an intermodal system rather than as discrete modes demands that transportation operations and decision making be more integrated across modes, and it must be more expansive (global), considering routes that may span continents. For example, a container of tennis shoes may move by water from China to a port in the United States (New York), then by rail inland to a distribution center in Cleveland, and ultimately by truck to the retail store and consumer in St. Louis, Missouri. To make sustainable transportation decisions, environmental analyses and management must move toward being systematic with the integration of impacts to determine the real consequences of maritime development projects and routine operations.

Until recently, the current paradigm typically consisted of separate planning, project design, maintenance, and operational decision making, with environmental considerations restricted to compliance with environmental regulations. Adversarial relations among the stakeholders, rather than cooperative relations, have characterized the process. Few, if any, incentives exist for cooperation, and many barriers exist to developing cooperative approaches to incorporating environmental considerations in transportation operations and decision making. This old paradigm resulted in transportation operations and infrastructure that were not meeting needed capacity or contributing to environmental objectives. This failure, in turn, affected the viability of community, environmental health, and transportation systems at the local, state, and national levels. The maritime industry is using new tools (environmental management systems, port environmental management systems, and others) to create improved environmental conditions for ports and the vessels that call. Issues such as sustainability, stewardship, corporate social responsibility, triple bottom line, and environmental streamlining are new values being discussed by the industry as desirable aspects and outcomes of the marine transportation environmental management process.

17.1.3 GREEN PORTS

A green port or a green ship is difficult to define, but in general it uses the applicable laws and regulations as a baseline for its environmental performance. Further, it is considered a port or ship that not only meets all these environmental standards in its daily operations, but also has a long-term plan for continuously improving its environmental performance. The task of developing a program that accomplishes environmental integration and improvement across media and modes is a challenging one. The issues facing ports (which are land based) and ships (which are water based) are obviously different, but the intention to improve environmental performance is the same. The complexity, lack of appropriate information, and apparently conflicting public policy goals often exacerbate the difficulty in making effective public and business decisions. However, methodologies are emerging to address these issues from the American Association of Port Authorities (AAPA),[2] the European Sea Ports Organization,[5] and other international organizations.

One of the ways to advance the goal of becoming a model green port is to implement an environmental management system (EMS). An EMS is a systematic process to align organizational and environmental objectives throughout the organization using, for example, the International Organization for Standardization (ISO 14001) or Eco-Management and Audit Scheme (EMAS) as the guide. By delineating roles and responsibilities an EMS provides a framework for a structured approach to managing environmental responsibilities as well as defining a consistent measure of environmental performance. Its comprehensive approach to managing the significant environmental aspects of an organization's activities will enable a port to find and fix root causes of potential problems, prevent pollution, and conserve energy and resources. It allows the port to be proactive instead of reactive on environmental issues related to its facilities.

For future port and maritime transportation activities to be successful, environmental considerations must be part of a systemic management and decision framework that encourages sustainability and environmental stewardship. This can and, indeed, must be done worldwide. The approach to an environmental management framework (e.g., EMS) presented herein is inclusive of all interests. It is formulated so that effective, informed decisions can be made that represent the environmental aspirations of the wide range of stakeholders that constitute the green ports and green ship community.

17.2   KEY ENVIRONMENTAL ISSUES FOR PORTS AND SHIPS

There are a wide variety of potential environmental impacts associated with maritime vessel operations and port development activities. They can range from waterside impacts such as the introduction of invasive species to landside impacts from contaminated soils or releases of caustic substances. These impacts can have significant financial consequences too. The two facets can converge to create a significant environmental mitigation program with major cost implications. For example, the ports of Los Angeles and Long Beach have embarked on a multi-million-dollar air emissions control program to improve port and intermodal transportation-related air quality.[6] Improved performance of this type of program can only come from a better understanding of the environmental interactions between ports and ships and the management measures available for the port manager to respond to their associated ecosystem and community impacts.

There are multiple media and factors that should be considered in any assessment of the environmental impacts of ports and port-related activities. A list of principal impact factors is presented in tables 17.1 to 17.3. The tables and summaries are

---

TABLE 17.1
Key Environmental Issues: Waterside

Types of Issues

A. Dredging/Disposal Operations
1. Water column impacts
   a. Physical—turbidity, suspended solids
   b. Chemical—releases, DO, nutrients
2. Benthic impacts
   a. Burial—smothering
   b. Chemical—availability and uptake
3. Beneficial uses
   a. Beach nourishment
   b. Construction
   c. Habitat development

B. Habitat and Resource Protection
1. Essential fish habitat protection
   a. Habitat
   b. Spawning grounds
   c. Reef construction
2. Fish windows (environmental windows)
3. Threaten and endangered species
   a. Marine mammals (whale strikes)
   b. Fish species
   c. Sea turtles
4. Wetland protection and restoration

C. Spills and Vessel Discharges
1. Oily and gaseous substances
2. Solid wastes
3. Hazardous materials
4. Ballast water management

TABLE 17.2
Key Environmental Issues: Landside

Types of Issues

A. Brownfields/Contaminated Lands

1. Containment of pollution
2. Cleanup of hazardous and toxic materials
3. Restoration of property

B. Discharges to Land and Waterways

1. Spills
2. Industrial sources (liquid and solid)
3. Storm water

C. Habitat Areas

1. Bird nesting areas
2. Erosion control
3. Wetlands filling
4. Endangered species

D. Operation-Associated Pollution

1. Light towers
2. Noise
3. Vibration
4. Dust

---

organized into three media categories: waterside, landside, and air quality issues. A search of the environmental literature will provide a complete description of the multiple facets of each potential impact.[7]

17.2.1 WATERSIDE

Table 17.1 presents environmental impacts associated with waterside activities. Many, but not all, waterside impacts result from activities that are performed in

---

TABLE 17.3
Key Environmental Issues: Air

Types of Issues (NOx, SOx, Particulates, and Greenhouse Gases)

A. Ports and Terminals

1. Oceangoing vessels (e.g., cold ironing)
2. Harbor craft (tugs, pilot boats, etc.)
3. Cargo handling equipment (cranes, yard equipment, etc.)

B. Intermodal Connections

1. Highway vehicles
2. Locomotives

---

support of maritime shipping. They may be either the direct result of construction of facilities, such as navigation channels, or waterway operations by vessels, such as the introduction of invasive species in ballast water discharges. Landside impact may be translated to the waterside of course, for example, discharges. A brief description of selected major issues is presented below.

17.2.1.1 Dredging and Dredged Material Disposal

Dredging activities in a port include both the removal of sediments from channels and berths in a port and the disposal of the excavated sediments. Dredging can be undertaken for new construction or maintenance purposes. Channel and berth dimensions and depths are determined by the size of the vessels they are designed to service. As vessel size is increased because of economies of scale, channel and berth design must be modified to accommodate these larger vessels.[8] Hence, dredging is frequently an integral part of port operational and capital investment activities.

Many industrialized harbors now face difficulties in performing new construction and maintenance dredging because of historic contamination of their sediments. For many years the polluted sediments were considered benign, but research in the 1960s and 1970s revealed that contaminants could magnify in the food web and impact human health.[9] Why is dredging and disposal of contaminated sediment a problem? First, there is potential release of contaminants during dredging when sediments are resuspended.[10] Mobilized contaminants can be biologically available for uptake and accumulation in edible aquatic species. Second, after these dredged materials are excavated, they must be placed in a location that will contain their toxic constituents and protect the environment from degradation and human health from impairment.[11] Disposal of maintenance material from berths and channels is considered to be potentially the most environmentally harmful because it typically is composed almost exclusively of fine-grained sediments. This size fraction contains the greatest percentage of contaminants.[12]

Disposal sites are becoming increasingly difficult to find. Existing sites are filling up or are being converted to other uses that provide a higher return, particularly in urban areas. New sites are difficult to permit. Aquatic sites are being managed more tightly because of regulatory changes, which mean greater quantities are no longer suitable for open-water disposal. In short, disposal of dredged material is difficult. Beneficial uses of the processed material, as well as clean dredged material, have become an important consideration in locating disposal options.

17.2.1.2 Resource Protection

In the United States, federal and state resource agencies identify and designate Essential Fish Habitats (EFH) for all life stages of federally and state-managed species of marine and anadromous fish. These include all threatened and endangered species. The EFH obligations focus on resource habitat protection. This action was taken in recognition of the depleted nature of many of the managed fisheries stocks. Restrictions are typically placed on navigational dredging or port construction activities in one of two forms. When there is a potential conflict between these activities and an aquatic species, a project or seasonal restriction is established. When no threat to a species or habitat is identified, an environmental window is defined and activities may proceed.[13] The environmental windows' regulation continues to pose a significant challenge nationwide because of the need to complete essential construction on time while complying with environmental mandates for the protection of endangered and threatened species.[13]

17.2.2 Landside Issues

Table 17.2 presents various types of landside environmental impacts. Landside impacts primarily stem from contaminant releases and petroleum spills. Other impacts may be associated with terminal operations. A broader category of impacts may arise from port-generated traffic, including congestion and air quality impacts, which are treated in the next section.

17.2.2.1 Discharges and Spills

Ports must also be concerned with the potential sources of pollution generated landside.[14] Toxic or harmful substances can be released to groundwater or the atmosphere during transfer onto the terminal or transport operations when spills and discharges accidentally occur. Causative agents, including cargo, equipment, or personnel, can categorize these sources. Cargo can be spilled during loading or unloading activities, fugitive dust generated during dry bulk handling, or leakage emitted from pipelines transferring liquid bulk. Equipment problems generally stem from petroleum products, including fuel or lubricant leakage or air emissions. Personnel can be the source of pollution because of sloppy operations and careless handling or movement of cargo.

17.2.3 Air Quality Issues

Table 17.3 presents the general categories for vessel- and terminal-associated air resources impacts. Because of the growing volumes of vessel and intermodal traffic at ports, there is increasing concern about air pollution from ships, terminal cargo handling equipment, and local truck movements on surrounding air quality.[14] Automobile emissions have been under the regulators' control for years. This has not been the situation for port-associated activities. As the levels of smog and particulates increase around ports, there is mounting evidence that maritime and port activities are causing health impacts. Air pollutants include nitrogen oxides (NOx) that contribute to smog formation and particulate matter (PM) that poses health risks, particularly to children.

The ports of Long Beach and Los Angeles, with the participation and cooperation of the staff of the Environmental Protection Agency (EPA), California Air Resources Board, and South Coast Air Quality Management District, have developed a cooperative strategy, the San Pedro Bay Ports Clean Air Action Plan,[6] to significantly reduce the health risks posed by air pollution from port-related sources. The region has some of the worst air quality in the United States. The SPB Clean Air Action Plan addresses every category of port-related emission sources (including ships, trucks, trains, cargo handling equipment, and harbor craft) and outlines specific actions to reduce emissions from each category. The plan's measures are expected to eliminate more than 50% of diesel PM emissions from port-related sources within the next 5 years. Smog-forming NOx emissions will be also be reduced by more than 45% under the plan. In addition, actions under the plan are anticipated to result in reductions of other harmful air emissions, such as sulfur oxides (SOx).

17.3　REGULATORY ASPECTS OF ENVIRONMENTAL CONTROL FOR PORTS AND SHIPS

Maritime transportation is regulated by many different international, national, state or regional, and local agencies. In the United States, several government entities may regulate specific portions of the transportation sectors. The U.S. Coast Guard and U.S. Environmental Protection Agency (EPA) primarily regulate the maritime transportation industry. In addition, there are several international treaties and conventions that also impose regulations on the maritime transportation sector.

The U.S. Coast Guard regulates all seagoing vessels and ensures that they comply with U.S. law, as well as international treaties and conventions. EPA has responsibility for regulating the marine facilities. EPA has traditionally relied on delegation to the states to meet environmental standards, in many cases without regard to the methods used to achieve certain performance standards. This has resulted in the states having varied requirements. Their requirements can be more stringent for water, air, land, and hazardous waste than the federal requirements.

17.3.1　INTERNATIONAL LAWS AND REGULATIONS

Ports must provide assistance to ships in complying with internationally agreed to rules for protecting the marine environment. The primary regulatory framework for vessels is contained in the 1973 Convention for the Prevention of Pollution from Ships and its amendment, the 1978 Protocol (MARPOL 73/78). This agreement is the only regulation with worldwide applicability for protection of the marine environment from ship-based activities. Specifically, ports are responsible for providing reception facilities for residue from ship operations and cargo transport.

MARPOL consists of basic obligations for contracting parties, rules for ratification, amendments, and denunciation, as well as protocols for reporting and arbitration. It is designed to address the problem of marine pollution from vessels under six annexes, each of which addresses a different type of marine pollution.

Annex I: Forbids the discharge at sea of oil in certain "special areas" and limits other discharges to 1/30,000 of the cargo. Annex I requires that all parties to the convention ensure that adequate facilities are provided for the reception of residues and oily mixtures at marine facilities.

Annex II: Contains regulations for discharges of noxious liquid substances (e.g., bulk chemicals). To date, more than 250 substances have been evaluated and regulated. Such substances can only be discharged to reception facilities, unless certain requirements are met.

Annex III: Requires the issuing of detailed standards on packaging, marking, labeling, documentation, stowage, quantity limitations, exceptions, and notifications for preventing or minimizing pollution by harmful substances.

Annex IV: Deals with prevention of pollution from sewage. Annex IV states that vessels are not permitted to discharge sewage within 4 miles (6.4 km) of the nearest land, unless they have an approved treatment plant. Between 4 and 12 miles (6.4 and 19.3 km) from land, sewage must be comminuted and disinfected before discharge.

Annex V: Seeks to prevent pollution from garbage and establishes specific minimum distances for the disposal of garbage at sea. The most important component of this annex is the complete prohibition on the disposal of plastics into the sea.

Annex VI: Relates to the prevention of air pollution from ships. The regulations in this annex set limits on sulfur oxide and nitrogen oxide emissions from ship exhausts and prohibit deliberate emissions of ozone-depleting substances.

These annexes are mandatory and all signatory nations, including the United States, are subject to them. The U.S. Coast Guard has published regulations imposing requirements implementing these annexes.

17.3.2 U.S. Laws and Regulations

It is important to remember that there is no one, specific definition or design for a port or marine facility. Each consists of various operations and will be subject to regulation based on those specific activities. The following discussion focuses on some of the regulatory programs that may be applicable to a port or maritime facility to protect water, air, and land resources, and transport hazardous materials. The laws and regulations are organized around the media they control: waterside oriented, landside oriented, and air quality.

17.3.2.1 Waterside Controls

17.3.2.1.1 Clean Water Act

There are three aspects to this law concerning ports: discharge of wastewater, discharge of storm water, and spills. The discharge of wastewater to surface waters from marine facilities is regulated under the National Pollutant Discharge Elimination System (NPDES). Under NPDES, ports must obtain a permit to discharge wastewater into navigable waters. In some cases, the individual facilities within a larger structure (e.g., within a port) may not have an individual permit but may discharge to a larger, port-wide system that is permitted. Although EPA develops effluent limitation guidelines for certain industrial wastewater discharges, there are no specific effluent limitation guidelines established for marine operations.

Storm water discharges are also regulated under the NPDES program, and permits are required to be obtained by certain dischargers of storm water to surface waters. Transportation facilities (e.g., marine terminals) with vehicle maintenance shops or equipment-cleaning operations are considered to have a storm water discharge associated with their industrial activity. Direct discharges must be permitted. Storm water discharges through municipal separate storm sewer systems are also required to obtain NPDES storm water permit coverage. Discharges of storm water to a combined sewer system or to a public wastewater treatment plant are excluded.

The Clean Water Act requires facilities to develop spill prevention, control, and countermeasure (SPCC) plans for petroleum products, such as oil or any substance that causes a sheen on water, if they are stored in large quantities at a particular site. The SPCC program requires reporting spills to navigable waters and the development of contingency plans that must be kept on site. SPCC plans document the location of storage vessels, types of containment, dangers associated with a major release of material from the tanks, types of emergency equipment available at each site, and procedures for notifying the appropriate regulatory and emergency agencies.

17.3.2.1.2   Ocean Dumping Act

The basic purpose of the Ocean Dumping Act is to regulate intentional ocean disposal of materials. The act basically prohibits all ocean dumping, except that allowed by permits, in any ocean waters under U.S. jurisdiction, by any U.S. vessel, or by any vessel sailing from a U.S. port. The dumping of certain materials is exclusively banned, including radiological, chemical, and biological warfare agents, any high-level radioactive waste, medical wastes, sewage sludge, and industrial waste.

Four federal agencies have authority under the Ocean Dumping Act: the U.S. EPA, U.S. Army Corps of Engineers, National Oceanic and Atmospheric Administration, and Coast Guard. EPA has primary authority for regulating ocean disposal of all substances except dredged spoils, which are under the authority of the Corps of Engineers. Permits for dumping other materials may be obtained from EPA if EPA determines there is no unreasonable danger to human health or the environment. Permits issued under the Ocean Dumping Act specify the type of material to be dumped, the amount to be transported for dumping, the location of the dump site, the length of time the permit is valid, and any special provisions for surveillance.

The act requires the EPA to make binding the 1972 Convention on the Prevention of Marine Pollution by Dumping of Wastes and Other Matters. This convention prohibits the dumping of mercury, cadmium, DDT, PCBs, solid wastes and persistent plastics, oil, high-level radioactive wastes, and chemical and biological warfare agents.

17.3.2.1.3   Oil Pollution Act

The 1990 Oil Pollution Act (OPA) establishes liability against facilities and maritime activities that discharge oil or which pose a substantial threat of discharging oil to navigable waterways. OPA imposes contingency planning and readiness requirements on certain facilities defined to include motor vehicles. These requirements affect port and maritime transportation activities that use diesel and other petroleum-based fuels for cargo handling equipment and intermodal operations.

17.3.2.2 Landside Control

17.3.2.2.1 Coastal Zone Management Act

The Coastal Zone Management Act of 1972 (CZMA) is the law that regulates activities at the water-land interface. CZMA established a program for states and territories to voluntarily develop comprehensive programs to protect and manage coastal resources (including the Great Lakes). There are twenty-nine federally approved state and territorial programs. Despite institutional differences, each program must protect and manage important coastal resources, including wetlands, estuaries, beaches, dunes, barrier islands, coral reefs, and fish and wildlife and their habitats. It also is the regulatory basis for seaport planning in coastal regions. Resource management and protection are accomplished in a number of ways, through state laws, regulations, permits, and local plans and zoning ordinances.

17.3.2.2.2 Resource Conservation and Recovery Act

Maritime facilities can generate a variety of regulated wastes under the Resource Conservation and Recovery Act (RCRA). The wastes are produced during the normal course of operations and the use of underground storage tanks for fuel storage. Vessel refurbishing and maintenance activities generate hazardous wastes such as spent solvents and caustics, and paints and paint sludges. Additional common materials from marine facilities that may be hazardous include lead-acid motor vehicle batteries, vehicle maintenance fluids, used oil, scraps of metals (cadmium, chromium, lead, mercury, selenium, and silver) and materials containing these metals (e.g., high-grade stainless steel or paint waste), waste solvents, empty paint cans and spray cans, and paint residue.

Note that petroleum products and petroleum-containing wastes (e.g., waste oil, contaminated fuel, or fuel spill cleanup wastes) are specifically exempted from RCRA regulations, unless they exhibit any of the hazardous waste characteristics. Many maritime facilities qualify as hazardous waste generators under RCRA law. Under RCRA, it is the facility's responsibility to determine whether a waste is hazardous.

17.3.2.2.3 Hazardous Materials Transportation Act

The U.S. Department of Transportation (DOT) under the Hazardous Materials Transportation Act (HMTA) regulates the transport of hazardous materials. Materials covered by the act include all RCRA listed wastes and some additional materials deemed by DOT to be dangerous to transport. The HMTA covers several activities, including packaging, labeling, shipping papers, emergency planning, incident notifications, and liability insurance. There is some overlap between the DOT regulation under HMTA and EPA regulations under RCRA for tank residues.

17.3.2.3 Air Quality Control

17.3.2.3.1 Clean Air Act

Air quality standards are important for the protection of human health and control releases of ozone and greenhouse gases. The Clean Air Act provides the regulatory basis for the U.S. air emissions standards. The EPA sets limits on how much of a pollutant can be in the air anywhere in the United States. This ensures that the entire nation has the same basic health and environmental protections. The law allows individual states to have stronger pollution controls, but states are not allowed to have weaker pollution controls than those set for the whole country. On the other hand, states may have stricter requirements (e.g., California).

National Emission Standards for Hazardous Air Pollutants (NESHAP) attempt to control several hundred compounds, the most notable being asbestos. All marine facilities must comply with the NESHAP requirements for asbestos when demolishing, or significantly remodeling, a building or vessel containing asbestos. The EPA is also concerned with reducing emissions of air toxics and volatile organic compounds that result from marine tank vessel loading operations.

States are required under the Clean Air Act to issue state implementation plans (SIPs) that regulate stationary sources, such as buildings and other permanent installations, and mobile sources, such as automobiles. Typical marine facilities and activities that may be subject to stationary source regulations include heating and refrigeration plants, fueling and fuel storage facilities, maintenance facilities, garages and parking lots, building demolition and construction, and redevelopment projects. Port facilities can handle significant traffic and are oftentimes parking areas for imported and exported vehicles. SIPs may also control mobile sources such as fleet vehicles and other vehicles using the marine facility.

These laws along with their accompanying regulations are templates for environmental groups to assess port environmental performance. Some environmental publications suggest that the industry is "dirty" even when it is in regulatory compliance.[14]

17.4   EMS: CERTIFICATION AND MONITORING

The period when port authorities were more concerned with the effect of environmental elements on their activities than with the impacts of port operations on the quality of the environment has long passed. It is now imperative that the environmental issues of a port's activities are managed effectively in terms of both day-to-day operations and long-term development.[15] By definition, environmental issues are often transboundary; involve several systems in terms of aspects, significance, and pathways; and their effects may impact on air, soil, sediment, water, and ecosystems. When a ship is within the port area it becomes a significant component of the management program.

17.4.1   COMMON CHALLENGES: DIFFERENT RESPONSES

Traditionally, shipping and port organizations appear to have pursued separate agendas when it came to considering safety and environmental management issues. Strategically, the close interrelationships between their respective interests were acknowledged, but in practice, the organization and implementation of policies was very much split across the land-sea divide.

Shipping interests are ubiquitous with strong international perspectives to their operation and organization. Shipping is largely subject to international conventions for environmental protection of the sea. Safety, health, and environmental considerations are industrial in their maritime application, with enforcement, inspection, certification, and financial provisions focused on the ship itself. Maritime regulations target general ship safety standards and the measures required to prevent and control pollution. The flag state has the responsibility for exercising jurisdiction and control over ships on its register[16] because it has jurisdiction over its vessels on the high seas and because it can enforce standards by implementing control over the ships placed on its registers and allowed to fly its flag. States are required to take measures necessary to ensure the safety at sea of ships flying their flag,[17] and the laws and regulations for the prevention, reduction, and control of pollution of the marine environment must be adopted by flag states.

The international port sector is comprised of member ports that are overtly diverse in form, function, and local characterization from each other to the extent that each port may justifiably be considered unique. Their commercial profiles, socioeconomic links with their communities, ownership, and even cultural and legal regimes show stark contrasts between individual ports. Environmental law is a complex of international, European, and national law, much of which has some relevance to ports. In terms of management, shipping and ports may appear to be treated as separate entities.

The shipping industry's management response to its environmental responsibilities has largely been driven by statutory regulations and, in particular, compliance with the ISM Code (International Safety Management) and the implications of the STCW Convention (International Convention on Standards of Training, Certification and Watchkeeping of Seafarers). The port sector's response is currently to encourage compliance through voluntary schemes of self-regulation.

In terms of management, shipping and ports may appear to be treated as separate entities. However, when a vessel is in a port area, the ship and its cargo are highly significant components of the overall environmental aspect. The importance of being able to produce evidence of environmental awareness, compliance with legislation, and implementation of best practices through effective management is increasingly regarded as a justifiable component of the business strategy for both shipping and port interests.

Prosecutions for environmental negligence and the high-profile attention given to liability and compensation regimes have focused the attention of shipping and port managers on the options available to provide quality assurance of their declared policies. With so many environmental aspects, impacts, and risks in common, and the transboundary nature of many potential incidents, it can reasonably be suggested that there is mutual advantage to be gained from a more harmonized approach to environmental management between the two interests.

The dynamics, inputs, effects, and issues of the various activities and operations may be different, but they all have the potential to impinge on each other and to contribute to the overall environmental quality. Ship emissions and cargo handling may well impact on the physical, chemical, and biological characteristics of the port area, and port expansion plans through dredging and reclamation schemes may impact on the wider city community through noise and habitat loss. It may be suggested that integrated environmental management that takes a holistic approach is far more likely to produce efficient and effective responses in support of the environmental imperative rather than a purely industrial approach.

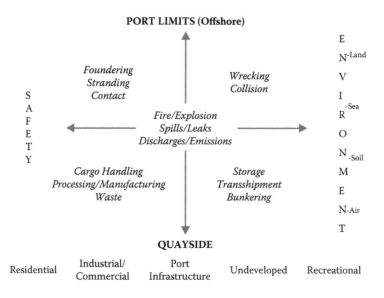

PORT LIMITS (Offshore)

Foundering
Stranding
Contact

Wrecking
Collision

S
A
F
E
T
Y

Fire/Explosion
Spills/Leaks
Discharges/Emissions

E
N-Land

V

I

R -Sea

O

N -Soil

M

E

N-Air

T

Cargo Handling
Processing/Manufacturing
Waste

Storage
Transshipment
Bunkering

QUAYSIDE

Residential Industrial/Commercial Port Infrastructure Undeveloped Recreational

FIGURE 17.1 Environmental risks in shipping and port operations.

As environmental planning and management takes on an even higher status in port development (driven by legislation, regulation, and stakeholder pressure), so the need to take a genuine systems approach to both daily operations and development projects becomes paramount.

Figure 17.1 shows just how intimate and compressed the close juxtaposition of the coastal and marine environment, shipping activities, port infrastructure, city development, and land-based natural environment can be in terms of the whole complex. To consider one component in isolation is to ignore the significance of interactions between the constituents and thereby lose the ability to manage effectively.

There is a close interrelationship between the environmental risks, aspects, and impacts of shipping and port operations. In terms of the functional organization of the port area, it is increasingly relevant to take a holistic view of safety, health, environmental, and security issues, and to integrate the interests of shipping and port operations and activities.

Each port may be considered as unique in terms of its geography, commercial profile, bathymetry, and operations, although it is acknowledged that ports often share common environmental challenges in terms of the growing requirement to demonstrate sustainable development and environmental protection. The traditional definition of a port as "the area where traffic changes between land and sea modes of transport" has been modified to demonstrate the indispensable role that ports play in the logistic chain. Ports are now "a mixture of industry and services that serve specific production and distribution processes."[18] In addition, ports are coming under pressure from the global market forces and the local political forces for redevelopment.[19] These new challenges make commercial ports subject to environmental protection and sustainable development by putting in place the functional organization necessary to effectively discharge a port authority's environmental responsibilities in a practicable and cost-effective manner. In this framework, port sustainable development can be defined as the situation in which the port is able to meet its own needs without endangering its own future. Nowadays, the concept of sustainable development is incorporated as the major component of the environmental policy statement of many port authorities, and environmental performance is a key component of the corporate social responsibility[20] and triple bottom line[21] concepts.

The apparent dilemma of protection of the environment in the port area or development of the port has been a recurrent theme. In recent years, experiences worldwide prove in practice that this dilemma does not make any real sense if ports adopt the following principles into their corporate policy[20] or the triple bottom line:[21]

1. Efficient economic performance
2. Ecological sustainability
3. Social responsibility

The largest challenge between policy objective and the attainment of environmental targets has been the availability of practical tools and methodologies for implementation and the framework within which to deliver the required results. The framework has evolved as an environmental management system that may be defined as the functional organization required in order to achieve compliance with environmental legislation, to create continuous improvement of environmental quality, and to prevent pollution. In addition, the demand for quality assurance (QA), the positive declaration of principles for the management of activities and functions aimed particularly at attaining high grades of safety and environmental standards, has emerged as another component of the environmental management of the activities and operations concerning the shipping industry and ports sector. Such an approach is rapidly becoming an increasingly significant component of the business plan and policy document of many major companies and organizations—and an influential driver for ports to consider implementing an environmental management system in order to demonstrate compliance with legislation.

17.5   EMS PRACTICE

An EMS covers the organizational structure, responsibilities, and ways and means of implementing environmental management. It is designed to ensure that the activities of the port authority, and their effects, conform to the environmental policy and associated objectives and targets. It includes the preparation and implementation of a documented system of procedures and instructions providing the basis for a program of continuous improvement.

There are two internationally recognized standards of EMS available:

1. **ISO 14001 (International Organization for Standardization):** ISO 14001 is a generic, international, voluntary standard that can be adapted to different types and sizes of organization in both industrial and service sectors, and which is based on the principle of continued improvement in environmental performance.

2. **European Union's EMAS:** The Eco-Management and Audit Scheme (EMAS) is promulgated under EC Regulation 1836/93. It is primarily of importance for the activities falling within the European market.

Essentially EMAS provides a model for the design, implementation, and maintenance of an EMS in the same way that ISO 14001 does. However, there are certain significant differences between the two. Under EMAS, a full environmental statement in nontechnical language should be published and independently verified, whereas under ISO 14001 only the environmental policy needs to be made public. Moreover, ISO 14001 does not require a preliminary review, whereas for EMAS it is both mandatory and assessable.

A useful summary of guidelines for the implementation of EMS and compatibility between standards is given by Paipai,[22] ABP (1999),[15] and in "ISO 14001 & EMAS."[23] However, as stated above, formal certification or registration under these schemes is neither mandatory nor necessarily appropriate for many ports at the present time for reasons of scale or nature of operations, significance of environmental effects, or, in some cases, arguably, cost. Nevertheless, an increasing number of ports are electing to adopt a phased approach on a voluntary basis that allows adaptation of EMS to local circumstances with the option to evolve toward formal registration under an accredited scheme in the future. Other, less formal models and approaches are available.

17.5.1 RESPONSE OPTIONS: THE EUROPEAN PERSPECTIVE

The port sector can demonstrate a positive and proactive program of environmental management initiatives and has particularly done so over the last 12 years.[24] The port sector's policy in Europe is currently to encourage compliance with legislation to high standards through voluntary schemes of self-regulation. In terms of development, the impacts of the ports activities and operations on the environment are coming under growing scrutiny from a wide range of stakeholders, including local communities and international legislators.[25] Port authorities are increasingly aware of the costs in terms of both poor public relations and publicity, and financial penalties, of neglecting their environmental duties and of failing to put effective environmental strategies in place. Ports are working toward a level playing field in terms of enforcement of environmental standards.

Ports have a wide range of reasons to respond to the new demands of environmental management and sustainable development. Significant drivers are represented in the following list:[4]

Compliance	Investor and shareholder
Port development	Director's liability
Risk management	Cost and risk reduction
Customers	Market opportunity
Community	Positive image
Insurance and banks	Influence policy

In this context, the *Green Paper on Seaports and Maritime Infrastructure,*[26] directive concerning environmental impact assessment,[27] the habitats directive,[28] and water framework directive[29] raise, on the one hand, the important role of ports in environmental protection and, on the other, the ports' need to incorporate the environmental dimension into their planning and development. While expansions of port facilities can make a significant contribution to economic and transportation development and the growth of a port, they may also create a wide range of potential adverse effects on the surrounding environment.

Self-regulation in the sector's policy refers to the various forms of EMS that many ports throughout Europe and the world are developing and implementing to assist in fulfilling their environmental responsibilities and duties. As stated above, the greatest challenge that lies in the gap between an environmental statement of intent or policy and the delivery of actual environmental protection is implementation. This is best achieved by the development and application of a structured EMS.[30] Internationally, the trend is that many of the more progressive and proactive port authorities are moving the environmental imperative to the heart of their business plan and high up the agenda for action plans. Successive surveys, conferences, workshops, and media reports confirm the growing status of environmental management and sustainable development as a major consideration and activity for a rapidly growing number of port authorities worldwide.

The European Sea Ports Organization (ESPO) has been remarkably consistent in its policy and attitude toward its environmental liabilities and responsibilities over the last decade.[31] In the face of increasing legislation and stakeholder pressure, it has placed a strong emphasis on the achievement of high standards of environmental quality through voluntary self-regulation.

In 1996, the factors quoted by respondents in a survey as being effective in motivating a small- to medium-sized enterprise to adopt an environmental policy were determined. The results are displayed in table 17.4.

TABLE 17.4
Reasons to Adopt an Environmental Policy

Factor or Driver	Percent of Respondents Identifying Factor
Cost savings	73%
Legislation and regulatory pressure	64%
Director's liability	62%
Market opportunity	61%
Positive company image	60%
Customer pressure	59%
Employee's concerns	42%
Local community concerns	33%
Feel-good factor	30%
Investors/shareholders	28%
Banks and insurance companies	28%

These same factors struck a chord with the port sector. Compliance with legislation and the quest for cost-effective operations were strong incentives to develop environmental programs appropriate to port authorities. With the rise in the public's perception of the status of the environmental imperative and the need to integrate environmental factors into the management of risk, there have been increased pressures from customers, investors, insurance companies, and local communities for port authorities to demonstrate their competence through some form of quality assurance. It would be incorrect, however, to suggest that the sector was always reactive. Ports can justifiably claim a proactive role in several of the most effective environmental initiatives aimed at protection and sustainability.[32]

Periodically repeated surveys within the sector helped to identify the key issues as identified by port professionals. The impacts are mixed among waterside, landside, and air resources. Programs of research and development were pursued through joint-funded activities between the EC and port authorities in order to develop practicable tools and methodologies with which to address their environmental challenges.[31] The ranking of the issues has changed over the years but in general is similar. Table 17.5 demonstrates the influence of European directives, such as 2000/59/EC on port reception facilities for ship-generated waste and cargo residues and the sustained significance of dredging activities. Priorities change with industry, public, and regulatory requirements.

Table 17.5 also reflects the continuity of issues that are of common interest between ports and shipping. The latter point is particularly significant. The port sector and shipping industry have much in common in terms of environmental management. A wealth of knowledge and best practice has been developed in both areas, yet it may be suggested that there is still scope for greater integration of respective environmental management programs to mutual advantage.

The port sector has a well-established policy that environmental issues should not be considered as competitive components between ports. Insofar as environmental issues are often transboundary and common to all ports, this approach is pragmatic and has certainly encouraged cooperation and collaboration. Indeed, the development

TABLE 17.5
Top 10 Environmental Issues

Environmental Issue (Ranking)	1996	2003
1	Dust	Garbage and waste
2	Dredging disposal	Dredging disposal
3	Port development (land)	Dredging operations
4	Dredging operations	Dust
5	Garbage and waste	Noise
6	Port development (water)	Bunkering
7	Noise	Air quality
8	Water quality	Port development (land)
9	Traffic volume	Ship discharge (bilge)
10	Hazardous cargo	Hazardous cargo

of shared-cost, practicable solutions to environmental challenges, and the free exchange of knowledge and experience have become key concepts of the sector's environmental research and development strategy. Environmental considerations may, of course, have an impact on competitiveness. A port that cannot be developed or expanded because of planning restrictions imposed on the basis of environmental considerations may well be placed at a competitive disadvantage. However, there is wide recognition of the fact that both the sector and individual port authorities benefit from a positive environmental policy based on an effective environmental management program.

17.5.2 ADAPTING TO CHANGE

More than 10 years of port-inspired initiatives aimed specifically at protection of the environment through appropriate policy and implementation of best practices can be demonstrated by reference to such benchmark events as displayed in figure 17.2.

The sector's response to its environmental liabilities and responsibilities has developed in reaction to changing legal and socioeconomic conditions and has evolved from early phases of merely raising environmental awareness to proactive, innovative programs and best practices implementation. Although compliance is still the prime objective, other significant drivers also steer policy and practice, such as sustainable development, cost and risk reduction, and issues related to corporate social responsibility and triple bottom line strategic frameworks.

Environmental programs activated within the maritime sector have been developed and adapted as the environmental remit itself has expanded in scope and status. Port authorities increasingly find themselves involved as major players in managing the whole port area, rather than just the aspects related to their immediate activities and operations. Environmental management necessarily has implications for landlords, tenants, and operators. Another significant development has been the closer integration of safety, health, environment, and security considerations where an effective, generic management model based on a systems approach can provide a cost-effective option for the port authority.

1994 - ESPO Environmental Code of Practice
1997 - Eco-information Project: (first, dedicated R&D program)
2001 - ESPO Review: (sector-wide overview of environmental status)
2002 - EcoPorts Project: (Major EC/Ports R&D initiative)
2003 - ESPO Environmental Code of Practice: (Objectives/guidelines)
    - Environmental review standard: (Preliminary EMS)
2004 - ESPO Conference, Rotterdam: (Benchmark performance)
    - First EPF Conference, Barcelona: (EcoPorts Foundation,
        Not-for-profit organization for exchange of environmental experience)
2005 - EPF Secretariat: (Centralized organization of services & products)
    - Second EPF Conference, Marseille
2006 - ESPO Conference, Stockholm—profile of EMS achievements
2006 - EPF Conference, Genoa—multiple workshops on key issues

FIGURE 17.2 Benchmark events in EU port environmental management.

The EcoPorts Foundation (EPF) has played a key role in encouraging collaboration on environmental issues among port authorities. It was established as a not-for-profit organization with the role of encouraging membership from as many ports as possible, identifying new topics for collaborative research, and acting as a focal point for port environmental managers to exchange ideas. Its major participants include representatives from the ports of Amsterdam, Antwerp, Associated British Ports, Barcelona, British Ports Association, Genoa, Gdansk, Goteborg, Hamburg, Rotterdam, and Valencia. Associated partners for purposes of research and development include the universities of Cardiff (United Kingdom), Amsterdam, Catalunya, Gdansk, and the World Maritime University, Malmo, as well as the specialists from ANPA and SOGESCA (Italy).

The EPF approach is based on voluntary collaboration and exchange of experience among seaports, inland ports, and dry ports throughout Europe. Port authorities themselves identify the environmental challenges faced by the port professional and specify the development required in terms of appropriate tools and methodologies to assist in effective management action. Existing products available to all port members of the EPF include two useful methodologies:

Self-Diagnosis Methodology (SDM): SDM is a validated procedure that assists ports in establishing their baseline performance in terms of comparison with best practices guidelines and assessing their environmental management status against the European benchmarks. The feedback provides an analysis that assists ports to identify priorities for action. Further, port authorities and other marine operators can use the analyses as a checklist to identify issues and level of management response in the port area and within the logistic chain.

Port Environmental Review System (PERS): PERS is another validated procedure that assists ports in implementing the first stages of a credible EMS. It can be used as a preliminary step to a more comprehensive system such as ISO 14001 or EMAS, but allows phased development. By following PERS procedures, ports can establish the core elements of an EMS. The EcoPorts Foundation identifies PERS as a standard for port environmental management, and as such, it carries the voluntary option of independent review (currently carried out by Lloyd's Register, Rotterdam, on behalf of the foundation) with a certificate of validation for ports that reach the standard. This is an important new development for ports in Europe, as it demonstrates attainment of a benchmark standard. A special package on the monitoring of environmental performance and the identification of appropriate environmental performance indicators provides a useful compendium of up-to-date information.

17.5.3 Quality Assurance and Benchmark Performance

The maritime sector's collaborative approach to developing and implementing effective environmental management can be tracked in terms of benchmark performance. From baseline values determined from dedicated surveys in 1996, it is possible to

TABLE 17.6
Progress in the Development of Environmental Programs

Environmental Management Component	1996 %	1999 %	2004 %	2006 %	Percent Change 1996–2006
Does the port authority have an environmental plan?	45	52	62	82	+37
Does the plan aim for compliance plus?	32	41	48	72	+40
Does the plan aim to raise environmental awareness?	44	62	69	68	+24
Is environmental monitoring carried out in the port?	53	60	69	72	+19
Does the plan involve community and other stakeholders?	53	60	56	78	+25
Is the ESPO Code available within the port?	41	48	53	53	+12
Designated environmental personnel?	55	65	67	88	+33

demonstrate continuous improvement in key components associated with the basic elements of any credible EMS. Table 17.6 presents progress from 1996 to 2006.

Similar progress can be demonstrated for other aspects of the maritime sector's environmental activities. The surveys have been carried out in a collaborative exercise between ESPO and EPF. As with other industrial surveys, some cautions must be expressed in terms of the declared value. Often, it is the most positive, enthusiastic, or responsible authorities that respond to such surveys, and there are national differences across Europe in the number of respondents and levels of activity. Nevertheless, even if the tables are interpreted in terms of trends rather than absolute values, it suggests that collaborative programs have helped to deliver continuous improvement in several key areas. Table 17.7 gives an indication of the performance using the ESPO Environmental Code as a benchmark.

17.6   EDUCATION AND TRAINING

The environmental objectives for the shipping industry and port sector are similar in many ways in terms of the mutual interest in delivering sustainable development, improving environmental quality, reducing risks, and demonstrating compliance in

TABLE 17.7
Benchmark Performance in Support of ESPO Environmental Code

Management Component	2004	2006	Percentage + or –
Review environmental management program?	43	50	+7
Publish review?	33	79	+46
Policy made available to public?	64	89	+25
Environmental performance indicators?	50	62	+12
Environmental management system?	20	57	+37

a transparent manner. Shipping and port activities share several significant environmental aspects, face common challenges from high-priority issues, interact with each other's operations, and benefit from a structured, environmental management system. While a vessel is within port limits or transiting adjacent coastal state waters, there are juridical, operational, commercial, administrative, and monitoring reasons to justify at least some degree of integration and collaboration of their efforts to comply with their respective environmental imperatives. Integrated environmental management seeks to integrate with the processes of planning and decision making those aspects of management that impact on assessment and evaluation.[33] Education and training in environmental management can be one of the most time- and cost-efficient methods of raising awareness, developing in-house capability, enhancing individual skill competencies, and bridging the still glaring gap, in some cases, between policy objectives and actual implementation. There is also the argument that better understanding by their respective personnel as to how shipping and ports interact can also contribute to more efficient management. Even if ports are not directly responsible for the activities carried out in the port area, port administrations still bear a certain general public responsibility that will be strengthened by the environmental liability directive[31] in Europe.

17.7 PORTS AND SHIPS: THE CONTINUING CHALLENGE

Global commerce has expanded at a double-digit pace for the last decade. Most of the cargo that this expansion has created moves on massive container and bulk cargo ships into and out of ports. Intermodal moves of cargo have skyrocketed to move this cargo into the hinterlands. There is an industry and public desire to have these activities managed in a way that avoids or ameliorates potential environmental impacts. In fact, there is a growing demand by the public for sustainable development.

There is much debate about the meaning of the term *sustainable development*. It was perhaps best articulated by the Brundtland Report of 1987 and by Agenda 21, which was established at the 1992 Earth Summit at Rio de Janeiro as "development that meets present needs without compromising the future." The Brundtland Report's definition is a sweeping change in policy not only for the maritime industry but society in general. It entails a broad process of adaptation and change in lifestyles and in systems of production and consumption. Today the idea of sustainable development has also moved beyond environmental policy to advocate principles of democratic decision making, community culture, and social equity as the frameworks that are best suited to carry out balancing of development pressures and sustaining the environment over the long term.

For the maritime and transportation industry to adopt a sustainable development theme there is a need to attempt a shift in policies and practices that embrace environmental values.[34] The EMS programs established by many ports are important steps toward a sustainable business process, but more may be required. Indeed, a paradigm shift from unilateral decision making to shared decision making may be necessary to avoid problems (including political problems). To make a paradigm shift to a systematic framework that has integrative features (i.e., sustainability and stewardship) and is streamlined, several policy and management measures deserve consideration, including conceptual and strategic changes. Elements of these two shifts are outlined below.

17.7.1   CONCEPTUAL SHIFT

Three conceptual shifts are proposed that would promote the development of environmental policies that are proactive, capable of practicable implementation, and support sustainable port projects and ship operations.

Compliance to performance: Moving from an environmental compliance framework to an environmental performance-based framework will enhance project outcomes. Compliance often results in the lowest common denominator solution, on a project-by-project basis, and a performance-based solution yields outcomes that achieve system-wide goals.

Incentive-based framework: The stick, rather than the carrot, is most often employed in environmental/transportation planning. On the other hand, incentive-based programs are often used in environmental stewardship programs in other disciplines.

Cooperation rather than competition: There is a need to integrate competing interests (public, private, and community) in transportation and environmental decision making. Legal, regulatory, and systemic policies that are barriers to this approach need to be identified and corrected. Lastly, the disconnection among local, national, and international maritime transportation desires must be addressed and strategies developed that acknowledge "one size does not fit all."

17.7.2   STRATEGIC SHIFT

Policies must be successfully applied to gain the desired effect. Two strategies are described that may be considered useful in promoting sustainable policy utilization and feedback.

Provide useful tools to decision makers: Many environmental and maritime transportation policy decisions are made based on model outputs and strategies encompassed in the current state of the practice. These strategies/models lack comparison factors. The state of the practice does not provide the level of information necessary to make informed decisions. However, new tools have not yet been developed that allow comparisons across modes and environmental media. Research is needed on innovative tools and strategies. Often these methodologies have been implemented and deployed in other disciplines. As an example, the application of scenario-based planning could provide a new way to view transportation/environmental planning. A common set of assumptions and metrics that allow comprehensive and useful comparisons across transportation modes and environmental media are needed for informed decision making.

Development of environmental performance measures: Maritime transportation investments can be used more effectively in achieving port and ship environmental goals through a performance-based mitigation strategy than the current compliance strategy. This approach allows environmental objectives to be set, balanced throughout the planning process, and evaluated (measured) following project implementation and operations. Measurement is known to be crucial in establishing success of program implementation, and although it is difficult to formulate meaningful metrics, they are necessary.

17.7.3 SUSTAINABLE ENVIRONMENTAL PHILOSOPHY

The concept of sustainability in maritime transportation is adding another layer of complexity to the expanding global transportation situation. However, it also provides a potential solution to the present regulatory gridlock by introducing needed structure to the decision-making environment. The concept of sustainability for the maritime sector means that decisions need to be integrated across media and modes, be balanced and optimize the triple bottom line of economic, social, and environmental objectives over the long term. The underlying logic of this concept is intuitively appealing, and public support for it can be expected to continue to grow in the coming years. The philosophical concept of green ports and green ships fits easily into this new sustainability framework. It is an environmental philosophy that works for today and future generations.

REFERENCES

1. Johnson, D. 2007. *America's container ports: Delivering the goods.* Washington, D.C.: U.S. Department of Transportation, Research and Innovation Technology Administration, Bureau of Transportation Statistics.
2. Chase, T. 1998. *Environmental management handbook.* Washington, D.C.: American Association of Port Authorities.
3. Couper, A. D. 1983. *The Times atlas of the oceans.* London: Times Books Ltd.
4. Wooldridge, C. F. 2004. The positive response of European seaports to the environmental challenge. Ports and harbors. *Official Journal of the International Association of Ports and Harbors* 49(8):9–12.
5. ESPO. 2003. *Environmental code of practice.* Brussels: European Sea Ports Organization.
6. Anon. 2006. *San Pedro Bay Ports Clean Air Action Plan.* Prepared by Port of Los Angeles and Port of Long Beach in cooperation with U.S. EPA, California Air Resources Board, and the South Coast Air Quality Management District.
7. Anon. 2000. *Green ports, environmental management and technology at US ports.* Boston: Urban Harbors Institute, University of Massachusetts.
8. Vickerman, J. 1998. Next-generation container vessels. *TR News*, May-June.
9. Wakeman, T., and Themelis, N. 2001. A basin-wide approach to dredged material management in New York/New Jersey Harbor. *Journal of Hazardous Materials* 2652:1–13.
10. Marine Board. 1997. *Contaminated sediments in ports and waterways: Cleanup strategies and technologies.* Washington, D.C.: National Research Council, National Academy Press.

11. PIANC International. 2002. *Environmental guidelines for aquatic, nearshore and upland disposal facilities for contaminated dredged material.* Working Group 5, Environmental Commission, Brussels, Belgium.

12. Demars, K. R., Richardson, G. N., Yound, R. N., and Chaney, R. C. 1995. *Dredging remediation, and containment of contaminated sediments.* ASTM STP 1293, Philadelphia.

13. Marine Board. 2001. *A process for setting, managing, and monitoring environmental windows for dredging projects.* Washington, D.C.: National Research Council, National Academy Press.

14. Bailey, D., Plenys, T., Solomon, G., Campbell, T., Feuer, G., Masters, J., and Bella, T. 2004. *Harboring pollution: The dirty truth about U.S. ports.* New York: Natural Resources Defense Council.

15. ABP Research. 1999. *Good practice guidelines for ports and harbours operating within or near UK European marine sites.* UK Marine SACs Project, English Nature.

16. United Nations Conference on the Law of the Sea (UNCLOS). 1982.

17. United Nations Conference on the Law of the Sea (UNCLOS), Article 94.

18. Suykens, F. 1986. Ports should be efficient (even when this means that some of them are subsidized). *Maritime Policy and Management,* Volume 13, Issue 2 April, 105–126.

19. Beresford, A. 2004. Intermodal transport in Europe: The key issues. *Marine and Port Review* 15–16.

20. Naniopoulos, A., Tselentis, B. S., and Wooldridge, C. F. 2006. Sustainable development of port operations: The role of research-led education. Paper presented at Proceedings of international conference in memory of the late Professor Basil Metaxas, Shipping in the Era of Social Responsibility, Cephalonia.

21. Savitz, A., and Weber, K. 2006. *The triple bottom line: How today's best-run companies are achieving economic, social and environmental success.* New York: John Wiley & Sons.

22. Paipai, E. 1999. *Guidelines for port environmental management.* Report SR 554, Wallingford HR.

23. ISO 14001 & EMAS: Experience to date and future directions. *Institute of Environmental Management Journal* 5(4), 1998.

24. Wooldridge, C. F. 2004. Environmental management: Progress through partnership. *Marine and Ports Review* 33–37.

25. Beresford, A. K. C., Gardner, B. M., Pettit, S. J., Naniopoulos, A., and Wooldridge, C. F. 2004. The UNCTAD and WORKPORT models of port development: Evolution or revolution? *Maritime Policy and Management* 31:93–107.

26. Commission of the European Communities (COM). 1997. *Green paper on seaports and maritime infrastructure.* COM (97), final of 10.

27. Council of the European Communities (EEC). 1985. *The assessment of the effects of certain public and private projects on the environment.* Council Directive 85/337/EEC, Official Journal L 175/40.

28. Council of the European Communities (EEC). 1992. *The conservation of natural habitats and wild fauna and flora.* Council Directive 92/43/EEC, Official Journal L 206.

29. Council of the European Communities. Water Framework Directive 2000/60/EC.

30. Wooldridge, C. F., and Brigdon, A. 1999. Environmental management practices in selected UK and overseas ports and harbours. In *Guidelines for port environmental management,* ed. E. Paipai. Report SR 554.

31. European Sea Ports Organization (ESPO). http://www.espo.be.

32. Whitehead, D. 2001. Quoted in Select Committee on Environment, Transport and Regional Affairs, Minutes of Evidence. Evidence given on February 14, 2001, Stationary Office, London.

33. Born, S. M., and Sonzogni, W. C. 1995. Integrated environmental management: Strengthening the conceptualization. *Environmental Management* 19:161–81.

34. Anon. 2003. *Using and environmental management system to meet transportation challenges and opportunities: An implementation guide.* Washington, D.C.: American Association of State Highway and Transportation Officials.

Index